SPRINGER STUDY EDITION

H.M. Staudenmaier (Ed.)

Physics Experiments Using PCs

A Guide for Instructors and Students

With 162 Figures

Professor Dr. H. M. Staudenmaier
Institut für Anwendungen der Informatik, Universität Karlsruhe
Kaiserstrasse 12, D-76131 Karlsruhe, Germany

First Edition 1993
Second Printing 1995

ISBN 3-540-58801-9 Springer-Verlag Berlin Heidelberg NewYork

CIP data applied for

Printed in the United States of America

Typesetting: Data conversion by Springer-Verlag

SPIN 10492217 55/3144-54321 - Printed on acid-free paper

Preface

In many fields of physics the use of on-line computers in research experiments is daily routine and without computers scientific progress is often unthinkable in our times. Computer-aided experimentation and measurement have become important aspects in professional life in research as well as in industry. Nevertheless these subjects are not really covered in the physics curricula of most universities and we believe it to be an important task for professors and teaching staff to familiarize students with computer applications in their own and neighbouring fields. In general it is desirable that students also learn some of the useful concepts of the new science called "informatics" or "computer science". Successful attempts have already been made in the field of computer applications in theoretical physics and various universities now offer elaborate courses; excellent textbooks are also available.

In the field of teaching computer applications in experimental physics, however, the situation is far less favorable and one aim of this book is to give some helpful trigger pulses for this important subject. In many universities, polytechnical schools and colleges one encounters isolated computerized experiments installed in undergraduate or graduate laboratories. Such experiments have often been developed by the local staff making use of the specialized research interests of the institute or department. The students may well get the impression that on-line computing and data taking is a distinctive feature of this particular experiment, and they will not learn that computerized experimentation has become a common method in nearly all fields of physics. We hope that the examples given in this book will demonstrate the powerful potential of computers to control experiments, take data, and analyze and display the results nearly on-line. We further hope to encourage the academic staff responsible for laboratory courses to include more, and more interesting, computerized experiments in their laboratory classes.

The experiments contained in this book are described in sufficient detail that they may be copied — totally or in part — without major problems, but the authors have kindly agreed to give additional information concerning hard- or software on direct request. The material is broadly based and ranges from fairly easy undergraduate projects to quite demanding graduate laboratory experiments. The chapters are grouped into 5 parts and the complexity of the experiments increases within each part.

The book is also intended for students participating in laboratory courses, and to this end the authors have given comprehensive descriptions of their experiments and additional information that is not touched upon in regular laboratory manuals.

The editor is grateful to his colleagues for their valuable contributions to this book and extends his thanks both to them and to Springer-Verlag for the excellent cooperation.

Karlsruhe, March 1992 H. M. Staudenmaier

Contents

Part I Mechanics

1. Fourier Analysis of Some Simple Periodic Signals
By *R. Lincke* (With 17 Figures) 3
1.1 Apparatus 3
1.2 Programs 4
1.3 Experiments 5
1.3.1 Simple Harmonic Wave 5
1.3.2 Beats 6
1.3.3 Amplitude Modulation 9
1.3.4 Rectangles 12
1.4 Didactic and Pedagogical Aspects 14
References 16
Appendix 1.A 16

2. Point Mechanics by Experiments – Direct Access to Motion Data
By *R. Dengler* and *K. Luchner* (With 20 Figures) 23
2.1 Introduction 23
2.2 ORVICO 24
2.2.1 Principle 24
2.2.2 Hardware 25
2.2.3 Software 28
2.3 Examples 29
2.3.1 Ballistic Motion 29
2.3.2 The Rigid Pendulum 31
2.3.3 Frame of Reference 33
2.3.4 Statistical Motion on an Air Table 34
2.3.5 Spheric Pendulum 35
2.3.6 Two Point Masses Observed 37
2.4 Conclusion 38
References 39

Part II Thermodynamics

3. Application of PID Control to a Thermal Evaporation Source
By *B. D. Hall* (With 12 Figures) 43
3.1 Introduction 43
3.2 The System to be Controlled: An Inert-Gas-Aggregation Source 45
3.2.1 Background 45
3.2.2 The Inert-Gas-Aggregation Technique 45
3.2.3 A Description of a Real Inert-Gas-Aggregation Source 47
3.3 Description of the PID Control Algorithm 49
3.3.1 The PID Control Algorithm 54
3.4 Implementing the PID Algorithm on a Computer 54
3.4.1 Program Structure and the Use of Interrupts 58
3.5 Adjusting the PID 59
3.5.1 The Ziegler-Nichols' Methods 60
3.6 Possibilities Offered by the Léman Source 63
3.7 Conclusions 65
Acknowledgements 66
References 66

4. Computer Control of the Measurement of Thermal Conductivity
By *B. W. James* (With 17 Figures) 67
4.1 Thermal Conductivity 67
4.1.1 Measurement of Thermal Conductivity with Parallel Heat Flow 68
4.1.2 Measurement of Thermal Conductivity with Non-Parallel Heat Flow 70
4.2 Experimental Considerations 71
4.2.1 The Thermocouple as a Temperature Measuring Device 72
4.2.2. The AD595 Thermocouple Amplifier Integrated Circuit 73
4.2.3 Thermocouple Accuracy 74
4.2.4 Calibration of the Thermocouples 75
4.2.5 Thermocouple Selection Multiplexing Circuit 75
4.2.6 Multiplexor Control 76
4.2.7 The IEEE-488 Bus Interface Unit 76
4.2.8 The Control and Measurement Software 76
4.2.9 Discussion of the Experiment 84
4.3 The Computer Simulation 85
References 86
Appendix 4.A 88
Appendix 4.B 99

Part III **Solid State Physics**

5. Experiments with High-T_c Superconductivity
By *M. Ottenberg* and *H. M. Staudenmaier* (With 4 Figures) .. 111
5.1 Experimental Setup 111
5.1.1 The Apparatus 111
5.1.2 Electronics 113
5.1.3 Computer, Interface and Software 114
5.2 Measurements 116
5.2.1 Resistance Measurement 116
5.2.2 Tunnel Diode Oscillator Measurement 118
5.3 Results 119
5.3.1 Detailed Analysis of the Resistance and TDO Measurements 119
5.3.2 Thermodynamic and Calorimentric Results 122
5.3.3 Experience Within the Laboratory Course 124
References 125
Appendix 5.A: Electric Circuit Diagrams 125
Appendix 5.B: Spline Fit Program SPLFIT 126

6. Computer Control of Low Temperature Specific Heat Measurement
By *G. Keeler* (With 13 Figures) 129
6.1 Basic Physics 130
6.1.1 Specific Heat 130
6.1.2 Low Temperature Specific Heat 131
6.1.3 The Debye Model for the Specific Heat 133
6.1.4 Specific Heat Anomalies 134
6.2 Experimental Setup 134
6.2.1 Specimen 134
6.2.2 Apparatus 135
6.2.3 Electronics 136
6.2.4 Microcomputer Control 138
6.3 Measurements and Results 138
6.3.1 Measurement Principles 138
6.3.2 Using the Computer Program 142
6.3.3 Typical Results 146
6.4 Discussion 148
References 149
Appendix 6.A: Circuit Diagrams 150
Appendix 6.B: Program Listing 150

7. **Computer-Controlled Observations of Surface Plasmon-Polaritons**
By *A. D. Boardman*, *A. M. Moghadam* and *J. L. Bingham* (With 14 Figures) ... 172
7.1 Introduction ... 172
7.2 A Computer-Controlled ATR Experiment ... 178
7.2.1 Prism Geometry ... 179
7.2.2 Computer Control of ATR Measurements ... 183
7.3 Comments on the Mechanics Design and the Computer Interface ... 187
7.4 Conclusion ... 191
References ... 191

Part IV **Optics and Atomic Physics**

8. **Molecular Spectroscopy of I_2**
By *U. Diemer* and *H. J. Jodl* (With 8 Figures) ... 195
8.1 Introduction ... 195
8.2 Some Basic Physics of the Diatomic Molecule ... 196
8.3 Experimental Setup ... 200
8.3.1 The Classical Arrangement ... 200
8.3.2 Extensions: Online Use of a Computer ... 201
8.4 Measurements ... 204
8.4.1 Calibration of the System ... 204
8.4.2 Recording the Absorption Spectra ... 204
8.4.3 Recording the Fluorescence Spectra ... 206
8.4.4 Some Additional Features of the Program *LAmDA* 207
8.5 Analysis of the Spectra Using the Program JOD ... 208
8.5.1 Analysis of Absorption Spectra ... 208
8.5.2 Some Optional Exercises ... 214
8.6 Pedagogical Aspects ... 217
References ... 218

9. **Optical Transfer Functions**
By *H. Pulvermacher* (With 14 Figures) ... 219
9.1 Introduction ... 219
9.2 Mathematical Tools ... 220
9.2.1 Fourier Transforms ... 220
9.2.2 Theory of Transfer Functions ... 221
9.2.3 Imaging with Space Invariant Systems ... 223
9.2.4 Coherent Optics ... 223
9.2.5 Incoherent Optics ... 225
9.2.6 Exercises and Questions ... 227

9.3 Experimental Set Up 227
9.3.1 Preliminary Considerations 227
9.3.2 The Optics 228
9.3.3 The Test Object 229
9.3.4 The Electronics 230
9.3.5 The Adjustment 231
9.3.6 The Software for Experimentation and Evaluation 233
9.4 Evaluation 237
9.4.1 The Tasks 237
9.4.2 The General Procedure of Evaluation 237
9.4.3 Influence of the Detector Slit 238
9.4.4 Pure Defect of Focus 238
9.4.5 Diffraction and Defect of Focus 239
9.4.6 Quasi-Coherent Illumination 240
9.5 Didactic and Pedagogical Aspects 243
9.5.1 Goals 243
9.5.2 Interpretation of Data 243
9.5.3 Presentation of Data 244
9.5.4 Complications and Limitations of the Model 244
9.5.5 Applications of Fourier Optics 245
Appendix 9.A: Diffraction by a Sector Star 245
References 247

Part V **Nuclear Physics**

10. Nuclear Spectrometry Using a PC Converted to a Multichannel Analyser
By *J. S. Braunsfurth* (With 13 Figures) 251
10.1 Introduction 251
10.1.1 Hardware Concept 251
10.1.2 Target Group 252
10.1.3 MCA Design Alternatives 252
10.2 Basic Physics 254
10.2.1 Interaction of Electromagnetic Radiation with Matter 254
10.2.2 Absorption of Electromagnetic Radiation in Matter 255
10.2.3 Interaction of Particle Radiation with Matter 256
10.2.4 Bremsstrahlung 257
10.2.5 X-Ray Fluorescence 257
10.3 Detectors and Measuring Equipment 258
10.3.1 Scintillation Detectors for β and γ Spectrometry . 258
10.3.2 Signal Recording Equipment; the Multichannel Analyser 259
10.3.3 Energy Resolution of a Detector 259
10.3.4 Radiation Detection Efficiency 260

10.4 Experimental Setup 261
10.4.1 Hardware Setup 261
10.4.2 General Structure of the MCA Program; Program Kernel 262
10.4.3 MCA Program Menues 265
10.5 Experiments 269
10.5.1 General Considerations 269
10.5.2 γ-Ray Absorption; Radiation Intensity Buildup by Compton Interaction 271
10.5.3 β Spectrum; Energy Loss of Electrons in Matter . 276
10.6 Student Reactions 279
References 281

11. Parity Violation in the Weak Interaction
By *E. Kankeleit*, *H. Jäger*, *C. Müntz*, *M. D. Rhein*, and *P. Schwalbach* (With 7 Figures) 282
11.1 Introduction 282
11.2 Basic Physics 283
11.3 Experimental Setup 285
11.3.1 Electronics 286
11.3.2 Software 288
11.4 Measurements and Results 289
11.4.1 General Remarks 289
11.4.2 Energy Calibration 289
11.4.3 Background Measurement 289
11.4.4 Measurement of the γ Polarization 289
11.4.5 Results and Discussion 290
11.5 Didactic and Pedagogical Aspects 292
References 292

12. Receiving and Interpreting Orbital Satellite Data. A Computer Experiment for Educational Purposes
By *T. Kessler*, *S. M. Rüger* and *W.-D. Woidt* (With 13 Figures) 293
12.1 Introduction 293
12.2 The UoSAT Satellites 294
12.3 The Receiving System 296
12.4 Discriminating Valid Data from Noise and Interference .. 297
12.5 The Real Time Data Acquisition System 299
12.6 Whole Orbit Data Analysis 303
12.7 Practical Experience and Further Aspects 307
Acknowledgements (from the third author) 308
References 308

Subject Index 309

List of Contributors

Bingham, J.L.
Applied Optics Group, Department of Physics, University of Salford
Salford M5 4WT, United Kingdom

Boardman, A.
Applied Optics Group, Department of Physics, University of Salford
Salford M5 4WT, United Kingdom

Braunsfurth, J.
Institut für Experimentalphysik, Ruhr-Universität Bochum
D-44780 Bochum, Germany

Dengler, R.
Lehrstuhl für Didaktik der Physik, Universität München
Schellingstrasse 4, D-80799 München, Germany

Diemer, U.
Fachbereich Physik, Universität Kaiserslautern
Erwin-Schrödinger-Strasse, D-67663 Kaiserslautern, Germany

Hall, B.D.
Office Fédérale de Métrologie
Lindenweg 50, CH-3084 Wabern, Schweiz

Jäger, H.
Institut für Kernphysik, Technische Hochschule Darmstadt
Schlossgartenstrasse 9, D-64289 Darmstadt, Germany

James, B.W.
Department of Pure and Applied Physics, University of Salford
Salford M5 4WT, United Kingdom

Jodl, H.-J.
Fachbereich Physik, Universität Kaiserslautern
Erwin-Schrödinger-Strasse, D-67663 Kaiserslautern, Germany

Kankeleit, E.
Institut für Kernphysik, Technische Hochschule Darmstadt
Schlossgartenstrasse 9, D-64289 Darmstadt, Germany

Keeler, G.J.
Department of Pure and Applied Physics, University of Salford
Salford M5 4WT, United Kingdom

Kessler, T.
Fachbereich Physik der Universität Berlin
Arnimallee 14, D-14195 Berlin, Germany

Lincke, R.
Institut für Experimentalphysik, Universität Kiel
Olshausenstrasse 40, D-24118 Kiel, Germany

Luchner, K.
Lehrstuhl für Didaktik der Physik, Universität München
Schellingstrasse 4, D-80799 München, Germany

Moghadam, A.M.
Applied Optics Group, Department of Physics, University of Salford
Salford M5 4WT, United Kingdom

Müntz, C.
Institut für Kernphysik, Technische Hochschule Darmstadt
Schlossgartenstrasse 9, D-64289 Darmstadt, Germany

Ottenberg, M.
Interfakultatives Institut für Anwendungen der Informatik und
Fakultät für Physik, Universität Karlsruhe
Kaiserstrasse 12, D-76131 Karlsruhe, Germany

Pulvermacher, H.
Institut für Medizinische Optik, Universität München
Barbarastrasse 16, D-80797 München, Germany

Rhein, M.D.
Institut für Kernphysik, Technische Hochschule Darmstadt
Schlossgartenstrasse 9, D-64289 Darmstadt, Germany

Rüger, S.M.
Fachbereich Physik der Universität Berlin
Arnimallee 14, D-14195 Berlin, Germany

Schwalbach, P.
Institut für Kernphysik, Technische Hochschule Darmstadt
Schlossgartenstrasse 9, D-64289 Darmstadt, Germany

Staudenmaier, H.M.
Interfakultatives Institut für Anwendungen der Informatik und
Fakultät für Physik, Universität Karlsruhe
Kaiserstrasse 12, D-76131 Karlsruhe, Germany

Woidt, W.-D.
Fachbereich Physik der Universität Berlin
Arnimallee 14, D-14195 Berlin, Germany

Part I

Mechanics

1. Fourier Analysis of Some Simple Periodic Signals

R. Lincke

During our studies, most of us who are now teaching physics have experienced Fourier analysis as a purely mathematical subject confined to the analytically tractable cases of rectangle, triangle and sawtooth [1.1–5]. Varying parameters, however, studying special cases in detail, illustrating the theory with graphics and applying it to real physical problems became feasible only with the advent of microcomputers. If this tool is available in the undergraduate laboratory, then there are few topics more rewarding than an experimental and theoretical investigation of Fourier series. In this article we shall show how one can produce and study signals containing one, two, three and many partial waves, i.e. we shall devote ourselves to simple harmonic signals, beats, amplitude modulation and rectangles.

1.1. Apparatus

The hardware requirements for these experiments are very modest: A PC (even a Commodore 64 will do) with an attached analogue-to-digital converter, a microphone, a pair of tuning forks, a simple signal generator and a low voltage transformer (see Fig. 1.1). A printer for making hard copies would be a nice addition.

The A/D converter should permit a sampling rate of several kHz; 8 bit are sufficient. Musicians tuning forks are not loud enough and do not permit controlled detuning; the well-known pair on wooden resonance boxes with a fundamental frequency of 440 Hz is ideal; attaching two small masses to one of them will lower the frequency down to 408 Hz. The signal generator should provide audio frequencies and rectangles and permit amplitude modulation.

At Kiel we use the apparatus available from NEVA [1.6]; many other makes should work as well.

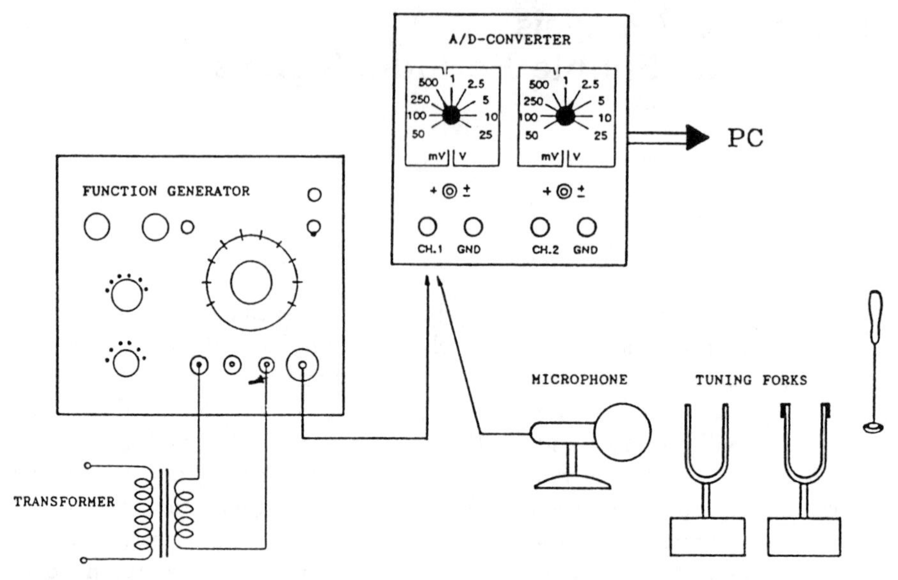

Fig. 1.1. Experimental setup

1.2 Programs

Although it is quite possible to do the experiments with simple self-written programs (see Appendix 1.A), we normally use the following more complete menu-controlled programs: SCOPE records 8-bit data at presettable speeds and permits saving the data on disk. It will also leave the data in the PC memory, so that a subsequent program can use them.

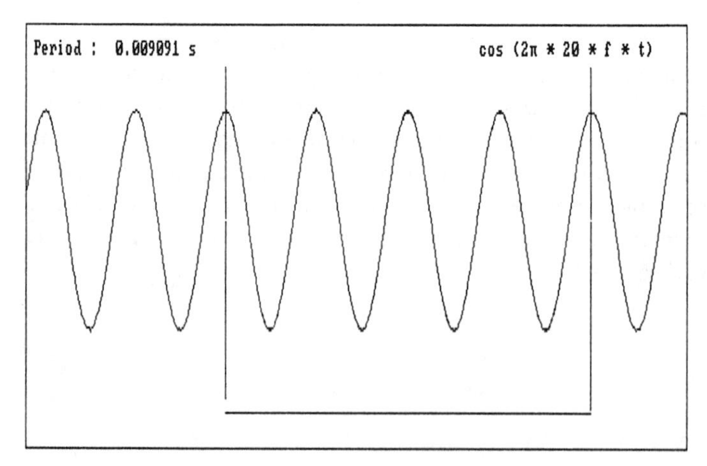

Fig. 1.2. Tuning fork. Amplitude plotted as a function of time. Four periods are selected as basis of the analysis

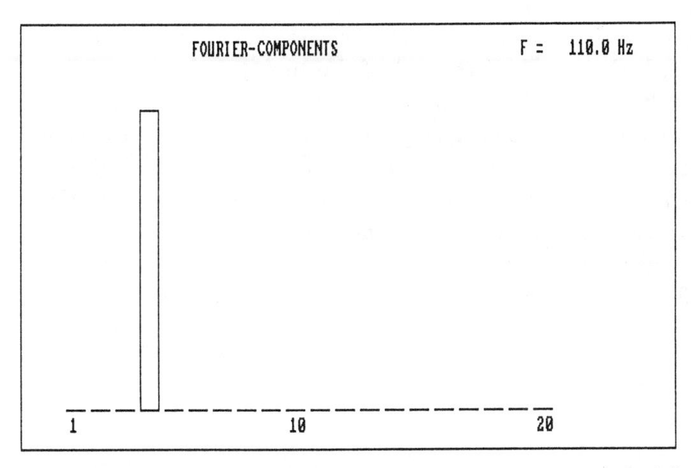

Fig. 1.3. Spectrum of the tuning fork. The spectral components of the Fourier expansion are plotted on an arbitrary scale for the first 20 harmonics

EVALUATION can read and display these data, determine amplitudes and time intervals and permit fitting envelopes etc. It may also be used for plotting functions.

FOUR-ANA performs the harmonic analysis: After plotting the data, the user has to define the fundamental period by setting markers (see Fig. 1.2). The program then calculates the expansion coefficients $a[n]$ for the cosines and $b[n]$ for the sines from the fundamental frequency $n = 1$ up to the 20th harmonic. As soon as a coefficient has been found to be significant (as compared to a preset cutoff) its partial wave is added to the present stage of expansion (see Fig. 1.3). When the agreement between this expansion and the measured data is perfect, only one trace will be visible (see Fig. 1.2). After completion, the result is presented as a table of coefficients $a[n]$ and $b[n]$ and the combined spectral components $s[n] = \sqrt{(a[n]^2 + b[n]^2)}$. Finally, the latter are displayed in a bar diagram (see Fig. 1.3). The fundamental frequency is shown in the upper right corner.

Our programs can be obtained from NEVA [1.7]; many similar programs should work as well. A minimal program in TURBO PASCAL 3 is listed in Appendix 1.A.

1.3 Experiments

1.3.1 Simple Harmonic Wave

Make a recording of the undisturbed tuning fork and measure its frequency. Make a Fourier analysis using first one and then several of its periods as fundamental period for the expansion.

In Fig. 1.3 only the 4th harmonic is present. Its frequency of 4*110 Hz agrees perfectly with the standard tone a^1 of the tuning fork.

Repeat the analysis with slightly misplaced markers for the period and also with 4.5 periods. It seems to be a frequent misconception that the result of the latter analysis should consist of equal contributions of the neighbouring frequencies $4f$ and $5f$. Here one can learn that setting the markers *defines* the period and thus the (infinitely long) signal to be analysed. With 4.5 periods repeated indefinitely, the signal shows discontinuities; its spectrum contains all harmonics from f to $20f$, not just the neighbours.

1.3.2 Beats

a) *Theory*

Most elementary discussions [1.4] treat beats as the superposition of

$$y_1 = A\sin(2\pi f_1 t) \quad \text{and} \quad y_2 = A\sin(2\pi f_2 t) \tag{1.1}$$

with equal amplitudes A and slightly different frequencies f_1 and f_2. Their superposition yields

$$y = y_1 + y_2 = 2A\cos\left[\pi\left(f_1 - f_2\right)t\right]\sin\left[\pi\left(f_1 + f_2\right)t\right] \ . \tag{1.2}$$

The envelope is a simple cosine with the beat frequency $f_1 - f_2$. Normally this phenomenon is only nearly periodic. For our Fourier analysis however, we need a strictly periodic function. Show that this is true only if f_1 and f_2 are commensurate:

$$f_1 = n\left(f_1 - f_2\right) \quad (n = \text{natural number}) \ . \tag{1.3}$$

Plot the above functions (either with EVALUATION or with an extra little program in BASIC or PASCAL, see Appendix 1.A) with commensurate and noncommensurate frequencies (see Fig. 1.4).

Repeat these plots with different amplitudes, keeping all other parameters unchanged (see Fig. 1.5).

Comparing the envelopes of these two phenomena reveals an apparent discrepancy: the cosine of the upper signal has a wavelength of about 14 units while the harmonic (?) envelope of the lower signal posesses a wavelength only half as long. This motivates a thorough analysis using a vector representation of the two oscillations; this yields

$$y = A_1\sin\left(2\pi f_1 t\right) + A_2\sin\left(2\pi f_2 t\right) = A\sin\left(2\pi f_1 - \Phi\right) \tag{1.4}$$

with the amplitude

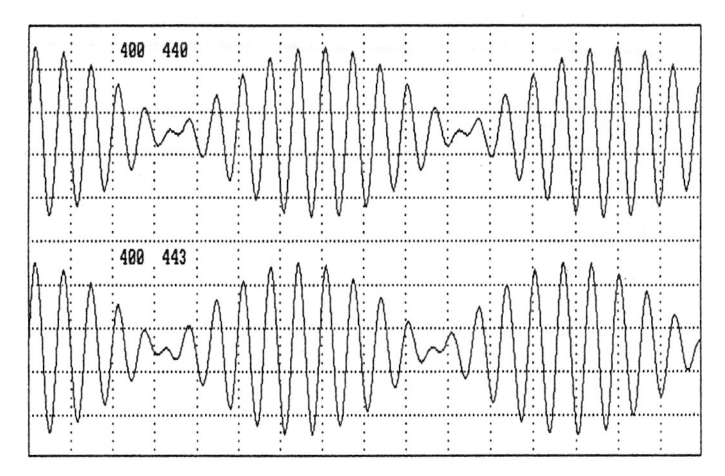

Fig. 1.4. Strictly and nearly periodic beats

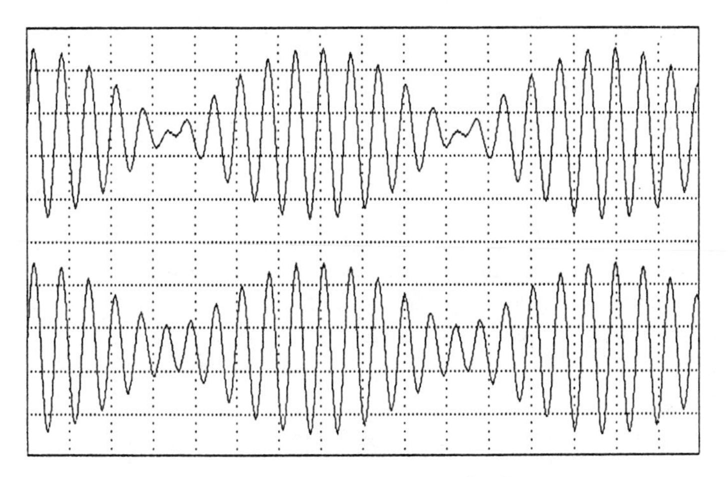

Fig. 1.5. Beats with equal and unequal amplitudes

$$A = \sqrt{(A_1^2 + A_2^2 + 2A_1A_2 \cos[2\pi(f_1 - f_2)t])} \tag{1.5}$$

and the phase

$$\tan \Phi = \frac{A_2 \sin[2\pi(f_1 - f_2)t]}{A_1 + A_2 \cos[2\pi(f_1 - f_2)t]} \,. \tag{1.6}$$

Since the phase is time dependent, the general beat phenomenon does not even have a well-defined frequency anymore!

Write a little program that plots this envelope for A_2/A_1=0, 0.2, 0.4, 0.6, 0.8 and 1. This shows how the general amplitude (1.5) reduces to the special case (1.2). Can you also show this analytically? Figure 1.6 shows the result.

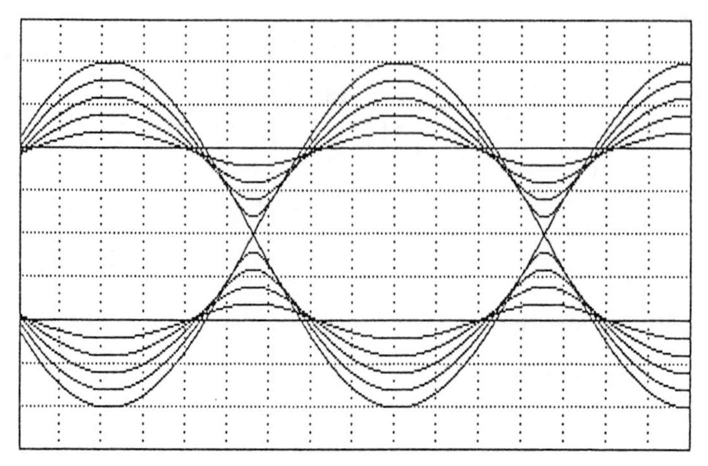

Fig. 1.6. Beat amplitudes for A_2/A_1=0, 0.2, 0.4, 0.6, 0.8 and 1

For small ratios A_2/A_1 the envelope resembles harmonic modulation. With growing A_2/A_1 the minima get sharper, and for equal amplitudes the upper envelope connects to the lower envelope to yield the elementary cosine function (1.2). This explains the apparent discrepancy.

b) *Experiment*

Reduce the frequency of one of the tuning forks as much as possible by placing both masses close to its ends. Record the beats. Move the masses carefully until the beats are strictly periodic. Repeat recording until you have some 'nice' beats (as in Fig. 1.7) with the maximum nearly filling the screen and the minimum less that 20 %. Save this data for later evaluation.

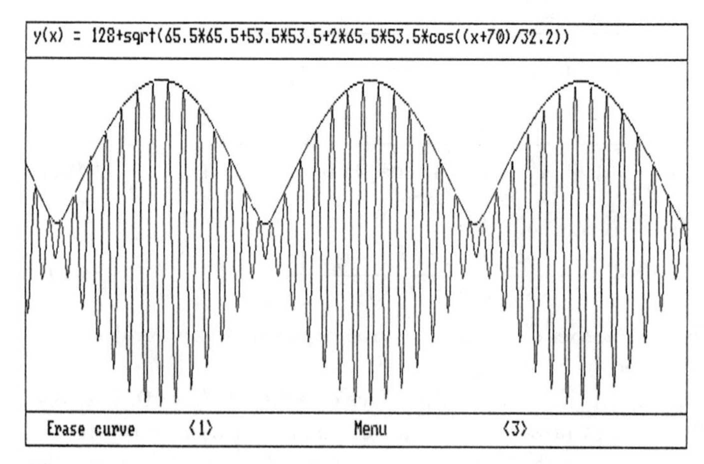

Fig. 1.7. Two tuning forks with envelope according to formula (1.5). The parameters in the line on top are best fits

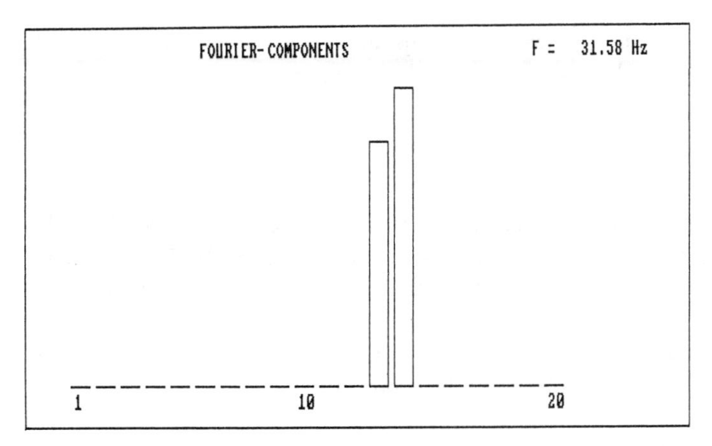

Fig. 1.8. Two tuning forks. The spectrum contains f_1 and f_2 only

Record the tuning forks separately and determine their frequencies. Calculate the beat period and compare the result with the recorded data set. Use the program EVALUATION to determine the largest and smallest amplitudes and calculate A_1 and A_2 from these values. Use the general form of the amplitude (1.5) to fit an envelope to the beats (see Fig. 1.7).

Perform a Fourier analysis now and compare the frequencies and the amplitudes with the values determined above (see Fig. 1.8).

Here are some typical results:

individual forks:	$f_1 = 440.0$ Hz $f_2 = 408.1$ Hz	
	calculated beat period: 31.4 ms	
recorded beats:	beat period:	31.7 ms, amplitudes: $A_1 = 66 \quad A_2 = 53$
Fourier analysis:	Frequencies:	$f_1 = 13{*}31.58$ Hz $= 410.5$ Hz
		$f_2 = 14{*}31.58$ Hz $= 442.1$ Hz
	amplitudes:	$A_1 = 65 \quad A_2 = 53$

1.3.3. Amplitude Modulation

a) *Theory*

Amplitude and frequency modulation are the most important methods of impressing information onto a high frequency carrier wave [1.8, 9]. Denoting the carrier amplitude and frequency by V_c and f_c and the modulation amplitude and frequency by V_m and f_m, the amplitude modulated carrier is described by

$$V = [V_c + V_m \sin(2\pi f_m t)] \sin(2\pi f_c t) \ . \tag{1.7}$$

The ratio $m = V_m/V_c$ is called the degree of modulation. Using well-known addition theorems, one can rewrite this as

$$V = V_c \sin(2\pi f_c t) + \tfrac{1}{2} V_m \sin[2\pi (f_c - f_m) t] - \tfrac{1}{2} V_m \sin[2\pi (f_c + f_m) t] \ . \tag{1.8}$$

Thus the carrier is accompanied by an upper and a lower side band with frequencies $f_c + f_m$ and $f_c - f_m$: the nonlinear process (1.7) generates new frequencies.

b) *Experiment*

Set the function generator to a sinusoidal voltage of 500 Hz. Modulate the signal with some volts from the transformer connected to the 50 Hz mains. Adjust the voltage until you have a nicely modulated signal as in Fig. 1.9. Save the data on disk.

Notice how well the spectrum (Fig. 1.10) agrees with the theoretical prediction (1.8): A carrier of 501.8 Hz with two side bands of equal amplitude

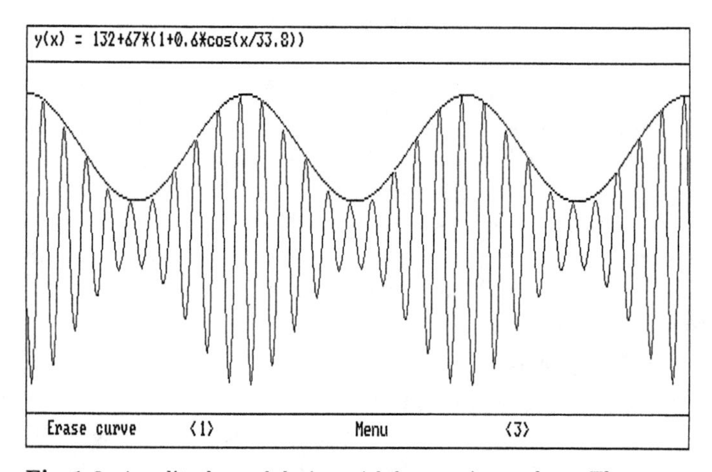

Fig. 1.9. Amplitude modulation with harmonic envelope. The parameters in the formula on top are best fits

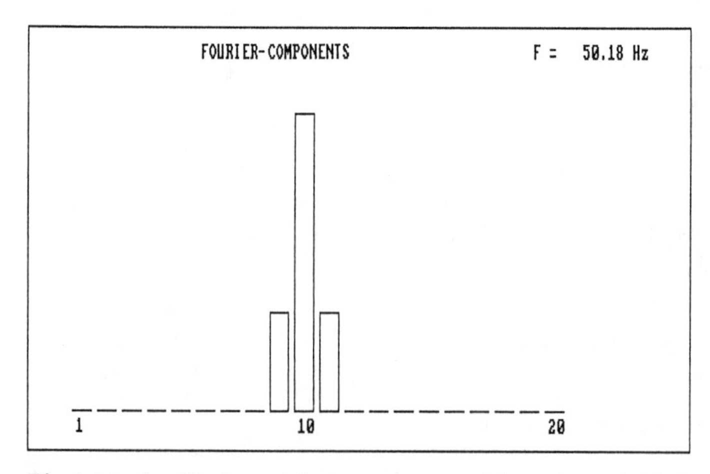

Fig. 1.10. Amplitude modulation, spectrum with carrier and side bands

separated by 50.18 Hz. (The frequency error of 0.4 % is, of course, due to the limited resolution).

If the modulating voltage is not harmonic, then a whole spectrum of side bands will appear. If a second function generator is available, you might repeat the experiment with rectangular modulation:

Figure 1.11 shows a strictly periodic amplitude modulation (2500 Hz modulated with a rectangular wave of 250 Hz). This modulating wave may be considered as a superposition of odd harmonics (250, 3*250, 5*250 etc., see Sect. 1.3.4). The spectrum of Fig. 1.12 therefore contains upper and lower side bands of just these frequencies. There are no even side bands present.

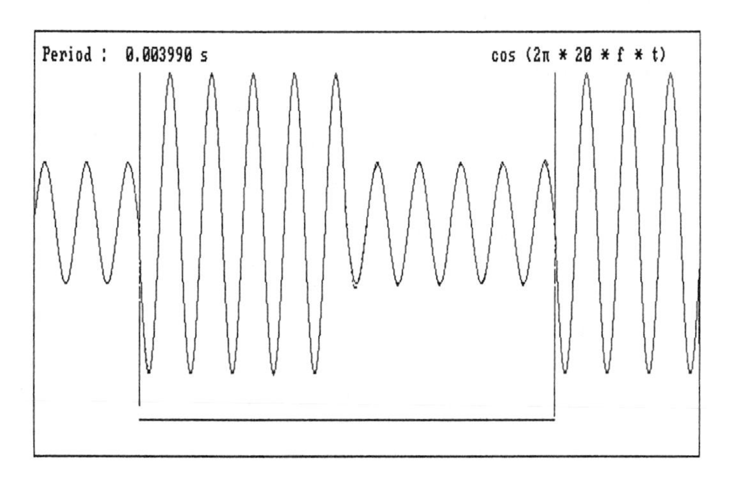

Fig. 1.11. Amplitude modulation (rectangle)

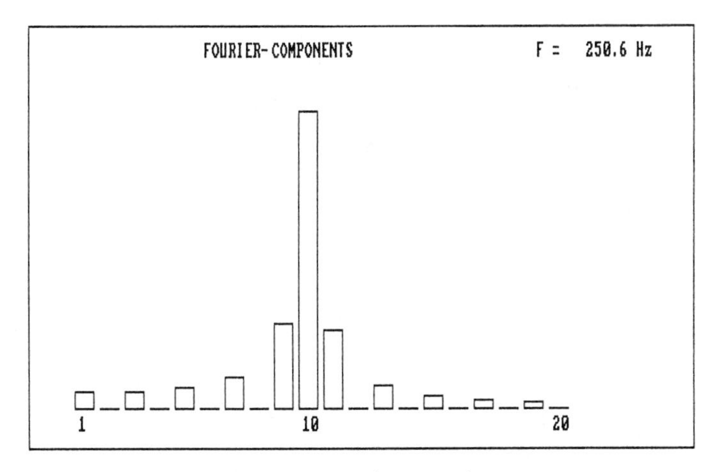

Fig. 1.12. Amplitude modulation (rectangle), spectrum

1.3.4 Rectangles

a) *Theory*

The Fourier expansion of a rectangular (or square) wave with equal on/off-times (50 % duty cycle) is a standard example in practically any book on mathematical physics [1.1–4]: only odd harmonics ($n = 1, 3, 5, 7 ...$) are present, and their amplitudes are proportional to $1/n$. If the signal is anti-symmetric (compared to the selected basis), then only the sines (i.e. $b[n]$) are present and all coefficients have the same sign. If the signal is symmetric, then there are only cosines in the expansion (i.e. $a[n]$) and their signs alternate.

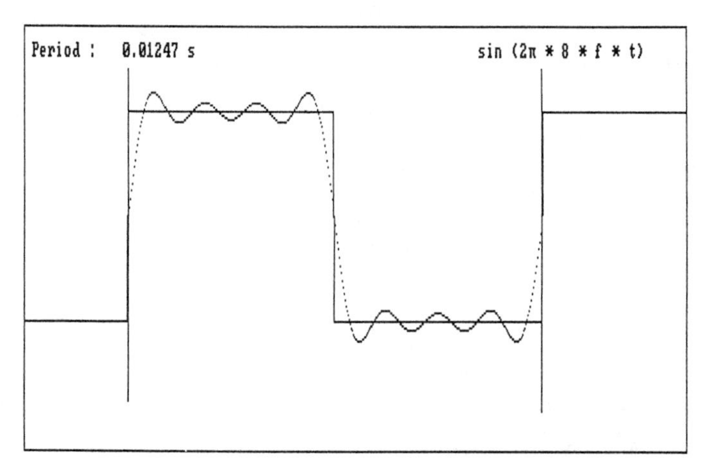

Fig. 1.13. Rectangle, 8 harmonics

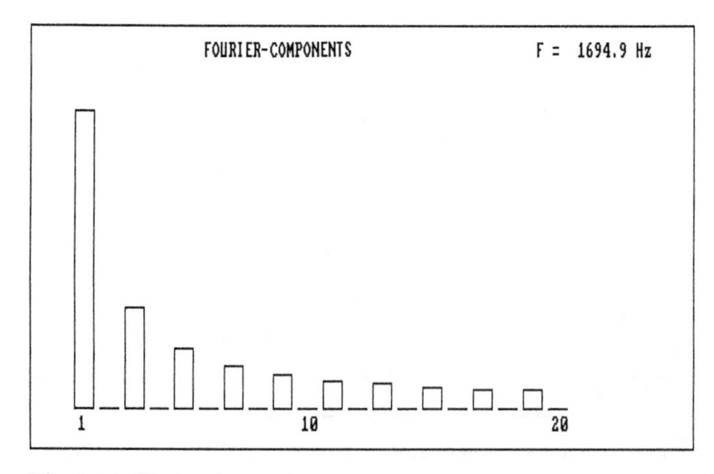

Fig. 1.14. Rectangle, spectrum

Use the function generator to record a nice rectangular wave. Save the data on disk. Make a Fourier expansion. Be careful to select equal on/off times! Compare the coefficients with the expected values. Figures 1.13 and 1.14 show the results.
The tabulated expansion coefficients are:

n:	1	2	3	4	5	6	7	8	9	10
$b[n]$:	89.1	0.0	29.7	0.0	17.8	0.0	12.7	0.0	9.9	0.0
n:	11	12	13	14	15	16	17	18	19	20
$b[n]$:	8.1	0.0	6.8	0.0	5.9	0.0	5.3	0.0	4.7	0.0

For this analysis, the numerical cutoff that differentiates between significant results and computational artefacts (see Sect. 1.2) has been set to 1.5. With this cutoff equal to zero there would have been cosine contributions of 0.7 at the odd frequencies. If you compare this tabulation with the predicted $1/n$ behaviour you see that even this exceedingly simple algorithm (compare Appendix 1.A) yields perfect results!

It is worthwhile to synthesize these values to regain the original signal. For this the program EVALUATION may be used (as in Fig. 1.15) or a small extra program in BASIC or PASCAL.

At discontinuities a Fourier synthesis displays a famous peculiarity, a marked overshoot known as Gibbs phenomenon [1.2]. This is clearly visible in an expansion containing harmonics up to 20 (see Fig. 1.16):

c) *Fourier Integral*

Only periodic functions can be represented by a (discrete) Fourier series. For non-periodic functions one needs a continuous sum of harmonics, the Fourier integral.

This transition from Fourier series to Fourier integral may be illustrated in the following way: record one narrow rectangular pulse, e.g. 30 pixels wide (out of the 640 of our CGA screen). If the basic period for the expansion is set twice as wide as the rectangle (n=2), then a square wave is defined as discussed in Sect. 1.3.4 a) and b). If the interval is doubled (n=4), then the fundamental frequency is only half as large. The spectral frequencies lie twice as close, and the spectrum contains also odd harmonics (see Fig. 1.17). If one continues this procedure (n=8 and n=12), then one separates the rectangle further and further from its neighbour; in the limit one approaches an isolated rectangular pulse with a continuous spectrum.

The spectra in Fig. 1.17 reveal an interesting feature: the nth harmonic is always missing! This can easily be understood: for n=8, e.g., the 8th harmonic will have 1 wave within the rectangle (and thus the contribution will average out) and 7 waves in the empty part of the interval (and thus contribute nothing).

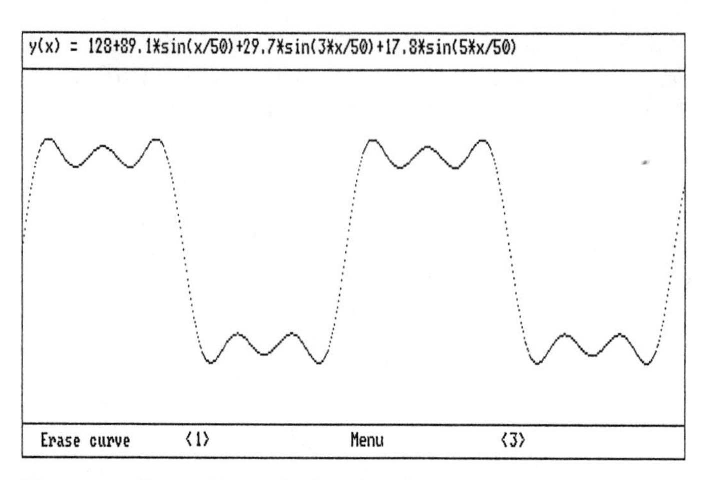

Fig. 1.15. Rectangle, synthesis using the coefficients determined by FOUR-ANA (see table above)

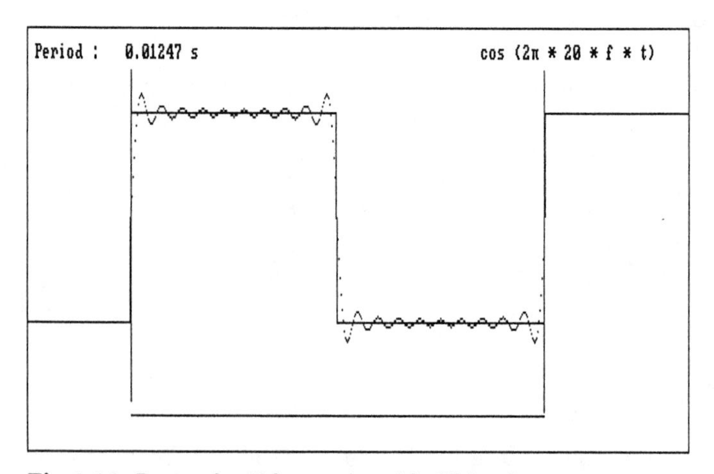

Fig. 1.16. Rectangle, 20 harmonics with Gibbs phenomenon

1.4 Didactic and Pedagogical Aspects

For the last eight years, we have treated these topics using different approaches:

a) Our undergraduate program contains a short interfacing course. Students write their own programs (originally with a C64 in BASIC, now with a PC in TURBO-PASCAL). Fourier analysis is a standard task for the better students; the application is mostly music.

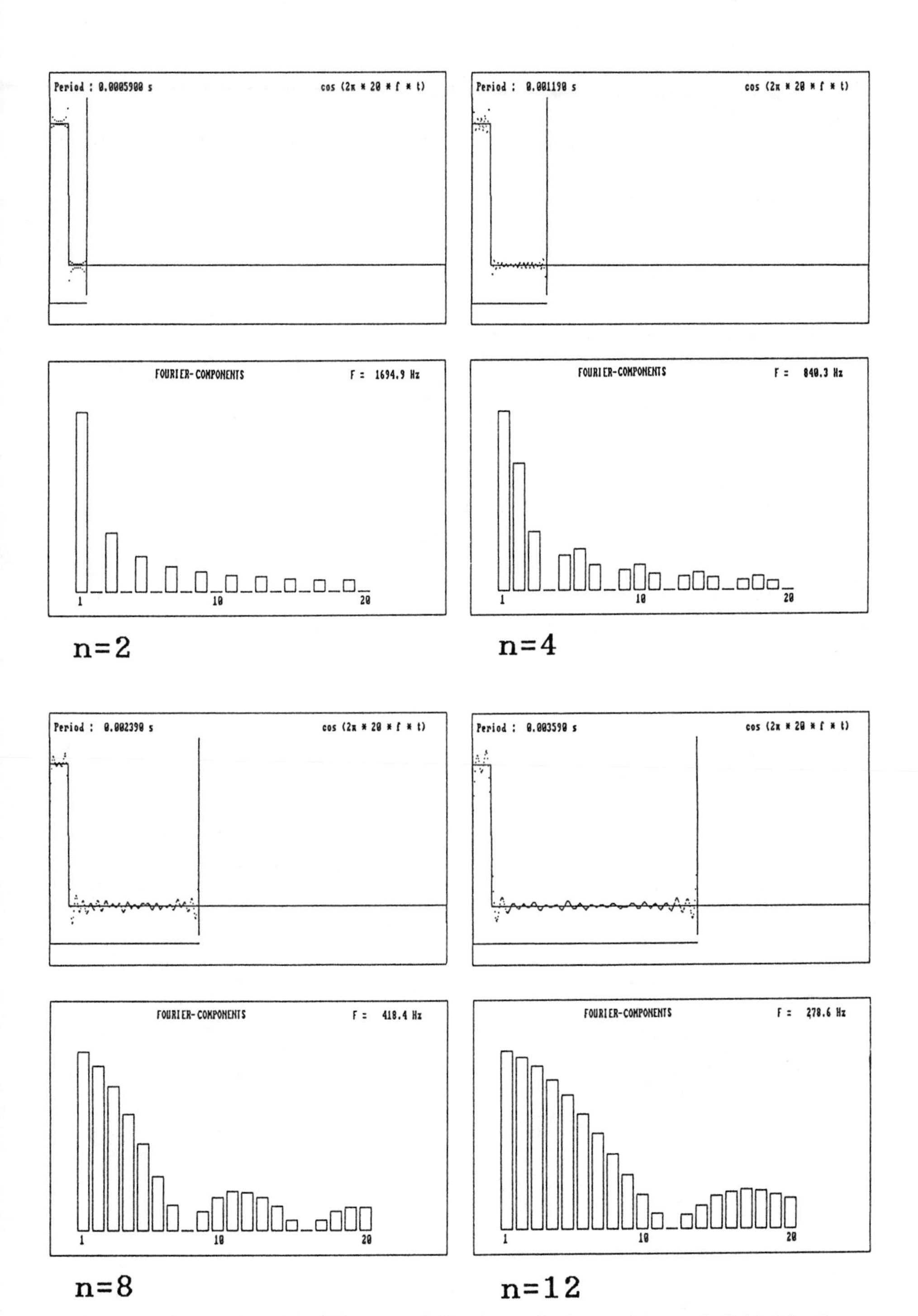

Fig. 1.17. Rectangles with different on/off-ratios. n is the total interval divided by the width of the rectangle

b) In the otherwise 'computer-free' part of the laboratory, beats as described above are a separate experiment.
c) In a special course for future physics teachers, we cover all the aspects described in this article using the menu-controlled programs.
d) The Fourier analysis program is regularly used in our experimental lectures (with a 3-colour video projector).

The student reaction has always been favourable to enthusiastic.

If one wants to let a larger number of students work on these subjects without being able to provide each computer station (used by 2, at most 3 students) with the necessary hardware, then it is quite feasible to record one set of data on disk and hand it to them for evaluation. It is even possible to work with artificial 'data' computed analytically and stored in the data format.

Some readers may wonder why we keep on using CGA graphics rather than the much finer resolution available with VGA. Our experience is that screens should contain coarse block graphics and not too much text to be effective in teaching. This is epecially true if several students have to watch the display (even if the data is being projected!). CGA is also the only graphic mode that is supported by GWBASIC and TURBO PASCAL 3 (see Appendix 1.A).

References

[1.1] G. Joos, E.W. Richter: Höhere Mathematik für den Praktiker (Barth, Leipzig 1968)
[1.2] E. Kreyszig: Advanced Engineering Mathematics (John Wiley, New York 1972)
[1.3] C.R. Wylie: Advanced Engineering Mathematics (McGraw-Hill, New York 1960)
[1.4] E. Hering, R. Martin, M. Stohrer: Physik für Ingenieure (VDI-Verlag, Düsseldorf 1988)
[1.5] M.R. Spiegel: Fourier-Analysis (Schaum's Outline) (McGraw-Hill, Düsseldorf 1976
[1.6] NEVA GmbH & Co, Postfach 1362, D-7340 Geislingen an der Steige
7890 Multifunktionskarte 6221 Kondensatormikrofon
3506 Satz von zwei Stimmgabeln 7217 Funktionsgenerator
[1.7] NEVA, Disks 7804,11 ,12 and ,13. Further descriptions in the LOSEBLATTSAMMLUNG 7801,80-,85.
[1.8] Bergmann-Schaefer: Lehrbuch der Experimentalphysik, Elektrizität und Magnetismus (de Gruyter, Berlin 1987)
[1.9] F.R. Connor: Modulation (Edward Arnold, London 1982)

Appendix 1.A

In this Appendix, two sample programs are shown for the theoretical discussion of beats (Fig. 1.4 and Fig. 1.6) and a minimal version of the program for doing the Fourier analysis. The more elaborate menu-driven programs described in Sect. 1.2 are very convenient; the physical problems, however, can be solved quite adequately with a much smaller amount of programming. For our work we use the language TURBO PASCAL 3. Except for one routine, everything could also be written in BASIC, but the data recording procedure (see the program FOURIER) is much too slow in BASIC – unless one could delegate this task to a routine written in machine code.

Version 3 is quite sufficient to tackle all problems and much less specialized than the later versions. Starting the graphics is already quite a formidable task in version 5! As a physicist who wants to use the computer as a tool rather than make it a hobby, one might define the quality of a language as 1/(thickness of the manual); from this point of view, TURBO PASCAL 3 is highly recommended.

1.A1 Programs for Sect. 1.3.2 a)

```
program FIGURE_4;                      { plotting theoretical beats }

var x,y,yold : integer; y1,y2 : real;
procedure grid;
 var n,m : byte;
 begin
   for n:=0 to 15 do                   { dotted line in y-direction }
      for m:=0 to 49 do plot(n*40,m*4,1);
   for m:=0 to 9 do                    { dotted line in x-direction }
      for n:=0 to 160 do plot(n*4,m*20,1);
   draw(0,0,639,0,1); draw(639,0,639,199,1); { border rectangle }
   draw(639,199,0,199,1); draw(0,199,0,0,1);
end;

begin                                              { main program }
hires; grid;
yold:=50;
 for x:=1 to 639 do begin                      { equal amplitudes }
   y1:=20*sin(x/4)+20*sin(x/4.4); y:=trunc(50-y1);
   draw(x-1,yold,x,y,1); yold:=y end;
yold:=150;
 for x:=1 to 639 do begin                  { different amplitudes }
   y1:=25*sin(x/4)+15*sin(x/4.4); y:=trunc(150-y1);
   draw(x-1,yold,x,y,1); yold:=y end;
repeat until keypressed
end.
```

```
program FIGURE_6;                                    { beat amplitudes }

var x,y1,y2,y1old,y2old,a1,a2,k        : integer;
    a,aold                             : real;

begin                                                  { main program }
hires; grid;                           { procedure grid from above }
a1:=40;
 for k:=0 to 5 do begin            { 6 different amplitude ratios }
   a2:=k*8; y1old:=100-a1-a2; y2old:=100+a1+a2;
      for x:=1 to 639 do begin
         a:=sqrt(a1*a1+a2*a2+2*a1*a2*cos(x/44));
          y1:=trunc(100-a); y2:=trunc(100+a);
           draw(x-1,y1old,x,y1,1); draw(x-1,y2old,x,y2,1);
          y1old:=y1; y2old:=y2;
      end;
 end;
repeat until keypressed
end.
```

1.A2 Minimal Program for Fourier Analysis

Some comments relating to the program FOURIER on the following pages:

const: The factor f=0.7804 scales the 8-bit reading of the ADC to the pixels y=0 to 199.

The address adc1=$104 depends on the hardware, of course. The time-per-pixel has to be determined with the actual recording procedure using the tuning fork for calibration.

Since we limit the Fourier expansion to the first 20 harmonics, the arrays are defined as [0..20] for the cosines (this includes the constant) and [1..20] for the sines.

Procedure recording: While recording and saving the data, the keyboard interrupts have to be turned off, otherwise there are discontinuities in the time scale.

Our ADC starts a conversion whenever it is being read. Thus one always gets the value at the time of the previous reading. We therefore discard the (uncontrollable) first value with a dummy reading.

The program is so fast that the ADC may be read before a conversion is complete. To avoid erroneous readings, a suitable counting loop has to be included.

For studying synthetic data (if real measurements cannot be made) one may just assign values to the elements $y[n]$: for n:=0 to 30 do $y[n]$:=200; for n:=31 to 639 do $y[n]$:=0;

Procedure interval: In the elaborate program FOUR-ANA the fundamental period is set with the CRSR-keys and a running marker. It is much simpler, however, to read the limits off the grid (40 pixels width) and enter them numerically.

Procedure A0,Ak,Bk: When calculating the Fourier coefficients, it is quite sufficient to approximate the integrals by sums. This saves time. One may introduce a cutoff criterium and regard a coefficient as real only if it contributes more than e.g. 1 pixel (out of the 200 on the screen).

Procedure coefficients: For most applications it is quite sufficient to write the table onto the graph. One could use a separate screen, of course.

Procedure save-on-disk: If one wants to fit envelopes etc. to recorded data one may want to use a trial and error method. In this case one has to make certain that the same set of data can be used over and over again.

program FOURIER;

```
const
     f=0.7804;                                  { scaling factor }
     adc1=$ 104;                    { address of A/D-converter }
     time_per_pixel=0.00001507;      calibrated by tuning fork }

var
     i,n,m,x,Xmin,Xmax,k,kmax : integer;
     period, frequency, omega, tempo : real;
     integral                        : real;
     y: array[0..639] of byte;
     a: array[0..20] of real;
     b: array[1..20] of real;

procedure recording;
begin
      inline ($FA);                             { no interrupts }
      y[0]:=port[adc1];                         { dummy reading }
      for i:=0 to 639 do begin
        for n:=0 to 0 do begin end { >>>>> try out !! <<<<< }
       y[i]:=port[adc1];
      end;
      inline ($FB);                    { interrupts allowed again }
end;

procedure display;
begin
hires; grid;                          { procedure grid from above }
 for i:=0 to 638 do
   draw(i,trunc(199-f*y[i]),i+1,trunc(199-f*y[i+1]),1)
end;
```

```
procedure interval;
begin
  gotoXY(2,2); write('Xmin: '); readln(Xmin);
   draw(Xmin,40,Xmin,160,1);
  gotoXY(16,2); write('Xmax: '); readln(Xmax);
   draw(Xmax,40,Xmax,160,1);
    period:=(Xmax-Xmin)*time_per_pixel;
   frequency:=1/period; omega:=2*Pi*frequency;
 gotoXY(40,2);
 writeln('fundamental frequency: ',frequency:6:1,' Hz')
end;

procedure A0;
begin
integral:=0;
for x:=Xmin to Xmax-1 do integral:=integral+y[x];
a[0]:=2*integral/(Xmax-Xmin)
end;

procedure Ak;
begin
tempo:=k*omega*time_per_pixel;
integral:=0;
for x:=Xmin to Xmax-1 do
   integral:=integral+y[x]*cos(tempo*(x-Xmin));
a[k]:=2*integral/(Xmax-Xmin)
end;

procedure Bk;
begin
integral:=0;
for x:=Xmin to Xmax-1 do
   integral:=integral+y[x]*sin(tempo*(x-Xmin));
b[k]:=2*integral/(Xmax-Xmin)
end;

procedure coefficients;
begin
gotoXY(2,4); writeln('a(0)= ',a[0]:5:1,'      ');
  for k:=1 to 20 do begin
    gotoXY(2,k+4); write('a(',k,')=',a[k]:5:1,'       ');
    gotoXY(16,k+4); write('b(',k,')=',b[k]:5:1);
  end;
readln
end;
```

```
procedure synthesis;
var s: array[0..639] of real;
begin
 for x:=Xmin to Xmax do s[x]:=a[0]/2;
 for k:=1 to 20 do begin
   hires; grid;
   draw(Xmin,20,Xmin,190,1); draw(Xmax,20,Xmax,190,1);
   gotoXY(2,2); write('up to ',k,'. harmonic');
   tempo:=k*omega*time_per_pixel;
   for x:=Xmin to Xmax do begin
     s[x]:=s[x]+a[k]*cos((x-Xmin)*tempo)+b[k]*sin((x-Xmin)*tempo);
     draw(x-1,trunc(199-f*s[x-1]),x,trunc(199-f*s[x]),1)
   end;
 repeat until keypressed
 end;
readln
end;

procedure spectrum;
var max:real;
    l: array [1..20] of real;
begin
 max:=0;
   for n:=1 to 20 do begin
       l[n]:=sqrt(sqr(a[n])+sqr(b[n]));
       if l[n]>max then max:=l[n]
   end;
hires;
  for n:=1 to 20 do
    draw(100+20*n,190-trunc(l[n]*160/max),100+20*n,190,1);
 gotoXY(40,2);
          writeln('fundamental frequency: ',frequency:6:1,' Hz');
   readln
end;

procedure save_on_disk;
const
  NMax = 640;                                 { number of databytes }
  blocks = 5;                            { number of 128-byte-blocks }
var
  f : file; jn : char;
begin
gotoXY(55,25); write('save data on disk? (y/n)');
  read(kbd,jn);
  if (jn='y') or (jn='Y') then begin
    assign(f,'DATA.640');
```

```
    rewrite(f);
     blockwrite(f,y,blocks);
     close(f)
  end
end;

begin                                                    { main program }
recording;
display;
interval;
gotoXY(2,4); writeln('please wait');
A0;
k:=0;
repeat
 k:=k+1; Ak; Bk;
until k=20;
gotoXY(2,4); writeln('continue with key');
repeat until keypressed;
coefficients;
synthesis;
spectrum;
save_on_disk;
clrscr
end.
```

2. Point Mechanics by Experiments – Direct Access to Motion Data

R. Dengler and K. Luchner

With all quantitative experiments on point mechanics, there are usually special gadgets necessary for data acquisition. Here we describe a method which enables the recording and evaluation of single-point motion data simply by using a video camera and a computer (online, or from stored information); with more sophisticated equipment it is possible to record the motion of up to three different points simultaneously. Typical examples are: linear accelerated motion, ballistic curve, accelerated frame of reference, statistical motion on air table, chaotic motion, collision of two bodies, coupled oscillations, rotation in addition to translation.

2.1 Introduction

Experiments on point mechanics play an important part in physics instruction; they appear on various levels and with various instructional intentions. The student's desire to obtain or to establish more detailed descriptions or recordings of any observation may be seen as an indication of a developing scientific attitude. In the many possibilities for observation, the description of the motion of bodies plays an important part. This has led to a variety of special experimental devices to obtain motion data; they are employed in schools and universities, by teachers and students, in order to demonstrate phenomena, to introduce new concepts, to inspire the student's creativity, etc.

In the following paper we describe a unique method to record, process and display motion data: It employs a video camera and a computer and is known as ORVICO (**O**bject **R**ecording by **Vi**deo and **Co**mputer), see Fig. 2.1. and [2.1]. This method, in our opinion, offers several advantages:

a) There are no other gadgets required to pick up data from the moving body, which might distract from the point of the experiment: The primary physics process is not covered by a tricky recording method. In a certain way, ORVICO corresponds to the role of the observers eyes and brain by using the camera and computer (see Sect. 2.2.1).

b) The same apparatus applies to many different types of experiment and allows various modes of observation and evaluation (see Sect. 2.3).

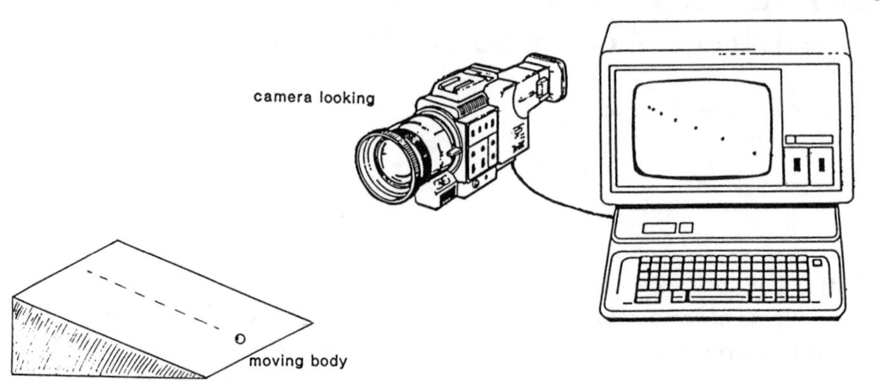

Fig. 2.1 The typical setup for ORVICO. The experiment (moving body) is observed by a video camera, and the video signal is fed to the computer. The experiment employed here (motion on an inclined plane) is only an example; a collection of other examples is given in Sect. 2.3

c) The data are directly produced for computer processing and, furthermore, the user may profit from the possibilities offered by the computer and software (see Sect. 2.2.2, 2.2.3).

2.2 ORVICO

2.2.1 Principle

As is well known, a video camera delivers electrical signals which contain the information necessary to reconstruct 50 pictures (frames) per second: For a scenic picture, this information consists of a great number of single dots of varying brightness (including colour) within each frame. For the field of single-point mechanics, however, the amount of information to be handled may be reduced considerably: In principle, the task is to observe only one single mass point, and nothing else! This leads to the idea that the "point mass" (the puck, the glider, the pendulum body etc.) is marked by a bright spot and is observed in front of a dark background. This considerable reduction of data – only two coordinate values have to be established and processed for each frame – enables fast and easy handling. Using a colour video camera, at least two separate channels (blue, red) are easily accessible, and thus the motion of two "point masses" (the one marked by a bright blue spot, the other by a bright red spot) may be recorded simultaneously.

With this simple idea, some requirements and possibilities for its practical application arise:

a) The motion to be observed has to take place within the focal plane of the camera (the method could be extended to also observe three-dimensional motion, but this requires a second camera and more sophisticated data processing; the thus widened scope, in our opinion, would not compensate for the effort).

b) It is advisable to have various modes of display, such as continuous trace of the moving body; trace as if stroboscopically illuminated; position–time diagram. In any case, the real values of time and position (calibrated optical imaging) are used.

c) The motion data thus recorded are available for further processing, such as calculating derivatives, showing distributions, finding Fourier components, etc.

d) Obviously, since the method can be instructional, the possibility of handling the data by software could also help to inspire personal ideas for application. Thus it is desirable for the user to have easy access to the software, in order to enable personal extensions.

In the following section, we firstly describe the hardware necessary to deduce the "single-point information" from the video signal, and later the handling of these data by software is briefly discussed.

2.2.2 Hardware

In order to pick up the video signal, to analyze it, and to form appropriate data for the computer, some hardware was developed. All of it is placed onto one card, the so called ORVICO card (commercially available); it is to be inserted into the computer (PC, XT or AT, plug-in unit with 62 contacts). The video signal is fed directly to this ORVICO card and the data transfer to the computer is controlled by software.

To understand the function of the ORVICO card, it is necessary to consider the structure of the video signal. As is well known, each frame consists of consecutive lines. Corresponding to the European Standard, 25 complete frames are delivered per second, each frame containing 625 lines. In order to produce pictures with suppressed flicker, the frequency of frames is doubled by simple trick: For one frame only the odd-numbered lines are used, for the next frame only the even-numbered lines are used, and so on. Thus, according to the supply frequency, the frequency of frames is 50 Hz. The ORVICO card also works with the video signal of NTSC standard; the different frequency of frames and number of lines are properly accounted for by software.

The lines are the carriers of information (voltage) on the local intensity within the frame; there are also special synchronizing pulses to mark the beginning of each line (line synchronizing pulse), and the beginning of a new frame (frame synchronizing pulse).

Let us now consider a typical signal delivered by the camera viewing one single bright spot in front of a dark background. Figure 2.2 shows an

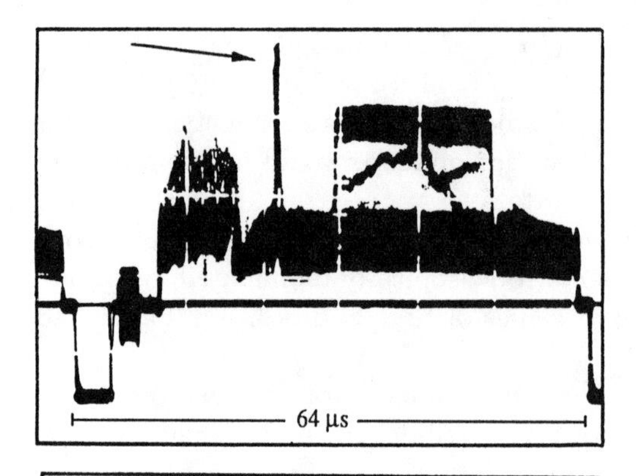

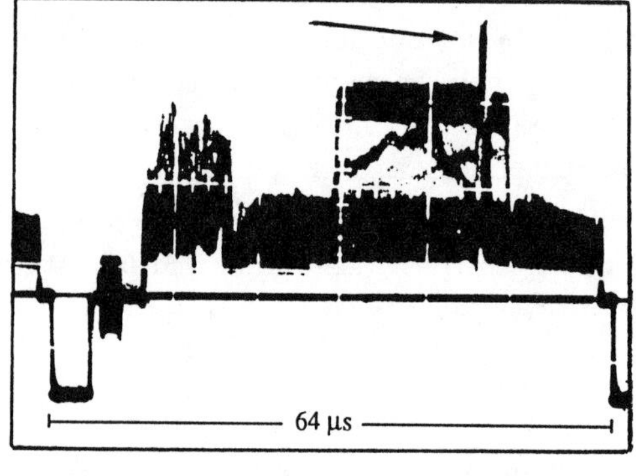

Fig. 2.2. Oscilloscopic traces of the video signal (on line) carrying the information on the position of one bright spot. A comparison of 'a' with 'b' shows that the spot has changed position (see arrows)

oscilloscopic trace of the electronic signal imposed on one of the 625 lines. The marked pulse arises from the bright object spot; in the following, it will be called the "object pulse".

From this video signal, the ORVICO card extracts the line synchronising pulse and the frame synchronising pulse, and also detects and localizes the object pulse: For each frame, it identifies the (x, y) coordinates of the bright spot (schematic diagram, see Fig. 2.3).

The line pulses and frame pulses are amplified and fed to an integrated circuit which directly separates them (standard procedure, as usually employed in TV sets). To identify the object pulse, a fast comparator is used. Its detection threshold value comes from a D/A converter controlled by software and keyboard. It is advantageous in that there is no potentiometer necessary and that the threshold value (which is displayed on the screen) is of high accuracy and can be reproduced easily.

Further processing is done by digital components.

A flip-flop called "status" is set to "low" by the frame pulse and to "high" by the object pulse. From the beginning of the frame, the "y counter" counts the line synchronizing pulses up to the very line carrying the object pulse; thus the y coordinate of the object is identified. This counter is reset by the frame synchronizing pulse. At the instant when "status" changes from

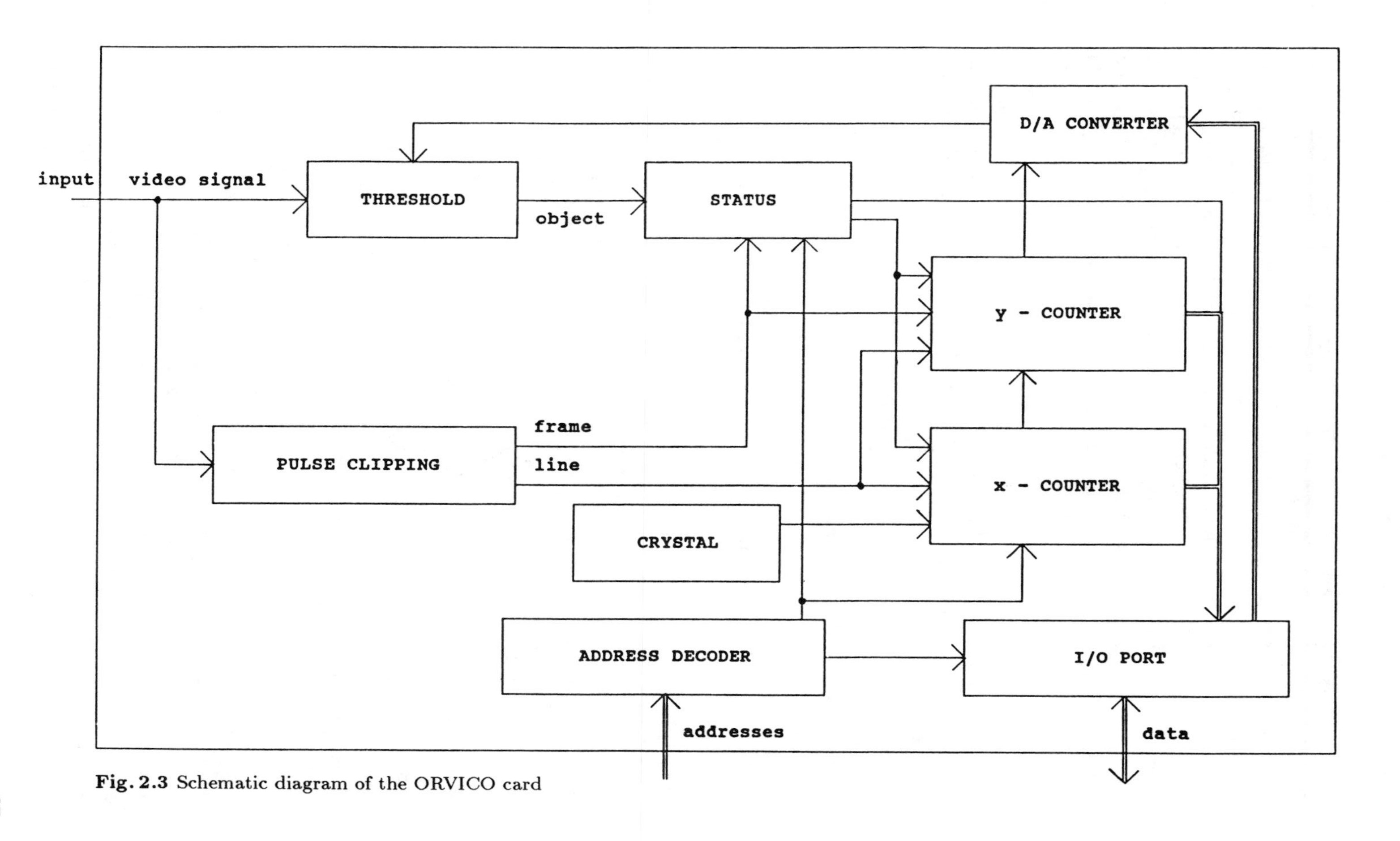

Fig. 2.3 Schematic diagram of the ORVICO card

"low" to "high", the actual value of the "y counter" is stored internally. As a 8-bit counter is used for this purpose, the upper 255 lines of the frame are handled. The remaining lower part (57 lines) is not recorded. Thus, in the resulting screen picture there is a free space, which is used to display additional information, such as working mode, threshold value, etc.

The x coordinate is identified by the "x counter" (10 bit), which measures the time between the line synchronizing pulse and the object pulse; it is reset by the line pulse, and it counts the pulses provided by an internal crystal oscillator. Only the counting result for the very line carrying the object pulse is important. For this reason, the x counter is also controlled by "status".

The flip-flop of "status" is not reset before the beginning of the next frame. Thus it is ensured that other bright spots in the lower part of the frame (the object has already been detected!) are ignored; they do not disturb the former recording.

The (x, y) coordinates found in this way are transferred to the computer when "status" is "high".

The complete data exchange (status, measured values, threshold value) is done via the "I/0 port" which is controlled by the address decoder. The accuracy of the data thus obtained is mainly limited by the quality of the camera and by the discrimination of the object against the background.

2.2.3 Software

The job to be done by the software is two-fold:

Firstly, it has to control and manage the data acquisition. There is a flow of data in two directions: From the computer to the ORVICO card, where the threshold level (to be adjusted by the keyboard) is transmitted in order to discriminate the object pulse from the background; the proper discrimination is adjusted by checking the screen. In the other direction, the computer receives the results of the x and y counter from the ORVICO card.

Secondly, it has to take care of data processing, storage and display. In writing up the program items, we have been led by the demands usually arising with instructional applications:

– The representation of the recorded motion in true coordinates. The proper calibration is obtained by recording two special positions of the point object and entering their real distance via the keyboard. The coordinates thus established are available for any further calculated displays.

– Various types of screen representation are possible "on line" (simultaneously) with the observed experiment, but also from earlier memorized recordings. These include the trace (x, y) of the moving body within the focal plane; due to the frame frequency this trace comes out as if the moving object was illuminated by a stroboscopic flashlight of the same frequency.

The representations $x(t)$ and $y(t)$ are also possible. In any case, the "stroboscopic" points may be connected by short straight lines, which leads to an almost "continuous" trace.

– These data can be processed to obtain and represent diagrams of certain functions which usually appear in context with point mechanics: $\dot{x}(t)$, $\dot{y}(t)$, $(\dot{x}^2 + \dot{y}^2)^{1/2}$, $\ddot{x}(t)$, $\ddot{y}(t)$, etc. As a nontrivial application, for a motion with change of direction the local acceleration (force acting) may be calculated from the recorded data, and the corresponding vectors may be shown overlain on the trace picture.

– More sophisticated graphs are also possible, such as polar coordinates $r(t)$ and $\varphi(t)$ with respect to an arbitrary centre, the Fourier spectrum (for a given time basis) of any periodic process, and the distribution function of statistical events.

– There are additional facilities concerning evaluation, handling and representation. They include enlarging details, simultaneous imaging of different sections, labelling of coordinates, help etc..

The menu structure guides the user to manage these possibilities and leads to various submenus. There also is a possibility to insert additional program parts (Turbo Pascal) according to individual ideas.

2.3 Examples

The examples described in this chapter are chosen to show the applicability of ORVICO to various instructional levels, and that it may be used by the teacher and by the student. It was not intended to give a complete collection of all possible applications in this chapter; instead, it is hoped that the reader feels inspired to continue with his own ideas.

2.3.1 Ballistic Motion

Ballistic motion is a typical example which can be treated at various levels. Starting with prescience experience, as a first step to a scientific treatment there is the need to characterize and quantify the trajectory of the thrown body. How is this usually done? One common way is to produce a photograph of the trajectory with long-time-exposure and stroboscopic illumination, and to evaluate it by hand.

With ORVICO, the same idea applies, but there are several advantages. The camera itself produces the stroboeffect (frequency of frames) and the data are available not only at the screen (as in the photograph) but also in a memorized form for further processing.

Figure 2.4 shows a "stroboscopic" trace of a small body, thrown by hand, observed and recorded by ORVICO. This experiment, obviously, may be performed as a lecture demonstration (either "on line" or several trials col-

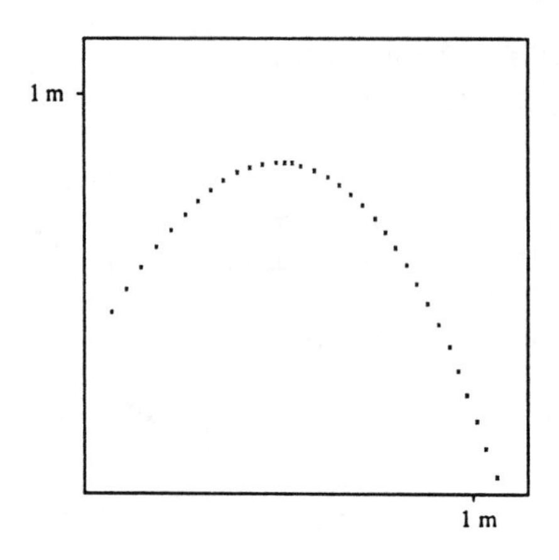

Fig. 2.4. A typical ORVICO recording: Trajectory (x, y) of a ballistic motion. The stroboscopic effect arises from the frame frequency; the time spacing between consecutive points is 20 ms. The scaling of the axes was obtained by previous calibration

lected and reproduced from the computer memory), or as a laboratory task (with the aim to have the student produce an evaluation or a verification of the laws of motion); since it is not necessary to receive the video signal directly from the camera, a portable video recorder may also be employed. Thus it is also possible to produce outdoor recordings (sports, vehicles) and take them to class for evaluation.

The evaluation of a stroboscopic trajectory as in Fig. 2.4 will firstly have to be made "by hand": How is $x(t)$, $y(t)$, $y(x)$, $\dot{x}(t)$, etc.? With this tedious activity, the desire for more comfortable handling arises: "All the necessary data are already stored in the computer. It should be possible to have the computer perform these tedious calculations ...". Thus access to software production or, if already present as in our case, for proper software application arises.

Figures 2.5, 2.6 and 2.7 show such software-produced evaluations of the trace shown in Fig. 2.4. One of the results is the order of magnitude for the vertical acceleration.

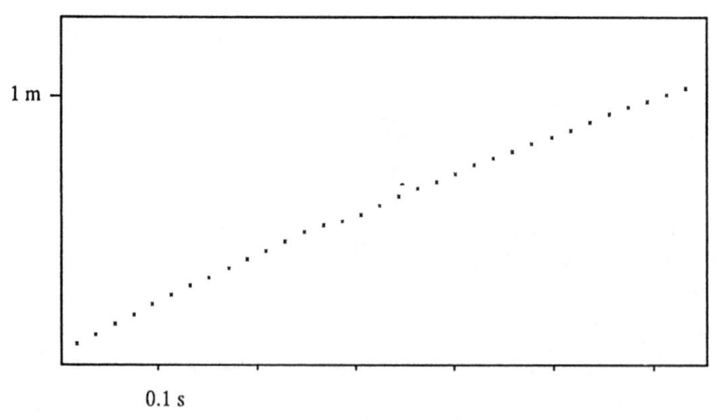

Fig. 2.5 The horizontal component $x(t)$, taken from Fig. 2.4

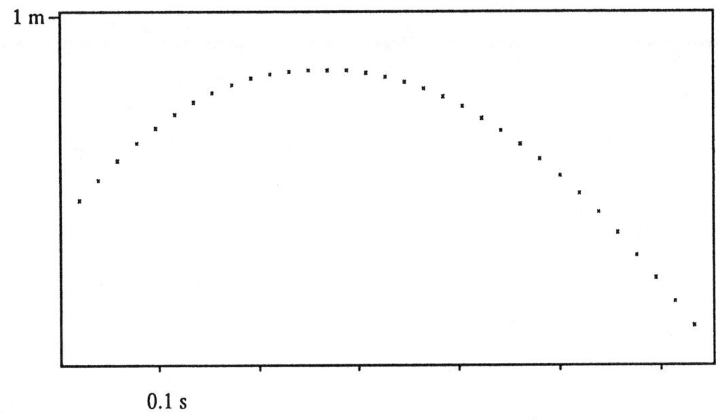

Fig. 2.6 The vertical component $y(t)$, taken from Fig. 2.4

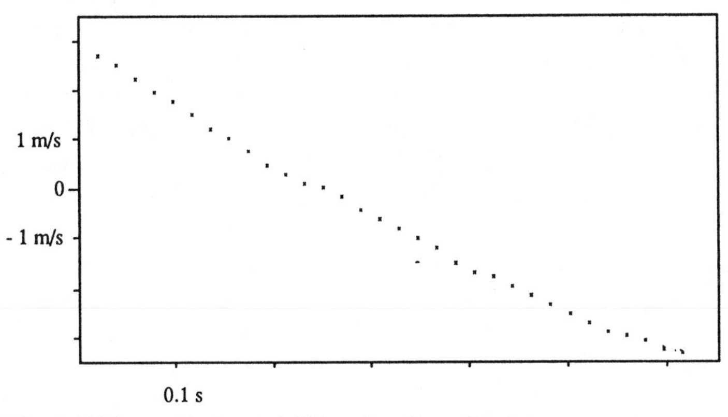

Fig. 2.7 The vertical speed $\dot{y}(t)$, taken from Fig. 2.4

In a similar way, several other experiments, not described here, may be recorded: motion on the air track, oscillations, a bouncing ball. Unusual applications seem to yield interesting information: motion of parts of the human body while walking, running or jumping.

2.3.2 The Rigid Pendulum

The mass point is forced to move along a circular path by a rigid rod, in a vertical plane (rotational axis is horizontal). In this example, again the stroboscopic picture gives a preliminary qualitative impression of the local speed. Fig. 2.8 shows a recording for a very specific initial condition. The pendulum is started at the point of instable equilibrium with almost vanishing initial velocity. At first, only the consecutive stroboscopic points should be considered: one recognizes the circular path and the changing speed. Obviously, this is an accelerated motion, and the total force producing it is

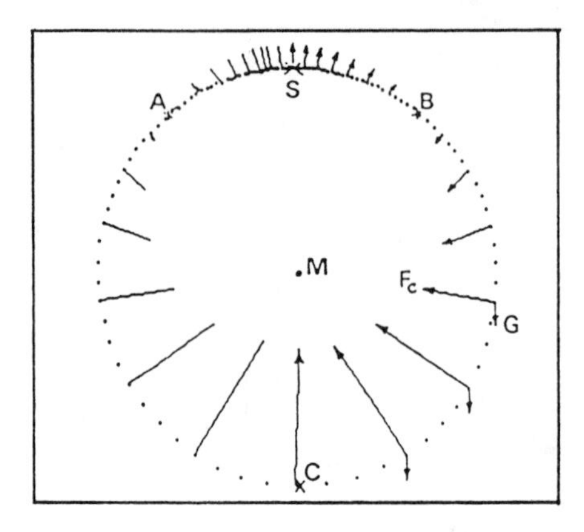

Fig. 2.8. Stroboscopic recording of the motion of a rigid pendulum. Arrows show the constraint forces F_c and the weight G

found by $\boldsymbol{F}_{\text{tot}} = md^2\boldsymbol{r}(t)/dt^2$, a calculation done by the computer. Subtracting the weight $\boldsymbol{G}$ of the body from $\boldsymbol{F}_{\text{tot}}$, the force exerted by the rod (constraint force $\boldsymbol{F}_c$) is obtained. The vectors in Fig. 2.8 indicate the direction and magnitude of $\boldsymbol{F}_c$ as obtained from the values $r(t)$ of the recording. There are several special cases to be seen: point S (the rod pushes upwards to compensate for the weight), point C (where the rod has to pull to provide the centripetal force and to compensate for the weight) and points A and B (no constraint force necessary; these are exactly the points where a free rolling body would jump off the track). It is a well-known exercise problem to calculate F_c as a function of the polar angle φ; the result is $F_c = G(2+3\cos\varphi)$. Figure 2.9 shows F_c as a function of $\cos\varphi$, derived from the measured values.

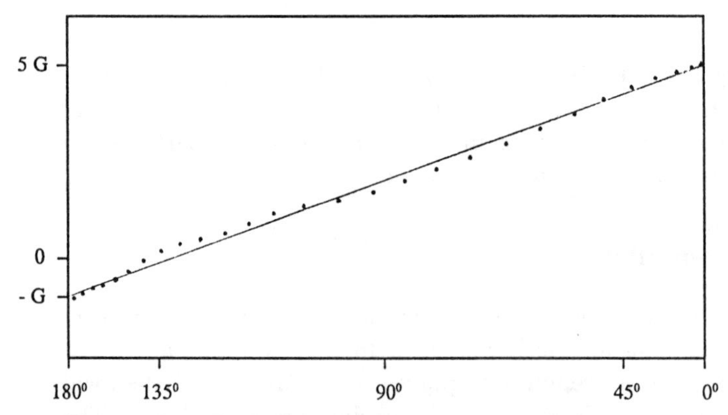

Fig. 2.9 Magnitude of constraint force F_c as a function of $\cos\varphi$, derived from Fig. 2.8 (points). $\varphi = 0$ is defined with point C from Fig. 2.8. (Solid curve: theoretically calculated)

2.3.3 Frame of Reference

The frame of reference in all ORVICO recordings obviously rests within the camera (in the examples described above, the camera was at rest in the laboratory). All consequences concerning moving frames of reference may be drawn from the data delivered by a moving camera. As an example, we show the fictitious forces derived from the data delivered by a rotating camera.

The camera is fixed to a rotating table (rotational axis is directed vertical; angular speed ω) looking vertically upward. Above the camera, a horizontal air track is mounted; the slider on it is marked with a bright spot visible to the camera, and it moves with constant speed along the air track. Figure 2.10 shows this motion seen by the rotating camera.

With this stroboscopic picture the interpretation towards the appearance of fictitious forces may performed, either point by point evaluating for apparent speed ($\boldsymbol{v}' = d\boldsymbol{r}/dt$), and force ($\boldsymbol{F}_{\text{tot}} \sim d^2\boldsymbol{r}/dt^2$), or directly by using the stored data and software. The result of the latter evaluation is displayed by the inserted arrows. The total force, $\boldsymbol{F}_{\text{tot}}$ (heavy arrows) in each instant can be represented by two components: A force $\boldsymbol{F}_{\text{cf}}$ which always points outwards from the centre of rotation (M), and a force $\boldsymbol{F}_{\text{co}}$ which always is perpendicular to $\boldsymbol{v}'$, the apparent speed. $\boldsymbol{F}_{\text{cf}}$ is the centrifugal force (note that its magnitude comes out to be proportional to the magnitude of $\boldsymbol{r}$; the dependence on ω only can be shown with an additional recording, employing another ω for the table; the total outcome is $F_{\text{cf}} \sim r\omega^2$); $\boldsymbol{F}_{\text{co}}$ is the Coriolis force (note that its magnitude comes out to be proportional to v'; again to show the dependence on ω another recording is required; the total outcome is $F_{\text{co}} \sim v'\omega$).

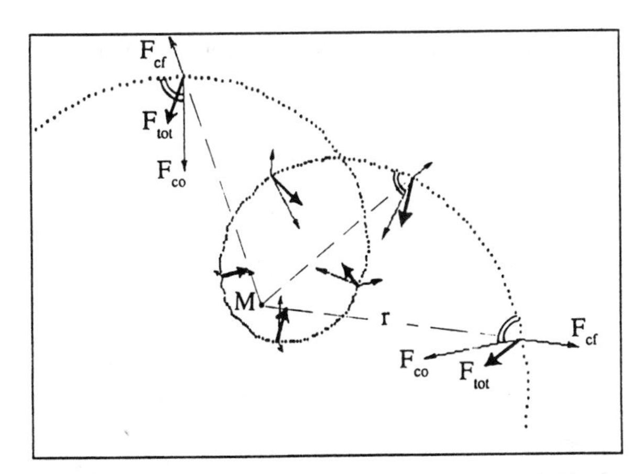

Fig. 2.10 Rotating camera: Stroboscopic trace of a body moving with $v = \text{const.}$ in an inertial system. Arrows are calculated by computer, but optically enhanced by hand (the screen display does this by using different colours). $\boldsymbol{F}_{\text{tot}}$ as derived from $d^2\boldsymbol{r}/dt^2$; $\boldsymbol{F}_{\text{cf}}$ (pointing outwards from centre of rotation, M) and $\boldsymbol{F}_{\text{co}}$ (perpendicular to trace) in their magnitudes are defined such that $\boldsymbol{F}_{\text{cf}} + \boldsymbol{F}_{\text{co}} = \boldsymbol{F}_{\text{tot}}$. The result is $F_{\text{cf}} \propto r\omega^2$; $F_{\text{co}} \propto v'\omega$

This inductive way of treatment cannot replace the usual theoretical considerations, but will help to induce a better type of physics intuition, which perhaps sometimes is not cultivated carefully enough.

2.3.4 Statistical Motion on an Air Table

One puck is marked with a white dot, and its motion amongst 10 other pucks on an air table ($1\,\mathrm{m}^2$) is recorded by ORVICO (Fig. 2.11). One can see various characteristics: straight line between collisions (hard sphere, no long-distance interaction), various spacings of trace points (various speeds), various lengths of free path. Distribution functions may be established, firstly by a tedious evaluation of many pictures by hand, but also by the computer (Fig. 2.12).

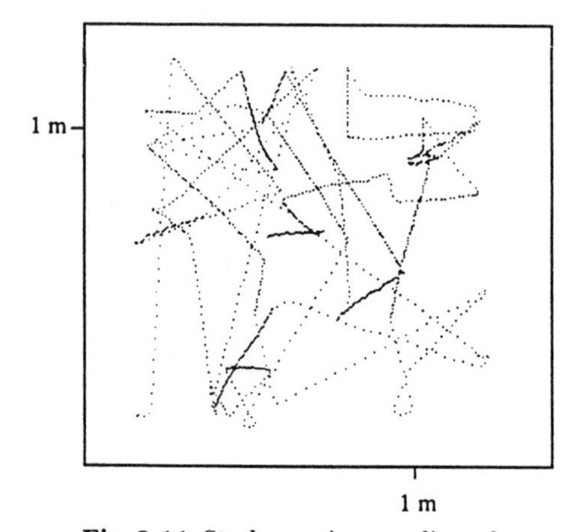

Fig. 2.11 Stroboscopic recording of one puck moving within 10 others on an air table. Type of interaction: Hard sphere elastic collision

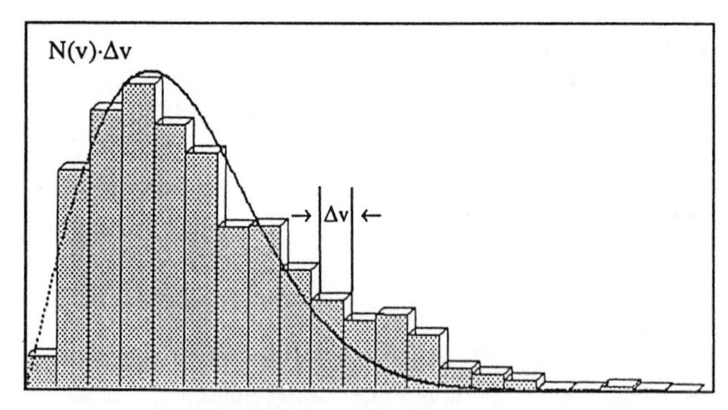

Fig. 2.12 Velocity distribution derived from Fig. 2.11. Solid line: Maxwell distribution for two-dimensional case. (Small deviation at high speed: Result of non-thermalized excitation by shaking from outside

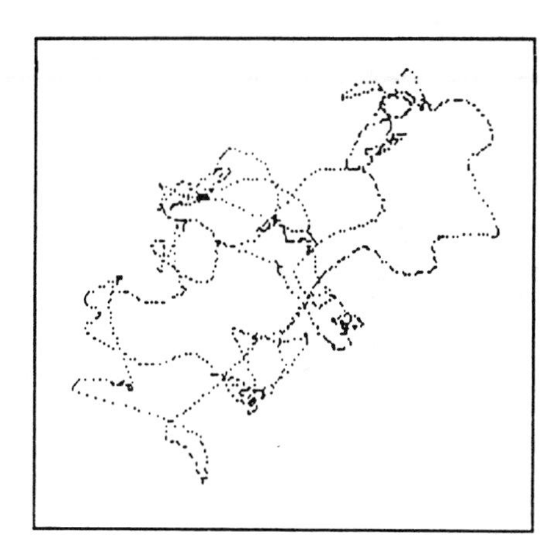

Fig. 2.13. Trace of puck moving among others on air table; repelling magnets

An interesting variation is shown in Fig. 2.13. Here a small air table ($0.1\,\mathrm{m}^2$) is filled with small pucks (repelling magnets). The trace of the one puck observed is no longer a straight line! This is a consequence of the magnetic field; the collisions are no longer of type "hard sphere".

2.3.5 Spheric Pendulum

The pendulum with two degrees of freedom offers many interesting observations. In the most simple case, the pendulum is exposed only to the gravitational field, where the trace of the pendulum body, according to the starting conditions, is either a straight line, or a circle, or a preceding ellipse (the precession arises from the two perpendicular linear oscillations, which slightly differ in period due to the differing amplitude). To take a record of the pendulum trace by ORVICO, the camera must be aligned vertically, and the pendulum body must move within the focal plane.

If, in addition to the gravitational field, another field of force is applied (for example, small permanent magnets are fixed to a glass plate placed in the focal plane, and also the pendulum is a permanent magnet), then many parameters for experimentation are open. In any case, after first viewing the motion of the pendulum, a desire arises to have it recorded [2.2]. Figure 2.14 shows a long-time-record of the pendulum trace, if there is one repelling magnet. The student may feel inspired to visualize the corresponding potential surface. Figure 2.15 shows a trace obtained with two repelling magnets. Again, by visualizing the corresponding potential surface, one can try to make predictions on the form of the trace, and its dependence on the starting conditions. Figure 2.16, with the same configuration of magnets, shows the trace obtained by a slight change in the starting condition. Both cases (Figs. 2.15, 2.16) are selected to show types of traces which can be considered periodic (neglecting damping): The starting conditions are reached again after a certain time. There are also traces, which are almost periodic

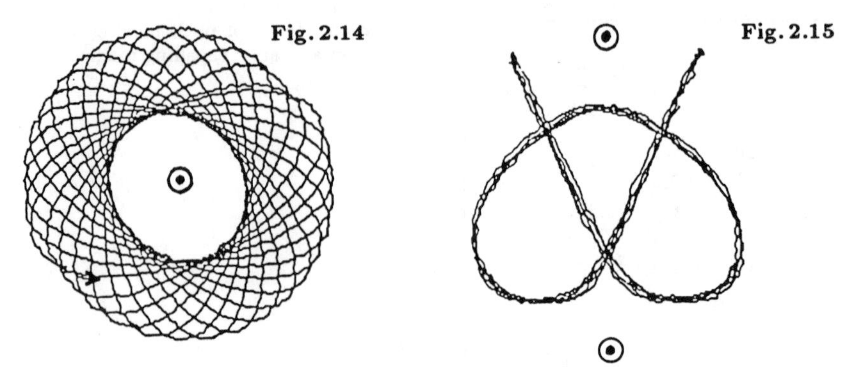

Fig. 2.14 Trace of spheric pendulum. One repelling magnet (centre) is fixed right below the equilibrium point of the free pendulum

Fig. 2.15 Trace of spheric pendulum; two repelling magnets

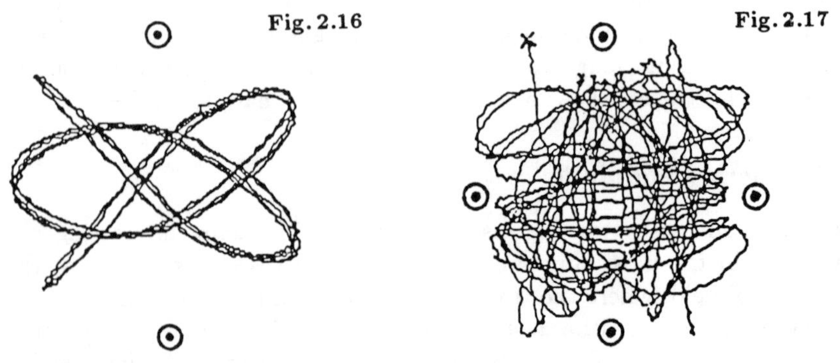

Fig. 2.16 Same arrangement as Fig. 2.15, but slightly changed starting position

Fig. 2.17 Trace of spheric pendulum; four repelling magnets. The starting conditions are chosen in order to obtain chaotic motion. (Other types of motion are also possible)

("quasiperiodic") and a previous section of trace can be reproduced, but slightly displaced, which leads to a kind of weaving pattern (Figure 2.14 is of this type). Finally, one can obtain a type of trace which is neither periodic nor quasiperiodic. Figure 2.17 (four repelling magnets) shows this very irregular trace. Typical for this type is not only the irregularity, but also the impossibility to reproduce this trace. If one tries, only in the very beginning will the reproduced curve match the first. This type of motion is called "chaotic". A representation of the pendulum motion in phase space is not directly provided by the software. However, this may be put up by the user according to the special selections to be made in order establish the Poincaré map. Although we do not intend to give a treatment on non-linear dynamics and chaotic motion here, it may be seen that, merely from the desire to obtain records of certain types of motion, a new field may be obtained.

2.3.6 Two Point Masses Observed

As mentioned before, with a colour video camera at least two independent channels (blue, red) can be employed. With the proper hardware (ORVICO plus card), the video signals of both channels can be processed and represented by software.

Figure 2.18 shows the well-known motion of coupled pendula. One of the pendulum bodies was marked blue, the other red.

Figures 2.19 and 2.20 are representations of the position–time function of two bodies colliding at the air track, elastic and inelastic. Again, it is obvious that recordings of this type can be used to establish or to test physics laws.

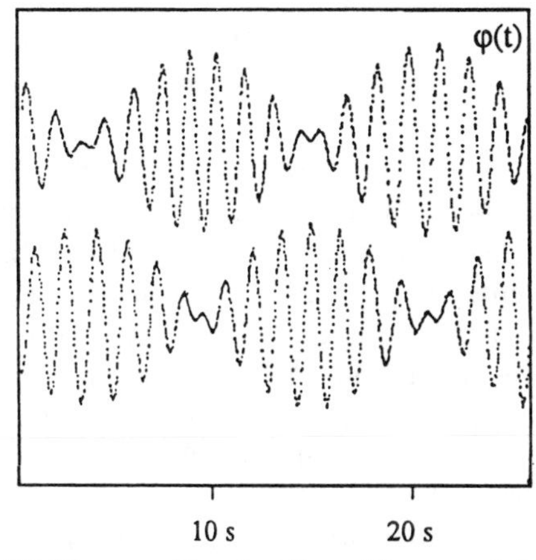

Fig. 2.18 Time–position function of two coupled pendula

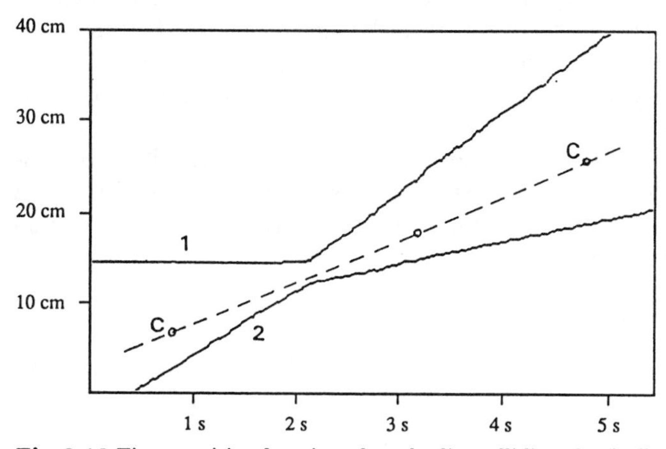

Fig. 2.19 Time–position function of two bodies colliding elastically on the air track. Body 1: 100 g; Body 2: 200 g. Inserted after display: Dotted line, position of centre of mass, *C*

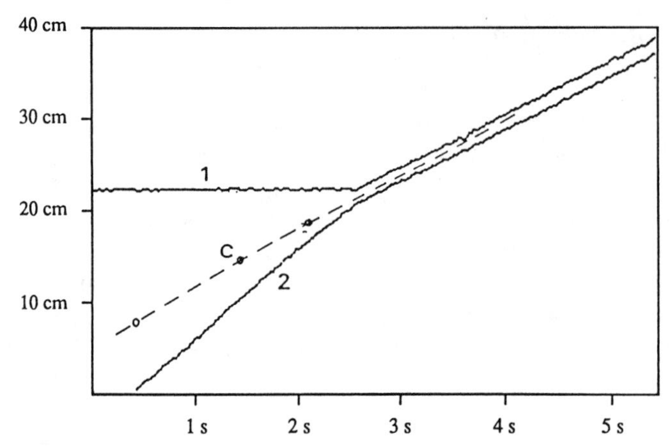

Fig. 2.20 Inelastic collision; otherwise same as Fig. 2.19

2.4 Conclusion

The typical users of ORVICO are persons engaged in physics instruction. The method exhibits attractive features both for the teacher and the student. Firstly, it employs components which usually are considered attractive, such as the video camera and the computer. Secondly, it follows the typical path one usually goes along when entering a subsection of science. The stages of qualitative observation, subsequent recording and quantitative evaluation may be realized. It seems especially appealing that all these stages can be covered by the same method.

Thirdly, perhaps the most essential advantage is that the motion process to be observed and discussed is not covered by any tricky experimental gadgets. Once the user has become familiar with ORVICO, there are no more problems with the recording process and attention can be paid to the physics. Finally, another advantage is the sensible use of the computer. Whereas in education the application of the computer can seem forced, with ORVICO it seems natural. There is the possibility to use or apply the method (without elaborating on the computer's role), or to introduce it as a natural help for data processing, and there is also the possibility to allow the user to apply his own ideas concerning software. Thus, ORVICO is a true example of an "on-line recording" and appropriate to the application of modern scientific methods.

References

2.1 R. Dengler, K. Luchner, R. Worg: "ORVICO – Eine neue Methode zur Auswertung von Bewegungvorgängen" in: PhuD **2**, 128–131 (1987);
R. Dengler, K. Luchner: "Object recording by video and computer: a new way to collect and analyse motion data" in Eur. J. Phys. **10**, 106–110 (1989)

2.2 K. Luchner, R. Worg: "Harmonische und chaotische Schwingungen" in MNU 40/6 (1987);
R. Worg: "Deterministisches Chaos – Wege in die nichtlineare Dynamik"; BI Wissenschaftsverlag Mannheim 1993

Part II

Thermodynamics

3. Application of PID Control to a Thermal Evaporation Source

B.D. Hall

3.1 Introduction

The fundamental reason for applying feedback control to a physical system is to obtain an *improvement* in the *performance* of that system. Applying feedback means that the input, or forcing function, of a system is derived, to some extent, from the system's output. By carefully choosing the way in which the output is fed back into the system, important gains in performance can be made. Of course, exactly what is meant by performance in any particular situation will need to be defined, within the specific context.

Control theory is a branch of engineering which has been given considerable impetus in recent years due to the rapid evolution of computing systems, especially microprocessors. However, the need to control processes and obtain the best performance possible arose much earlier, because of increasing mechanization in industry. The analytical methods originally developed by control engineers are based on feedback systems that operate continually to compensate for undesirable system behaviour; they can be described mathematically with systems of differential equations. Computer-based systems, on the other hand, must calculate their response before acting to compensate a change in the output. This takes time, and their operation is usually analyzed by considering the discrete instants at which the controller will act; they are described mathematically with difference equations. In many cases, well-established principles of control in continuous systems find close analogues in computer-based controllers. The PID algorithm, the subject of this chapter, is a classic example.

Thanks to the low cost of microcomputer systems, and to their wide acceptance in the laboratory, a scientist wishing to install a simple controller probably has the choice of either buying it as a ready made unit, or building it using a small computer and suitably chosen instruments. Many scientists, however, being unfamiliar with elementary control theory, tend to overestimate the difficulty involved in developing their own system; provided the associated computer interfaces are available, the implementation of simple control algorithms on a computer is not a complicated affair.

With this in mind, it would seem appropriate to give students, heading for research work in the laboratory, a simple introduction to the design and use of controllers. This chapter is intended to do that. By describing an experimental situation in which a microprocessor-based controller has been applied to a problem of temperature control, it is intended to emphasize how the simplest principles of feedback control can be of real service in practical situations. Although drawn from the context of a research – not a teaching – laboratory, the physics involved in the problem is within the grasp of senior undergraduate students. Furthermore, an equivalent apparatus for teaching purposes could be assembled at modest cost.

This chapter is intended to place emphasis on the implementation and performance of a standard PID (Proportional plus Integral plus Differential) control algorithm rather than the actual process to which it is applied. The importance is to convey, to the student, the elementary principles of a very widely applicable technique. Nevertheless, the next section will introduce the field of research known as cluster physics, from which this example has been drawn, and describe the processes occurring within an inert-gas-aggregation source, which is the physical system to which feedback control is to be applied here. Adequate bibliographic references will be given for the interested reader to develop the subject further, however, those wishing to concentrate on the controller alone may wish to skip this section. Following this, an introduction to feedback control and a description of the PID algorithm are given, before going on to describe its implementation on a microcomputer. Examples of the performance and tuning of the PID coefficients will then be given and the chapter will close with a discussion of the relative merits of this type of system.

A formal development of control theory will not be attempted here; mathematical treatments using Laplace and z transforms will be avoided, as will detailed performance analyses, such as frequency response or stability considerations. The concepts necessary to the implementation of PID control will be presented in an intuitive manner, leaving the formal aspects of control theory to a more methodical treatment of the subject.

Because digital controllers are necessarily interfaced with peripheral instruments, and because these interfaces are highly specific, it is of little interest, here, to describe the hardware and interface drivers of a particular system. For this reason, such details will not be discussed. On the other hand, these are likely to be the most intricate parts of a digital control project, and this fact should probably not be kept too well hidden from students arriving in the teaching laboratory, least it bias their impressions!

3.2 The System to be Controlled: An Inert-Gas-Aggregation Source

3.2.1 Background

In the field of study known as *cluster physics* the system of interest – the cluster – is so small as to be usually described by the number of atoms it contains. Clusters span a range of sizes: from just a pair of atoms (the dimer) to *ultrafine particles*, containing several thousand of atoms and measuring several nanometers in diameter. These small systems are remarkable because their study traces out the evolution of physical properties from those of the bulk solid to those of an isolated atom. In clusters, physical properties that in the bulk are considered to be independent of the amount of matter present, such as melting point, crystal structure and electronic structure, become a function of the system's size [3.1–7].

Among the methods used to create clusters for study, the technique known as *inert-gas-aggregation* has been extensively employed. Inert-gas-aggregation is best suited to the production of larger clusters, those with upwards of several hundred atoms, and it can be applied to a wide variety of materials with little difficulty. Evidence of this versatility is to be found in the rather beautiful observations of Uyeda and co-workers, made in the early days of small particle research. These researchers performed electron microscope and electron diffraction studies on a number of different elements in ultrafine particle form, produced by inert-gas-aggregation. The electron microscope photographs featured in their publications show a wealth of beautifully facetted microscopic crystals [3.8].

The inert-gas-aggregation source described below is a part of an experiment that has been designed to study the crystal structure of ultrafine particles, using electron diffraction [3.9, 10]. Although the experiment is rather specialized – measurements are made on clusters in free-flight in vacuum – the inert-gas-aggregation source itself is straightforward. The control problem that arises is that of stabilizing the operation of this source: it must generate ultrafine particles of a certain average size, and at a constant rate, over the time interval required to make a diffraction measurement, and it should be possible to change the conditions prevailing in the source that affect cluster production quickly and smoothly.

3.2.2 The Inert-Gas-Aggregation Technique

The method of inert-gas-aggregation works by producing the conditions necessary for homogeneous nucleation of clusters from their vapour. We begin, therefore, with a quick review of nucleation theory.

In the thermodynamic description of homogeneous nucleation, that is, in the absence of foreign nucleants which might serve as seeds for condensation, there is an energy barrier to the formation of a liquid phase which is due to the creation of a surface, separating liquid and vapour [3.11]. If we compare the Gibbs free energy of a pure vapour system with a system containing a spherical droplet, of radius r, in equilibrium with its vapour, then the difference in free energy, ΔG, is given by:

$$\Delta G = 4\pi r^2 \gamma - \frac{4}{3}\pi r^3 \rho RT \ln S \ , \tag{3.1}$$

where γ is the surface tension, ρ the liquid density, T the temperature, R the molar gas constant, and S the supersaturation ratio – the ratio of local partial pressure of the vapour to its equilibrium vapour pressure at temperature T. The first term in (3.1) represents the energy cost of creating a surface, while the second term represents the gain ($S > 1$) in free energy of the condensed phase over the vapour phase.

Other things being equal, we see from (3.1) that the stability of a droplet will depend on its size. More specifically, a droplet must be bigger than the so-called *critical radius* – the value of r for which ΔG in (1) is a maximum – if it is to avoid shrinking to oblivion. This is so because the energy of forming a droplet at the critical size is a maximum. It follows that any droplet larger than the critical radius will be able to reduce its energy by growing bigger still; continued growth will be energetically favourable. On the other hand, droplets smaller than the critical size find it energetically favourable to shrink, and will therefore tend to evaporate away. Of course, this does not explain how a droplet can ever reach the critical size in the first place, since in doing so it must spend some time in the size range in which growth is energetically unfavourable. The explanation for this lies in the kinetics of the nucleation process, which we will not go into here (see [3.11]).

By differentiating (3.1), we can obtain expressions for the critical radius and the critical Gibbs energy. These are, respectively,

$$r^* = \frac{2\gamma}{\rho RT \ln S} \tag{3.2}$$

and

$$\Delta G^* = \frac{16\pi\gamma^3}{3(\rho RT \ln S)^2} = \frac{4}{3}\pi {r^*}^2 \gamma \ . \tag{3.3}$$

The inert-gas-aggregation method works by creating a very high level of supersaturation, S, thereby lowering ΔG^*. In fact, the high values of S that are obtainable in an inert-gas-aggregation source are one of the reasons why this technique can be applied to a large number of materials, especially metals. Because ΔG^* is very sensitive to the surface tension, γ, other cluster production techniques tend to have difficulty with materials having a high surface tension, which appears to the third power in (3.3) (e.g.: $\gamma_{\mathrm{Pb}}/\gamma_{\mathrm{H_2O}} \sim 15$).

In our inert-gas-aggregation source, vapour is produced by evaporating the material to be studied from a joule-heated crucible. The rate of evaporation depends very strongly on temperature (Clausius-Clapeyron relation): the hotter the crucible is, the more vapour will be produced and the hotter the vapour will be near the crucible. However, an inert gas at room temperature is continually introduced to the chamber, and mixing of the two gases causes the vapour to cool. The drop in vapour temperature leads to supersaturation (the equilibrium vapour-pressure is a very strong function of temperature [3.12,§ 4.1.2]) and the conditions for nucleation are met.

3.2.3 A Description of a Real Inert-Gas-Aggregation Source

The inert-gas-aggregation source used in this chapter is shown in Fig. 3.1. We will refer to it as the Léman source, because it was designed and built on the shores of the beautiful Lake Léman in Switzerland.

The Léman source consists of a cylindrical vacuum chamber, 9 cm in diameter and 11 cm long. At one end (on the right in the figure), the chamber is closed off by a flange accommodating an inlet for the inert gas, electrical feed-throughs for the connections to the crucible and thermocouple, a pressure gauge and a small window for observation (it is possible to check the temperature measurements of the crucible using an optical pyrometer through this window). At the other end, the chamber is narrowed down, in a cone, to a small aperture, 4 mm in diameter, through which the gas and clus-

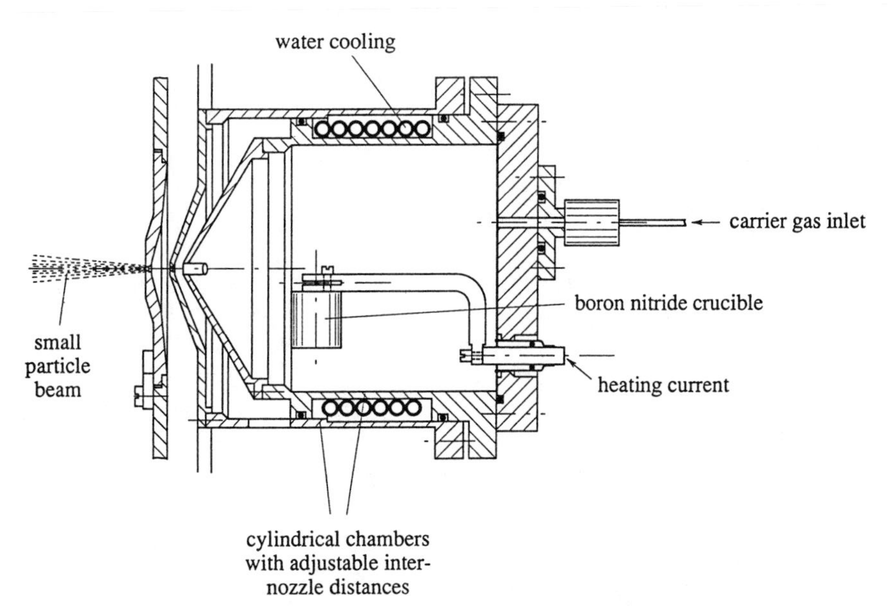

Fig. 3.1 The Léman source. Inert gas (Ar or He) is allowed into the chamber at the rear and pumped out through a nozzle in the cone on the left. Nucleation and growth occurs in a region close to the crucible (see [3.10])

ter mixture is continually pumped (mechanical roughing pump $\sim 12 \text{ m}^3/\text{h}$). The walls of the chamber are cooled by water circulating through a copper tube spiraling around the outside of the cylinder that is also in contact with the flange. A crucible, made of boron nitride, is situated in the middle of the chamber. Heating is provided by a tungsten wire, sandwiched between two boron nitride layers, and the crucible is shielded by a double layer of tantalum foil, in order to reduce radiation losses. The volume capacity of the crucible is $\sim 2 \text{ cm}^3$ and operating temperatures of 1200 – 1300°C are typical, although the system will sustain temperatures in excess of 1600°C (a type-C thermocouple is used, with a tantalum sheath).

During an experiment, the rate of gas flow into the chamber can be controlled by a valve in the gas supply line, and the current supplied to the heating coils of the crucible is determined by a programmable power supply. The crucible temperature and inert gas pressure are both monitored. Unfortunately, the quantity that is of real interest – the rate of cluster production and the distribution of their sizes – cannot be directly measured; in the complete experiment, variations in the rate of production of clusters are only observed as changes in the intensity of the diffraction signal. Such changes cannot easily be used in a feedback system, and it is desirable to try and stabilize the source using those parameters available, namely the temperature and pressure.

Heat losses from the crucible are dominated by radiation and convection. The difficulty in fixing the temperature of the crucible arises because different parts of the chamber, and its immediate environment, heat up at different rates. As heat losses are dependent on the local temperature gradients, the changes in temperature of the surroundings will influence the rate at which heat is lost from the crucible and therefore influence its temperature. To put it another way, the source chamber takes a long time to reach thermal equilibrium when the crucible is heated; any change in the power supplied will therefore take a long time to stabilize.

Typical gas pressure in the chamber is ~ 1 mbar (argon or helium), and, because at these pressures the gas state is viscous (the mean free path is considerably shorter than the typical dimensions of the chamber), the thermal conductivity of the gas should be pressure-independent [3.12, §2.7.3]. In fact, changes in chamber pressure do affect the heat losses from the crucible, although only slightly. This is due to the continual renewal of gas in the chamber and also to changes in the convective heat transfer. Increasing the gas flow through the chamber, which corresponds to an increase in the chamber's pressure, increases the rate of heat loss from the crucible and lowers its temperature (This will be illustrated later: see Fig. 3.12).

In contrast to the crucible temperature, the inert gas pressure is easily stabilized when the flow rate is set by a mechanical valve. There is, essentially, no time delay involved in the setting of the chamber pressure. The gas pressure does, however, depend slightly on the crucible temperature (heating the gas causes its pressure to increase), although this effect is inconsequential when compared to the variations in the evaporation rate due to fluctuations in the crucible temperature. The control problem for the Léman source is thus reduced to one of providing a stable, and manageable, crucible temperature.

3.3 Description of the PID Control Algorithm

The preceeding section argued that a means of stabilizing the temperature of the crucible in the Léman source is required and that the only acceptable way of doing this is to regulate the current supplied to the crucible heater filament. Although other factors do also affect the temperature (heat exchange with the buffer gas, diminishing the amount of material evaporating, the amount of heat lost through radiation, etc.), these cannot be manipulated in a useful way during an experiment.

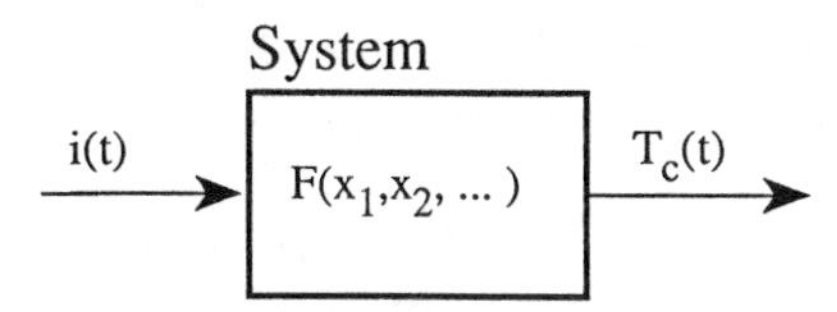

Fig. 3.2 Block diagram representation of the system to which feedback will be applied. The input is a current and the output is the temperature of the crucible. The relationship of $i(t)$ to $T_c(t)$ will depend on a number of factors, represented by the parameters $x_1, x_2, \ldots$

Control systems are often depicted in block-diagram form. In this chapter, the system, which consists of the crucible and its environment, will be represented by a single block (Fig. 3.2) labelled by a function, $F(x_1, x_2, ...)$. The input to the block is a current, $i(t)$, and the output is the crucible temperature, $T_c(t)$. Both input and output are continuous functions of time. The system's response (i.e. temperature) to a given input (i.e. current) will depend on a number of parameters, $x_1, x_2, \ldots$. Although Sect. 3.2 discussed some of the influences to the system response, the nature of $F(x_1, x_2, ...)$ is not very well known. This section will describe how a feedback control algorithm can be applied to an arbitrary system so that fluctuations in F, due to the x_i, can be compensated for by varying i, thereby maintaining the output in the desired state. In Sect. 3.5 the algorithm will be applied to the real problem posed by the Léman source.

Now consider the diagram in Fig. 3.3. A loop has been closed around the system, F, which is now in series with a control block. The contents of the controller will be developed progressively in the remainder of this section. An external control value, T_0, is applied to the loop (this is the desired

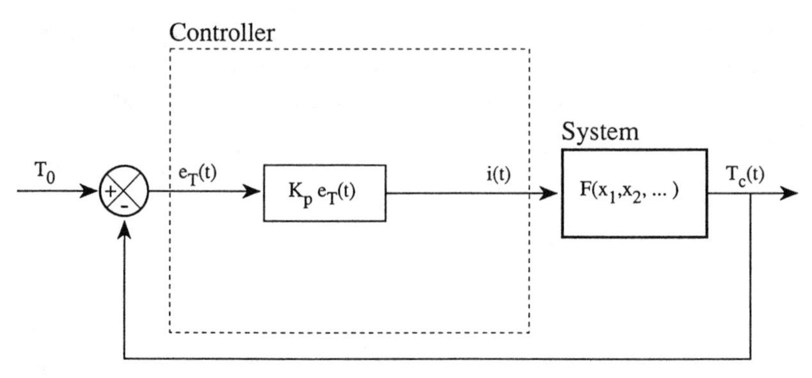

Fig. 3.3 Proportional control. The difference between the desired temperature, T_0, and the measured temperature, $T_c(t)$, is multiplied by a positive constant and used to set the output current. Proportional control results in a stable operating point at which the measured and target temperatures are not equal

temperature, or target) at a difference node where the crucible temperature, T_c, is subtracted from T_0. The difference, e_T, is then used as an input to the control block. Clearly, when $e_T = 0$ the system is in the desired state, however, while $e_T \neq 0$ the controller must act so as to return e_T to zero by changing the current supply.

As a first step in choosing the control algorithm, consider the effect of simply multiplying the input, $e_T(t)$, by a positive constant in order to set the current. This is called proportional control:

$$i(t) = K_p e_T(t)\,, \qquad K_p > 0\,. \tag{3.4}$$

Suppose now that the system happens to be in a state such that $e_T > 0$ and $i > 0$. The bigger e_T is, the higher the current will be, and so we would expect that the crucible should heat up and that e_T will diminish. The algorithm does, therefore, seem to reduce e_T. Unfortunately, it will only stabilize the crucible temperature at some value $T_0' < T_0$. This is because at the desired temperature, where $e_T = 0$, no current can be supplied to the crucible. Now, we have already seen that heat is continually removed from the crucible by a variety of processes. So, as the crucible temperature initially rises, the current is reduced until a point is reached at which heat losses equal the power supplied. Beyond this, any increase in the crucible temperature cannot be sustained by the diminishing current and the crucible will begin to cool down. The stable temperature, T_0', is that at which the power supplied to the crucible, by the electric current, is balanced by the sum of the various heat loss processes. The difference between T_0' and T_0 will depend on the value of K_p and can be reduced by increasing the proportional gain.

The difference between T_0' and T_0 is unavoidable in the proportional control algorithm and is referred to as the steady-state error. However, consider the modification to the control algorithm shown in Fig. 3.4. A constant is

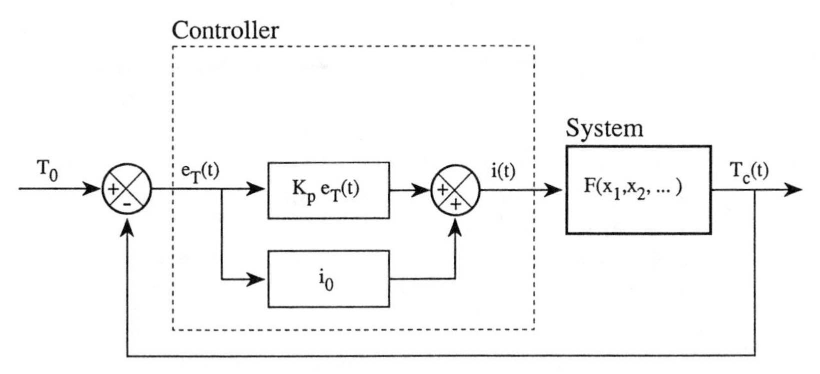

Fig. 3.4 This arrangement, though not a practical one, provides proportional control with no steady-state error. The constant term, i_0, is assumed to be the steady-state current value required by the system when $T_c(t) = T_0$

now added to the output of the proportional control algorithm. This value, i_0, is assumed to be just the current required to balance the heat losses from the crucible when $T_c = T_0$, which means that the system will now be stable when $e_T = 0$. If, for any reason, $e_T < 0$, meaning that $T_c > T_0$, then $i < i_0$ and the system will tend to cool. Similarly, if $e_T > 0$ then the system will heat up, due to an increase in the current supplied.

Of course, it is not possible to know what the correct value of i_0 to use is, because we have admitted that our knowledge of the system response is limited. Instead, we can avoid this difficulty, in a simple and elegant way, by introducing an integrator into the control algorithm. This is called proportional plus integral control (PI), for obvious reasons.

Figure 3.5 shows the output current from the control block as the sum of two terms: proportional and integral, so that the current correction is now given by

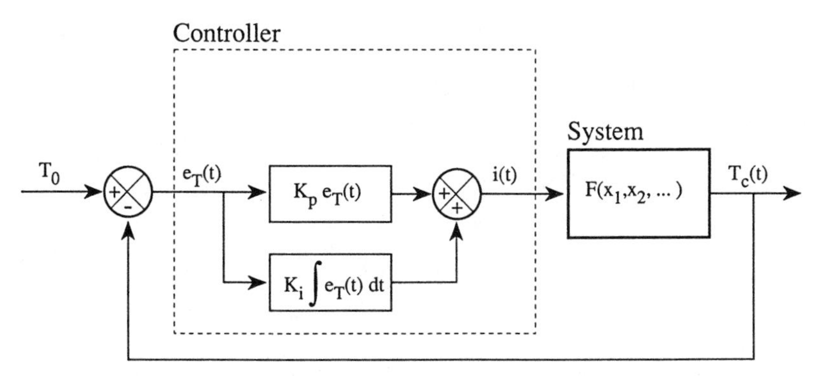

Fig. 3.5 Proportional plus integral (PI) control. By integrating the error signal, $e_T(t)$, it is possible to estimate the constant term of Fig. 3.4. When $e_T(t) = 0$ the contribution from the proportional block is zero and that from the integral block is constant

$$i(t) = K_{\mathrm{p}} e_{\mathrm{T}}(t) + K_{\mathrm{i}} \int e_{\mathrm{T}}(t) dt \,, \qquad K_{\mathrm{p}}, K_{\mathrm{i}} > 0 \,. \tag{3.5}$$

To see the role played by the integrator, imagine that the system has stabilized using only the proportional control, i.e. $K_{\mathrm{i}} = 0$. The steady-state temperature will then be $T_{\mathrm{c}} = T_0' < T_0$ and $e_{\mathrm{T}} > 0$. If the integrator is then suddenly included ($K_{\mathrm{i}} > 0$ but $K_{\mathrm{i}} \ll K_{\mathrm{p}}$) then the current supplied to the crucible will immediately increase, causing the crucible temperature to rise slightly. Now, this temperature rise causes the proportional term to reduce current to the crucible, however, provided e_{T} remains positive, the integral term will steadily *increase* its contribution to the total current with time. When the system again reaches a steady state, the output from the integrator will be equal to i_0, and the proportional term will only act when $e_{\mathrm{T}} \neq 0$.

The relative values of K_{p} and K_{i} are important when the stability of the system is considered. This will be shown by example later (Fig. 3.12). Clearly, if K_{i} is too small, it will take a long time for the integral term to settle when the control value, T_0, is changed or if a change to any of the system parameters occurs, leading to a change in the steady-state error. On the other hand, if K_{i} is too big then the integral term will change too quickly. In this case, there is not enough time for the proportional term to stabilize; the integral term will tend to drive the system too far in one direction and then too far in the other. This is known as *ringing* and can lead to instability if the oscillations become too large.

Let us leave aside the integrator for the moment and consider again the behaviour of the simple controller shown in Fig. 3.4. Imagine that the system is initially at $T_{\mathrm{c}} = T_0$ when suddenly a rapid change in T_{c} occurs, to some value $T_{\mathrm{c}} = T_1 < T_0$. The general behaviour of the system will be the following: e_{T} is initially positive and causes a strong reaction from the proportional term, tending to drive the crucible temperature back towards T_0. As soon as the temperature of the crucible starts to rise, the *extra* current, due to the proportional term, will decrease. Put another way; a sudden drop in crucible temperature causes a surge in the power supplied, initiating a compensating rise in crucible temperature.

The change in power surge accompanying a change in crucible or target temperature is the basis of the control strategy, and it is desirable to provide a reaction that is as large as possible in order to enhance the performance of the controlled system. The larger the value of K_{p}, however, the larger the amplitude of the reaction to any perturbation to the system; too strong a reaction and the system will become unstable. This instability arises when the system over-compensates, driving the crucible temperature beyond the desired temperature, and provoking successively bigger swings in temperature.

A way of acting against the tendency of a system to oscillate is to introduce a term in the control algorithm which is proportional to the derivative of e_{T}. Figure 3.6 shows the control algorithm as being composed of a pro-

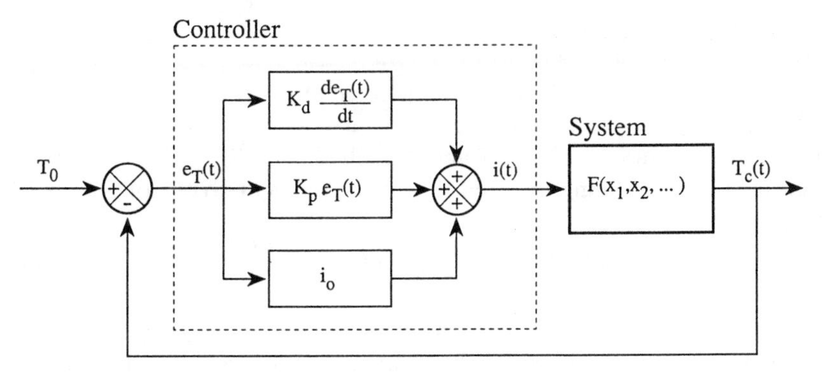

Fig. 3.6 Proportional and derivative (PD) control. Only when the derivative of $e_{\mathrm{T}}(t)$ has the same sign as $e_{\mathrm{T}}(t)$ does the derivative correction reinforce the proportional one. This will occur whenever there is a tendency for the temperature to move away from the target value. As soon as $T_{\mathrm{c}}(t)$ begins to return towards T_0, the derivative term acts against the proportional one, thereby reducing the tendency to overshoot and oscillate

portional and a derivative term, as well as a constant which defines the operating point. The current correction is now

$$i(t) = i_0 + K_{\mathrm{p}} e_{\mathrm{T}}(t) + K_{\mathrm{d}} \frac{de_{\mathrm{T}}(t)}{dt}, \qquad K_{\mathrm{p}}, K_{\mathrm{d}} > 0\,. \tag{3.6}$$

To see what the effect of the derivative term will be, consider, for a moment, that $K_{\mathrm{d}} = 0$ and that the proportional gain, K_{p}, is strong enough to cause the system to oscillate about the target temperature, T_0. As above, the proportional term increases power to the crucible when the temperature is too low and reduces it when it is too high. The system oscillates because too much power is supplied to the crucible during the time that $e_{\mathrm{T}} > 0$, causing the temperature to overshoot. Then, while $e_{\mathrm{T}} < 0$, too little power is supplied and the crucible temperature falls back below the target.

Now suppose that $K_{\mathrm{d}} > 0$. During the cycle in which $e_{\mathrm{T}} < 0$, the sign of de_{T}/dt will be first negative, then positive. That is, the derivative term will begin by reinforcing the proportional correction, while the tendency is for the crucible temperature to move away from the target temperature. However, as soon as the crucible temperature begins to move back towards the desired temperature, the sign of de_{T}/dt changes and the derivative term will oppose the proportional term. The combination of proportional and derivative (PD) control thus applies a strong power surge during the periods in which the crucible temperature is moving *away* from the control value. As soon as the tendency to move *back* towards the desired value is established, the PD controller reduces the current, and in so doing reduces the tendency to over-supply current to the crucible.

3.3.1 The PID Control Algorithm

The complete PID control algorithm is the combination of the control and correction strategies developed above; the controller consists of the sum of the three terms, as shown in Fig. 3.7 and described by (3.7), with the relative strengths of each term being adapted to suit the particular control problem.

$$i(t) = K_{\mathrm{p}} e_{\mathrm{T}}(t) + K_{\mathrm{i}} \int e_{\mathrm{T}}(t) dt + K_{\mathrm{d}} \frac{de_{\mathrm{T}}(t)}{dt} . \qquad (3.7)$$

As suggested by the development above, it is possible to consider the PID algorithm as a simple control strategy (proportional control) which has been refined by including two corrective terms: an integral, to eliminate steady-state error, and a derivative, which tends to damp oscillatory behaviour in the output. Although only an intuitive guide, thinking about the PID algorithm in this way can provide a useful basis for interactive tuning of the three coefficients $K_{\mathrm{p}}, K_{\mathrm{i}}$, and K_{d}. This will be discussed in Sect. 3.5. A more complete description of the PID algorithm can be sought in standard texts on control theory (e.g. [3.13, §8.10]). The next section will describe how a PID control algorithm can be approximated on a digital computer.

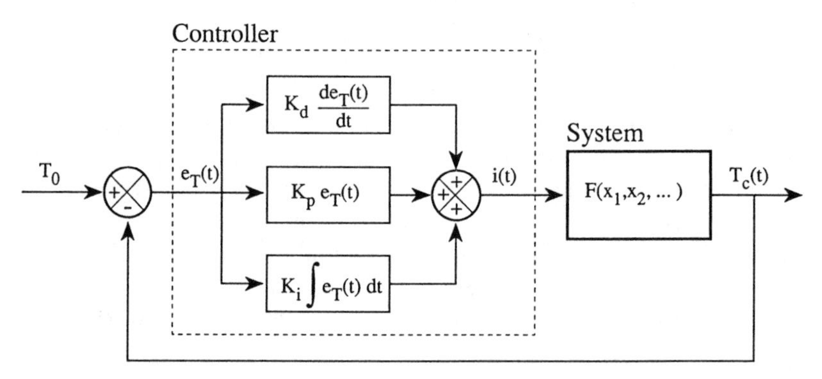

Fig. 3.7 The PID controller

3.4 Implementing the PID Algorithm on a Computer

The discussion of control algorithms so far has paid no attention to any practical limitations that real controllers might have. In particular, algorithms are assumed to act *continuously* in the feedback loop, according to the respective differential equations. Indeed, continuous control systems are designed and implemented by engineers regularly. However, in this chapter, we are concerned with finding a control algorithm that can be programmed on a microcomputer.

When a computer is interfaced with measuring equipment and used to make a series of measurements of some quantity, the representation of that quantity is inevitably discrete, both in amplitude and in time. This cannot be avoided; there will always be a finite interval of time between successive measurements, and the computer will always measure and record a quantity with finite precision.

Although we will not be concerned with the question of precision, it is necessary to show how the continuous PID algorithm, described by (3.7), can be approximated using only discrete samples of the difference signal $e_{\mathrm{T}}(t)$. We will assume that the instants at which sampling occurs are quite regular, and that they are separated in time by an interval, Δt. The crucible temperature will therefore be measured as a series,

$$T_{\mathrm{c_n}} , \qquad n = 0, 1, 2, \dots , \tag{3.8}$$

where we have written

$$T_{\mathrm{c_n}} \equiv T_{\mathrm{c}}(n\Delta t) , \qquad n = 0, 1, 2, \dots . \tag{3.9}$$

Let us now develop, term by term, a sampled-data approximation to the PID algorithm.

The proportional term, $K_{\mathrm{p}} e_{\mathrm{T}}(t)$, can be represented by

$$K_{\mathrm{p}} e_{\mathrm{T_n}} . \tag{3.10}$$

That is to say, the current supplied, by proportional control, during the interval Δt after the nth sample, is proportional to the difference signal, $e_{\mathrm{T_n}} = T_0 - T_{\mathrm{c_n}}$, evaluated at the nth sampling instant. Note that we are assuming that the current supply will be held constant during the interval Δt between samples.

The integral, $\int e_{\mathrm{T}}(t) dt$, can be represented by the discrete sum

$$I_{\mathrm{n}} = \Delta t \sum_{\mathrm{m=0}}^{\mathrm{n}} e_{\mathrm{T_m}} , \tag{3.11}$$

which is a rectangular approximation to the desired integral. The integral term is then written as

$$K_{\mathrm{i}} I_{\mathrm{n}} . \tag{3.12}$$

Finally, the derivative, $de_{\mathrm{T}}(t)/dt$, can be approximated with a backward difference, that is, the difference between the two most recent samples, and the derivative term written as

$$\frac{K_{\mathrm{d}}}{\Delta t}(e_{\mathrm{T_n}} - e_{\mathrm{T_{n-1}}}) . \tag{3.13}$$

A sampled-data approximation to the PID algorithm can thus be written as

$$i_{\mathrm{n}} = K_{\mathrm{p}} e_{\mathrm{T_n}} + K_{\mathrm{i}} \Delta t \sum_{\mathrm{m=0}}^{\mathrm{n}} e_{\mathrm{T_m}} + \frac{K_{\mathrm{d}}}{\Delta t}(e_{\mathrm{T_n}} - e_{\mathrm{T_{n-1}}}) . \tag{3.14}$$

It is important to realize that even though (3.14) is an *approximation* to (3.7) for a continuous system, a detailed analysis of this difference equation is possible using the mathematical techniques of sampled-data control theory. In general, however, this will not be necessary and, because the continuous PID algorithm is easier to visualize than (3.14), it is useful to have it in mind when working with a sampled-data controller.

Now, to implement a sampled-data PID, one possibility is just to write a computer program that will explicitly calculate the terms in (3.14). There are, however, several draw backs to this, in terms of the smooth running of the controller (see [3.14,15]). On one hand, designing a system which can easily be switched between manual and automatic control is not obvious using (3.14); the problem being to pass some knowledge of the operating point from the automatic system to the manual, and vice versa, when the control is switched. Furthermore, problems can also arise when the PID coefficients are changed during controller operation, and if the controller drives its associated instrumentation beyond their working limits. These considerations make it worth while considering an equivalent expression to (3.14), which will provide a more practical implementation.

The expression we will use is written as

$$i_{\mathrm{n}} = \sum_{\mathrm{m}=0}^{\mathrm{n}} (ae_{\mathrm{T_m}} + e_{\mathrm{T_{m-1}}} + ce_{\mathrm{T_{m-2}}}) \,, \tag{3.15}$$

which is represented schematically in Fig. 3.8. In this figure, the algorithm is shown as an assemblage of units corresponding to the operations to be performed on the sampled data, the output is i_{n} and the input, $e_{\mathrm{T_n}}$. Internally, samples are manipulated by units which multiply, add or delay them one sample period.

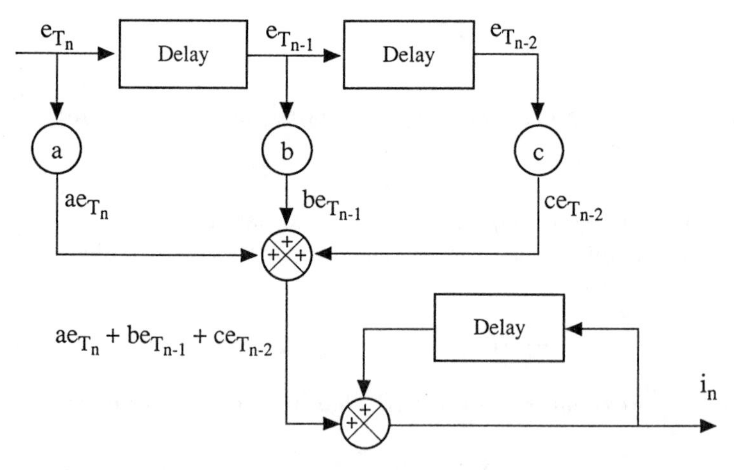

Fig. 3.8 An alternative, equivalent, sampled-data PID implementation

In order to show the equivalence of (3.15) and (3.14), it helps to notice that the final stage of our machine implements the sum in (3.15). Therefore, if we can find three terms that, when applied to the input of a summing unit, will produce the same output as the desired discrete PID form, then these equations can be compared with the second-order difference (3.15) to obtain the coefficients a, b, and c. Thus we have, for the proportional term,

$$K_p(e_{T_n} - e_{T_{n-1}}) , \tag{3.16}$$

which has the effect of setting the most recent proportional correction value and at the same time removing the previous one. For the integral term, we have

$$K_i \Delta t e_{T_n} , \tag{3.17}$$

and for the derivative term,

$$\frac{K_d}{\Delta t}[(e_{T_n} - e_{T_{n-1}}) - (e_{T_{n-1}} - e_{T_{n-2}})] . \tag{3.18}$$

Taking the sum of these three expressions and equating it with the bracketed term in (3.15), one obtains

$$a = K_p + \frac{K_d}{\Delta t} + K_i \Delta t , \tag{3.19}$$

$$b = -\frac{2K_d}{\Delta t} - K_p , \tag{3.20}$$

$$c = \frac{K_d}{\Delta t} . \tag{3.21}$$

The advantages offered by an implementation of (3.15) are the following:

- If the PID coefficients are changed, while the system is operating, the new coefficients will be used only with those samples arriving after the change was made, and not those previously accumulated.
- Inclusion of a manual operation mode, that can be interchanged smoothly with automatic control, is easily provided by inserting a switch at the input to the summing stage. The summing stage can then be connected to either the second-order difference equation in (3.15) or to a manual control which can provide incremental changes to the output.
- When the controller drives the output stage beyond acceptable limits (e.g. overload protection) an error will accumulate in the integral term (wind-up effect). Wind-up occurs when the output stage cannot deliver the correction required by the PID algorithm and, as a result, the error signal tends to persist rather longer than it would if the controller were operating within its normal range. In such cases, if (3.14) is used, the integral term continues to grow, regardless of whether or not the output can deliver the desired correction. Wind up changes the effective steady-state value of the controller unnecessarily and leads to unstable

behaviour when the output stage returns to operation within its normal range again.

The advantages of (3.15) are a direct result of placing the summing unit after the calculation of the incremental correction, instead of inside the PID integral term. In fact, in a practical controller, the summing unit may even be a part of the output stage instrumentation, thus requiring only incremental corrections from the control algorithm.

3.4.1 Program Structure and the Use of Interrupts

Now that we have a convenient discrete PID algorithm, we will look at the program structure needed to implement the complete controller. Figure 3.9 shows the sequence of operations that must be performed during a single control cycle, that is, once every Δt seconds. Some of the operations shown will probably be useful elsewhere in the program too, such as *read the temperature* and *set the current*, other operations will also be needed, for example: *change the target temperature*, *change the PID coefficients*, etc. The exact choice will depend on the problem at hand, however, we can safely assume that the sequence of Fig. 3.9 will be available as a single routine call, so that it can be activated easily at regular intervals.

It is important to consider how the controller routine will be activated. There are two possibilities: either the routine is placed inside an infinite

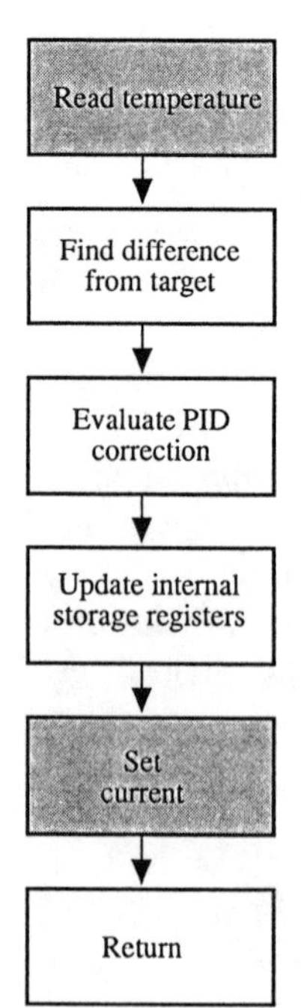

Fig. 3.9. Flow Diagram for a PID Controller. These operations will be executed once every cycle of the controller. The operations shown shaded will probably also be useful as independent routines as well

loop, possibly with other operations, that will cycle at the desired rate; or, the control routine can be used as an interrupt handler and be activated by regular interrupt signals. The latter is likely to be a more versatile approach, although it certainly demands more delicate programming.

The purpose of interrupts in a computer is to allow for a rapid, pre-programmed, response to a given event, without suspending program activity by waiting for that event to occur. Interrupts allow microcomputers to give priority to certain events, temporarily suspending normal program execution, while the interrupt handling routine is allowed to run. All microcomputers are equipped with hardware and software for interrupt handling, although, not all programming languages provide support for these.

If the control routine, is implemented as an interrupt handler then either an internal clock (if possible) or an external time source can be used to generate interrupts every Δt seconds and thereby activate the control cycle. Provided the time required for the control routine to execute is less than Δt, there will be spare time available for another program to run. This program need have little relation to the process being controlled, e.g. a graphical display of data.

Some care is needed when using interrupts, however. Anywhere that interrupt programming is provided for, that is in any particular programming language, there will be a risk of interference between the main program and the interrupt routine. For this reason, programming languages always provide instructions for disabling and enabling a response to interrupts. It is important to make use of these instructions to avoid interference between routines when designing a program that uses interrupts. For example, consider what might happen if a subroutine in a program began to change the PID coefficients and was interrupted by the control routine itself. Conceivably, some but not all of the coefficients could have changed and the PID algorithm would therefore calculate a current correction based on an inconsistent set of values.

3.5 Adjusting the PID

Once the hardware and software have been assembled, it is necessary to adjust the coefficients of the sampled-data PID algorithm, in order to achieve acceptable performance from the system. Fine tuning is possible interactively, with some understanding of the role of each coefficient in the controller. However, some initial values are required as a starting point, so as to get the system up and running.

One possibility is to proceed by trial and error from the outset: increasing the proportional gain until a reasonable step response is obtained, then applying an integral term to remove the steady-state error, and finally increasing the derivative gain to damp down any oscillatory tendency. Having obtained an initial set of values this way, they can be refined until an acceptable response is obtained (e.g. [3.13, §15.5]).

3.5.1 The Ziegler-Nichols' Methods

An alternative approach, referred to as the Ziegler-Nichols' (Z-N) methods, is widely employed. There are two Z-N methods, either of which can be used to derive a set of PID coefficients for a system to which feedback control is to be applied. In both methods, a reasonably simple measurement of system response is required, from which the calculation of suitable coefficients can be made. The Z-N methods are intended to provide a closed-loop system response that is slightly under damped; that is, the system's response to a step at the input will be a rapid change towards a new steady-state value, although the response will oscillate several times before settling.

In what is known as the closed-loop Z-N method, the system is placed in a feedback loop with a continuous proportional controller (i.e. as in Fig. 3.3). The gain of the controller is then increased until the system oscillates continuously. The value of the critical gain, K'_{p}, and the period of the oscillations, T_{Osc}, are then used to calculate PID coefficients according to the following relations [3.13, §15.3]:

$$K_{\mathrm{p}} = 0.6\,K'_{\mathrm{p}}\,, \qquad K_{\mathrm{i}} = \frac{2K'_{\mathrm{p}}}{T_{\mathrm{Osc}}}\,, \qquad K_{\mathrm{d}} = 0.125\,T_{\mathrm{Osc}}\,K'_{\mathrm{p}}\,.$$

There are several drawbacks to using this method with the Léman source. Firstly, it is difficult to incorporate a continuous proportional controller in this system, because the instruments already installed are intended for use with a computer. Secondly, the oscillations that occur when the gain is very high induce rather large current swings. This means that the heater filament tends to be overworked by brief bursts of high current, and this leads to fatigue and eventual break-down of the tungsten wire. The large amplitude of the current variations also means that the non-linear relationship between the current and power supplied to the crucible will be quite important, this will cause incorrect estimates to made using the Z-N method, which is based on a linear approximation to the system response.

The second, open-loop, Z-N method is more appropriate for use with the Léman source. It requires an abrupt change to be applied to the input of a system (a step), and the response of the system to this change is then recorded as a function of time. No special instrumentation is required as the feedback loop is opened for this test, and the step and measurement of the response can be made with the computer-based instruments already available. Furthermore, the amplitude of the step at the input can be quite small, avoiding the problems of non-linearity and overheating mentioned above.

The method is illustrated by example in Fig. 3.10 where the response of the crucible temperature to a sudden increase in the current supply is

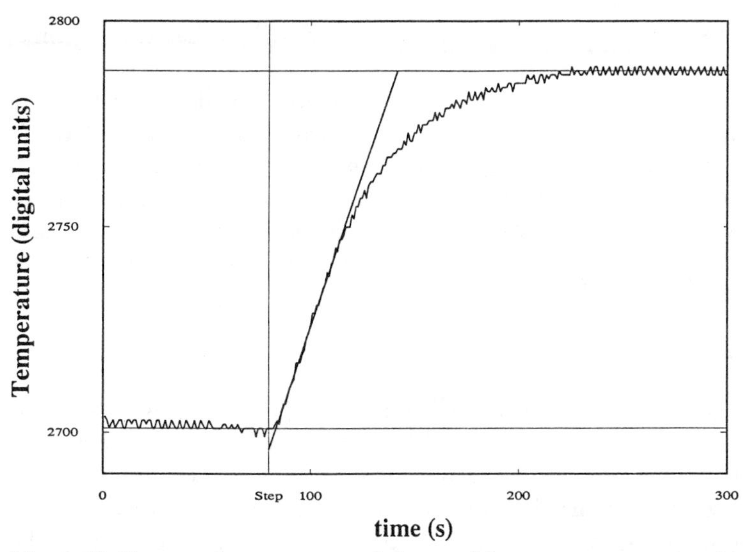

Fig. 3.10 The open-loop response of the crucible temperature to a sudden increase in the heater current supply. The input current increase came after 80 s (vertical line) and then, after a delay of ~ 6 s, the temperature began an exponential rise to a new value. The initial temperature was ~ 1064°C and the final temperature ~ 1098°C. The increase in current supply was ~ 0.5A

shown. The crucible temperature was initially ~ 1064°C, with the rapid fluctuations of ±0.5°C in Fig. 3.10 being due to noise in the measurement process. In Fig. 3.10, after 80 s, the current supply was incremented by 1000 digital units (~ 0.5 A). The temperature remained stable for about 6 s, after this change of input, before beginning an exponential climb to a new steady value.

The parameters of the response needed for the calculation of the Z-N coefficients are in this case: the lag time, L, between the application of a step at the input and any appreciable change at the output, and also the greatest rate of change in the output with time, R, during its rise to a new stable value (see Fig. 3.10).

In order to apply the Z-N method, we require the following values from the step response shown in Fig. 3.10:

- lag time, $L = 6$ s
- input step, $\Delta = 1000$ digital current units (~ 0.5 A)
- maximum slope in the rising output, $R = 1.5$ digital temperature units per second (~ 0.6°C s^{-1})

The PID coefficients are found according to the Z-N relations [3.15, p. 284]

$$K_{\mathrm{p}} = \Delta/(LR) = 111, \qquad K_{\mathrm{i}} = K_{\mathrm{p}}/(2L) = 9, \qquad K_{\mathrm{d}} = K_{\mathrm{p}}L/2 = 333 .$$

The time constant, τ, of the rising output was ~ 30 s which is a useful guide to the choice of sampling rate for the PID controller. In general, $\Delta t < \tau/10$

and in this case Δt was chosen to be 2 s, somewhat less that the lag time, L.

As mentioned above, the Z-N coefficients are intended to provide a slightly underdamped system response, in which a step input will result in an initial overshoot of the output followed by 2–3 diminishing oscillations around the final steady-state value. Such a response is not ideal for the Léman source, for which little or no overshoot is required. It is natural to look for gain values to somewhat below the Z-N values. The coefficients that were finally adopted are:

$$K_\mathrm{p} = 75\ , \qquad K_\mathrm{i} = 5\ , \qquad K_\mathrm{d} = 300\ , \qquad \Delta t = 2\ \mathrm{s}\ .$$

The system response to a change of target temperature down 10°C and then back up again is shown in Fig. 3.11. The initial temperature here was 1067°C and the gas pressure 1.5 mbar. The response shows no overshoot during the downward change in temperature and only slight overshoot, without oscillation, during the subsequent return to the initial temperature. The time required to change from one temperature to the other is $\sim$ 8 s which should be compared with the 2 - 3 minutes required in the open-loop system (Fig. 3.10).[1]

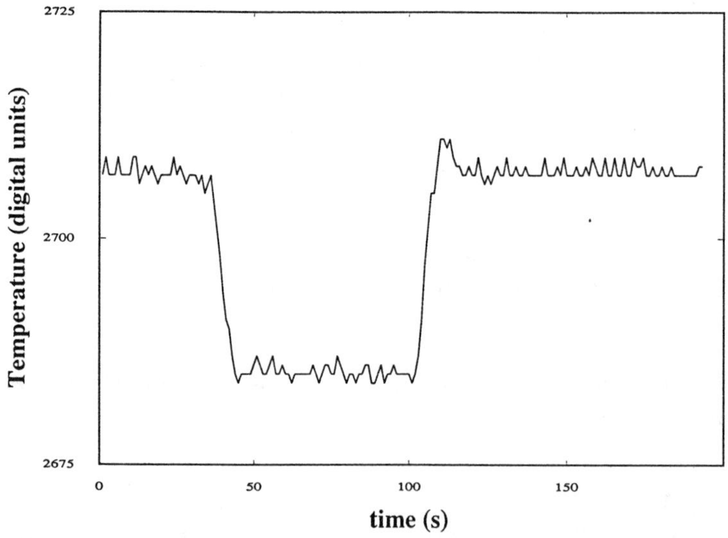

Fig. 3.11 System response under PID control. The target temperature was first stepped down 10°C and then, after about a minute, stepped back to its original value. The time for the temperature to change in each case was $\sim$ 8 s

[1] Incidentally, it is instructive to try and adjust the temperature manually, thereby playing the role of the controller. It is a laborious business trying to hold the temperature steady while achieving a reasonable rate of change between temperatures. This helps to bring home the practical advantages of the dedicated controller.

It is noteworthy that these final coefficients are in roughly the same relative proportions to each other as the initial Z-N estimates. As a general rule, when trying to refine the response of the system from the Z-N settings, one should start by changing all three coefficients by a common factor until the response cannot be improved upon. Only then should the individual coefficients be varied to make final adjustments to the response.

3.6 Possibilities Offered by the Léman Source

While by no means a difficult problem of control, the Léman source does provide a striking example of the performance gains available by application of feedback to a simple system.

From a pedagogical point of view, the source can be used to illustrate a variety of interesting properties of PID control.

- The steady-state error and its relationship to proportional gain can be easily demonstrated if $K_{\mathrm{i}}, K_{\mathrm{d}} = 0$. Furthermore, changing the background gas pressure will alter the heat losses at any given temperature and hence the steady-state temperature will vary.
- The tendency of the uncontrolled source temperature to drift can be exacerbated by reducing the circulation of coolant in the source walls.
- The choice of optimum PID coefficients will depend on the temperature at which the source is to be operated. The assumption of linear system response will be valid for the purposes of stabilization of temperature drift and fluctuations, and for moderate changes in operating temperature. However, if a large range of temperatures is covered, the non-linear relationship between current and power will become clearly apparent, and a corresponding set of coefficients will be required to span the temperature range.
- The relationship of sampling interval, Δt, to system performance and stability can be illustrated. It is easily shown, by starting with a very short interval, that by increasing Δt the system tends towards oscillation and instability, requiring a reduction in PID gain and a resulting degradation of performance.
- The roles of the derivative and integral PID terms can be illustrated separately. In the case of the integral term, this is shown in Fig. 3.12. The system's response to abrupt changes in the gas pressure are shown in Fig. 3.12 for three values of gain, K_{i}. In each case, the pressure is first increased from 0.5 mbar to 4.0 mbar and the temperature is allowed to regain the target value. The pressure is then reduced again to 0.5 mbar and the system allowed to settle. When the gas pressure is increased, cooling of the crucible is more efficient and the integral term works to raise the temperature. If K_{i} is rather small (as in Fig. 3.12.(c)) the recovery time will tend to be long. If K_{i} is large (as in Fig. 3.12.(a))

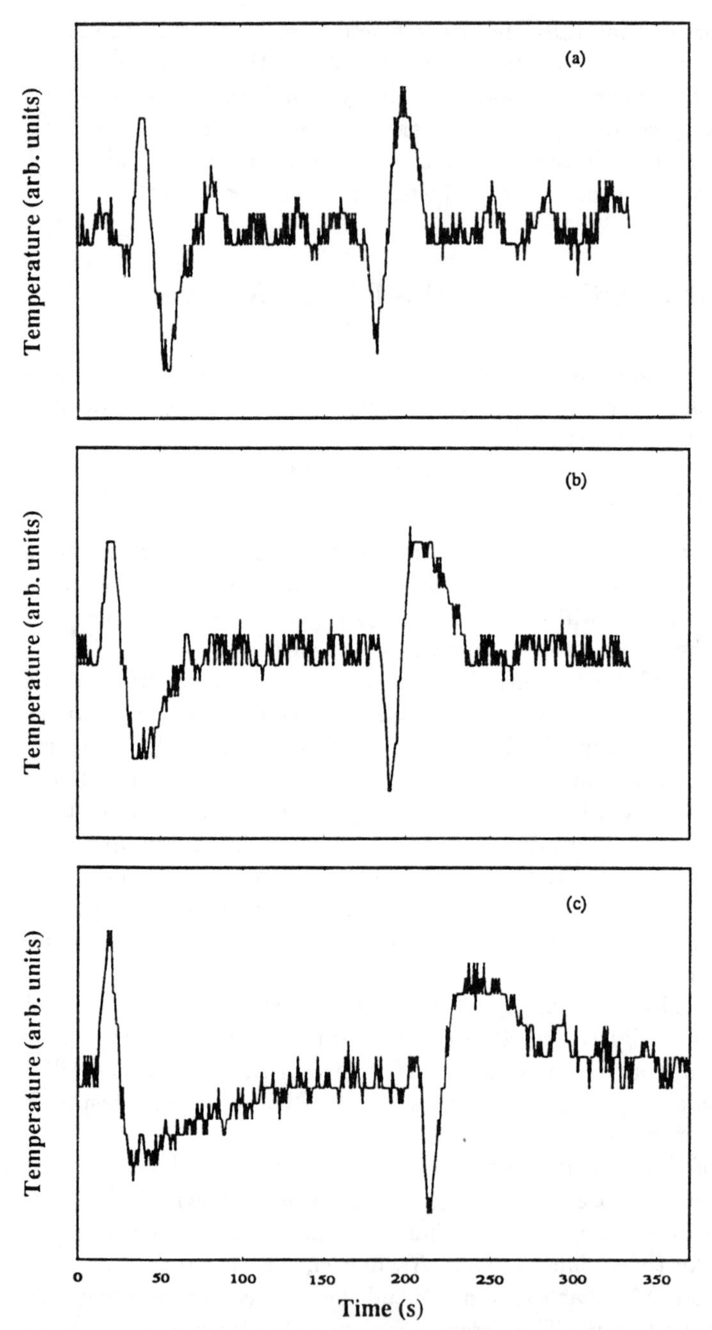

Fig. 3.12a–c. An example of how the integral gain coefficient, of the PID, affects the speed of adaptation to a change of steady-state conditions. In the figure, the integral gain, K_i is: (**a**) 8.5; (**b**) 5; and (**c**) 1.5. In each case, the (Ar) gas pressure was abruptly changed: first up, from 0.5 mbar to 4 mbar, then down again to 0.5 mbar. The target temperature and other PID coefficients remained the same throughout (T_c = 1064°C, $K_p = 75$, $K_d = 300$)

the system will tend to oscillate. An acceptable choice of K_i is shown in Fig. 3.12.(b).

For teaching purposes, and especially if it is not desired to produce small particles by inert-gas-aggregation, a similar system could be constructed with relative ease. In the Léman source, careful attention was paid to the design of the crucible and its heat shielding. This was to insure optimum performance in the experiment, but in a teaching experiment, when it is intended to illustrate the principles involved in stabilization of the temperature only, a rather crude heating element and thermocouple would be quite sufficient.

On the other hand, if an inert-gas-aggregation source is built, the experiment might usefully be combined with one on electron microscopy in an advanced teaching context. In this case, it might be possible to observe some of the beautiful small particle forms or, alternatively, investigate the distribution of particle sizes produced by the source as a function of changing gas pressure and crucible temperature [3.16,17]. Some means of collecting small particles will be required. This can be achieved by exposing suitably prepared electron microscope grids directly to the gas-particle flux, as it leaves the orifice in the source chamber. Because the density can be quite high, some form of shutter mechanism will be required in order to obtain reasonable coverage.

3.7 Conclusions

This chapter has developed, in an intuitive manner, the sampled-data PID algorithm and showed how it can be applied to a practical problem of temperature control arising in the context of a state-of-the-art experimental apparatus.

It is shown that considerable gains in system performance can be obtained with only a limited amount of knowledge concerning the response of the system. Nor is it necessary to refer to the detailed mathematical formalism of digital control theory in order to implement and tune a simple sampled-data PID controller. By bringing these points to the attention of students, and by giving them the occasion for hands-on experience with the adjustment of a PID controller, they will be able to make better use of commercial controllers, or be capable of tackling their own development problem with confidence.

The software implementation of the algorithm is straightforward. In practice, difficulties are more likely to arise in interfacing the computer with the instrumentation associated with the experiment.

For teaching purposes quite a simple set-up could be used to mimic the behaviour described. Alternatively, a complete inert-gas-aggregation source could be developed for use in an advanced teaching laboratory.

Acknowledgements

The inspiration for the original Léman source and the associated diffraction experiment came from Prof. G. D. Stein, and the source design and construction are due to Dr M. Flüeli. The author is very grateful to Dr Philippe Buffat and the members of the Institut Inter-départemental de Microscopie Electronique at EPFL, who gave their support during the writing of this manuscript. The examples of source performance were made in association with Dietrich Reinhard. Criticisms of the manuscript, from both D. Reinhard and Dr D. Ugarte, were also much appreciated.

References

3.1. G. D. Stein: Physics Teacher **Nov** 503 (1979)
3.2. F. Träger and G. Freiherr zu Putliz: Interdisciplinary Sci. Rev. **11** 170 (1986)
3.3. C. Hayashi: Physics Today **Dec** 44 (1987)
3.4. M. A. Duncan and D. H. Rouvray: Scientific American **Dec** 60 (1989)
3.5. M. L. Cohen and W. D. Knight: Physics Today **Dec** 42 (1990)
3.6. D. R. Huffman: Physics Today **Nov** 22 (1991)
3.7. A. Howie: Faraday Discuss. Chem. Soc. **92** *in press* (1991)
3.8. R. Uyeda: J. Cryst. Growth **24/25** 69 (1974);
Y. Saito, S. Yatsuya, K. Mihama and R. Uyeda: J. Cryst. Growth **45** 501 (1978);
S. Kasukabe, S. Yatsuya and R. Uyeda: Jpn. J. Appl. Phys. **16** 705 (1977);
T. Hayashi, T. Ohno, S. Yatsuya and R. Uyeda: Jpn. J. Appl. Phys. **16** 705 (1977)
3.9. B. D. Hall, M. Flüeli, D. Reinhard, J. -P. Borel and R. Monot: Rev. Sci. Instrum. **62** 1481 (1991)
3.10. B. D. Hall, M. Flüeli, R. Monot and J. -P. Borel: Phys. Rev. B **43** 3906 (1991)
3.11. J. E. McDonald: In *Homogeneous Nucleation Theory,* ed. by F. F. Abraham (Academic Press, New York, 1974)
3.12 A. Roth: *Vacuum Technology* (Elsevier North-Holland, New York, 1990)
3.13. C. L. Phillips and H. T. Nagle *Digital Control System Analysis and Design* (Prentice-Hall, Englewood Cliffs N. J., 1990)
3.14. P. A. Witting: In *Industrial Digital Control Systems* ed. by K. Warick and D. Rees (Peter Peregrinus, London, 1986) p.37
3.15. J. L. Min and J. J. Schrage: *Designing Analog and Digital Control Systems* (Ellis Horwood, Chichester West Sussex, 1988) p.276
3.16. B. G. de Boer and G. Stein: Surf. Sci. **106** 84 (1981)
3.17. C. G. Granqvist and R. A. Buhrman: J. Appl. Phys. **47** 2200 (1976)

4. Computer Control of the Measurement of Thermal Conductivity

B. W. James

If one end of a metal rod is heated while the other is held in the hand of an observer, that part of the rod that is held will become hotter, even though it is not itself in direct contact with the source of heat. Heat is said to travel along the rod by conduction through the material of the rod.

If initially the whole of the rod was at room temperature, the end of the rod which is held will gradually increase in temperature while the other end is heated. In this process heat will be conducted along the rod so as to increase the temperature at every position along the rod. After a long time a steady state is reached in which a temperature gradient exists along the rod, and the rate at which heat flows into the rod is equal to the rate at which it flows out. Before this steady state is reached, the rate at which heat flows out of the rod will be less than the rate at which heat flows into the rod because some energy is needed to increase the temperature of the rod. In order to measure the thermal conductivity of the material of the rod it is usual to wait for the steady state to be attained.

4.1. Thermal Conductivity

Thermal conductivity is a measure of how well a material will conduct heat. It also provides a way of comparing the rates of flow of heat in different materials under the same conditions. Consider a rod of some material connecting a heat supply at constant temperature to a heat sink at a lower temperature and for the present assume there is no flow of heat from or to the rod apart from that at the heat supply and the heat sink (see Fig. 4.1).

Experiments show that the rate of flow of heat through a thin section of the rod of length Δx (see Fig. 4.2) in the steady state is:

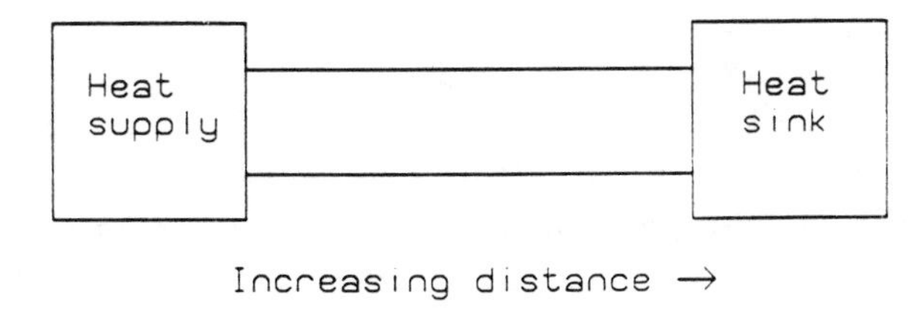

Fig. 4.1. A rod connecting a heat source to a heat sink

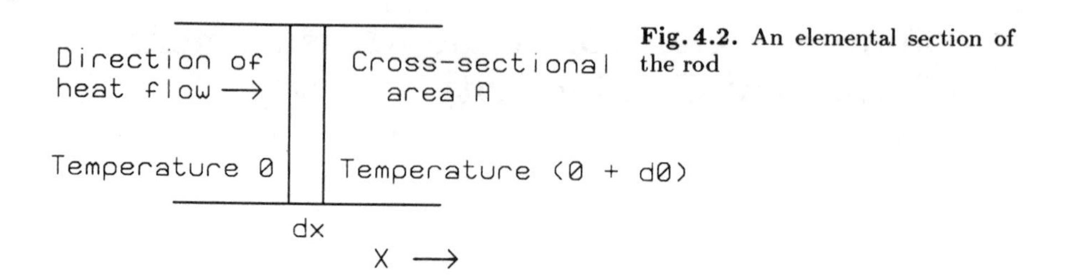

Fig. 4.2. An elemental section of the rod

(i) proportional to the temperature gradient for fixed cross-sectional area

$$\frac{\Delta Q}{\Delta t} \propto \frac{\Delta \theta}{\Delta x}$$

(ii) proportional to the cross-sectional area for a fixed temperature gradient

$$\frac{\Delta Q}{\Delta t} \propto A \, .$$

Thus the rate of flow of heat is given by the expression

$$\frac{\Delta Q}{\Delta t} \propto A \frac{\Delta \theta}{\Delta x}$$

or

$$\frac{\Delta Q}{\Delta t} = -KA \frac{\Delta \theta}{\Delta x}$$

where K is a constant of proportionality, called the thermal conductivity. (The negative sign arises because heat flows from a region of high temperature to a region of low temperature, or x increases as θ decreases.) The temperature gradient $d\theta/dx$ is negative. Thus the sign of K will be positive. If we take the limiting case of this expression when $\Delta x \to 0$, then

$$\frac{dQ}{dt} = -KA \frac{d\theta}{dx} \, . \tag{4.1}$$

This is the equation which defines thermal conductivity; the justification for it is based on experimental evidence.

4.1.1 Measurement of Thermal Conductivity with Parallel Heat Flow

It is assumed here that the rod along which the heat is flowing is perfectly insulated so that no heat is lost radially. The lines of heat flow (that is, the lines which indicate the directions in which the heat is being conducted) are parallel. This is illustrated in Fig. 4.3 in which a bar of length X and cross-sectional area A is placed between a heat supply at temperature θ_1 and a heat sink at temperature θ_2 ($\theta_1 > \theta_2$). In this case the temperature gradient $d\theta/dx$ is the same for all positions along the rod. Thus, after a long time, in the steady-state conditions

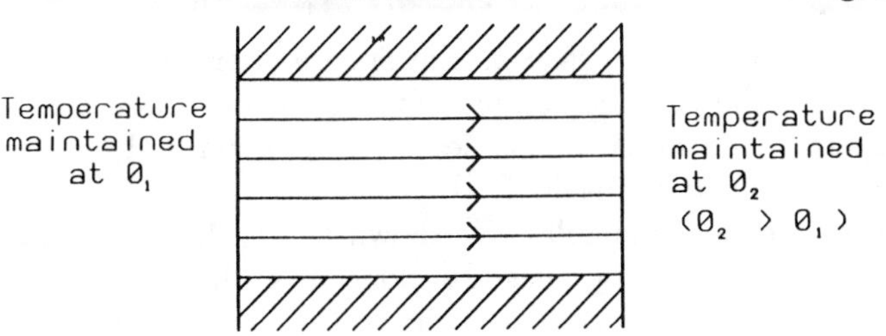

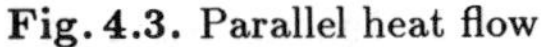

Fig. 4.3. Parallel heat flow

$$\frac{d\theta}{dx} = \frac{\theta_2 - \theta_1}{X} = \frac{-(\theta_1 - \theta_2)}{X}$$

and

$$\frac{Q}{t} = KA\frac{(\theta_1 - \theta_2)}{X} \ . \tag{4.2}$$

This is known as Fourier's equation of heat flow.

Thus one way to measure the thermal conductivity of a material is to supply heat, by electrical or steam heating, at a steady rate to one end of a rod and to measure the resultant temperature gradient in the steady state attained after a long time. The simplest arrangement is to ensure that the flow of heat through the sample material is parallel to its axis. This is the case when the material is in the form of a cylinder of uniform cross section which is well insulated. The dimensions of the apparatus used will depend on the expected value of the thermal conductivity. The Searle's apparatus [4.1], shown in Fig. 4.4, is designed so that the rate of the flow of heat can be measured conveniently.

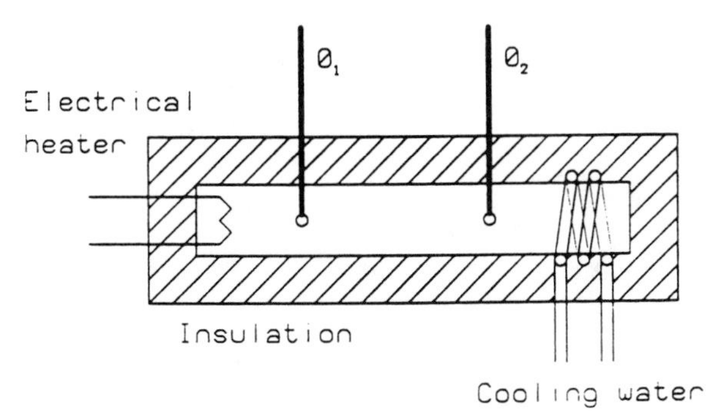

Fig. 4.4. Searle's experimental apparatus

4.1.2 Measurement of Thermal Conductivity with Non-Parallel Heat Flow

In practice it is difficult to ensure perfect insulation. An alternative experiment used to determine the thermal conductivity of a good conductor is that of Forbes [4.2]. In this a metal bar is heated at one end and heat is allowed to escape along its entire length.

Consider the section of a rectangular bar shown in Fig. 4.5 which has a heater at one end and is cooled by heat loss along its entire length. Let the heat flux at x be $F(x)$. For the element shown the net heat influx is $F(x)$-$F(x+\delta x)$, the rate of increase of heat in this element is dQ/dt and the rate of loss of heat from the surface of this element is $P\delta x f(\theta)$, where $P=2(a+b)$ is the perimeter of the element and $f(\theta)$ is a function of the temperature. From the principle of the conservation of energy we get:

$$\frac{\partial F}{\partial x}\delta x + \frac{\partial Q}{\partial t} + P\delta x f(\theta) = 0$$

where

$F = -KA\left(\frac{d\theta}{dx}\right)$
$Q = \rho A S \delta x \delta\theta$
$A = ab$
$K =$ thermal conductivity
$\rho =$ density
$S =$ specific heat capacity.

Therefore

$$-KA\frac{\partial^2\theta}{\partial x^2} + \rho AS\frac{\partial\theta}{\partial t} + Pf(\theta) = 0\,. \tag{4.3}$$

In the steady state ($\partial\theta/\partial t = 0$) and assuming $f(\theta) = h\theta$ where h is a constant, then

$$KA\frac{d^2\theta}{dx^2} - Ph\theta = 0$$

therefore

$$\theta = \theta_0 e^{-\alpha x} \tag{4.4}$$

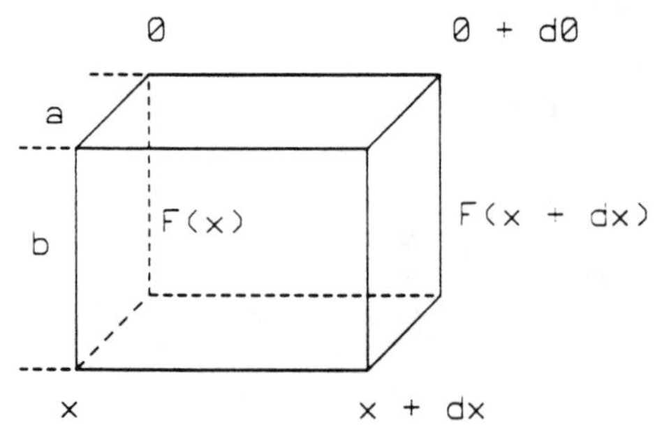

Fig. 4.5. An element of Forbes bar

where

$$\alpha = \sqrt{\frac{Ph}{KA}}\,.$$

In order to find the thermal conductivity K it is usual to carry out a subsidiary experiment to determine the constant h so that it may be eliminated from (4.4). This is done by heating a small piece of the rectangular bar to a higher temperature and allowing it to cool whilst observing its temperature. In this subsidiary experiment the cooling from the ends must be taken into account. For this subsidiary experiment $F = 0$, as there is no heat flow into the rod, thus

$$\rho AS\ell \frac{d\theta}{dt} + (P\ell + 2A)\, h\theta = 0$$

where ℓ is the length of the small piece of bar, therefore

$$\theta = \theta_0 e^{-\beta t} \tag{4.5}$$

where

$$\beta = \frac{\left(\frac{P}{A} + \frac{2}{\ell}\right) h}{\rho S}\,.$$

From (4.4) and (4.5)

$$K = \frac{\beta}{\alpha^2} \frac{\rho S}{\left(1 + \frac{2A}{P\ell}\right)}\,. \tag{4.6}$$

Thus in order to determine K by the method of Forbes two experiments are necessary. The temperature distribution along the bar in the steady state when one end is heated and heat is lost along the entire length yields α from the slope of a plot of $\log_e \theta$ against x. The second experiment leads to a value of β from the slope of a plot of $\log_e \theta$ against t. It is assumed that data for density ρ and specific heat capacity S are known for the material investigated.

4.2 Experimental Considerations

The general arrangement of equipment needed to carry out the Forbes bar experiment is shown in Fig. 4.6. The experiment is most conveniently carried out on a square steel bar of about 75cm length and 1cm^2 cross section, with thermocouple holes about 4cm apart along the bar. The heating element used is a soldering iron encapsulated cylindrical heater of 30W and about 5cm long by 4mm diameter.

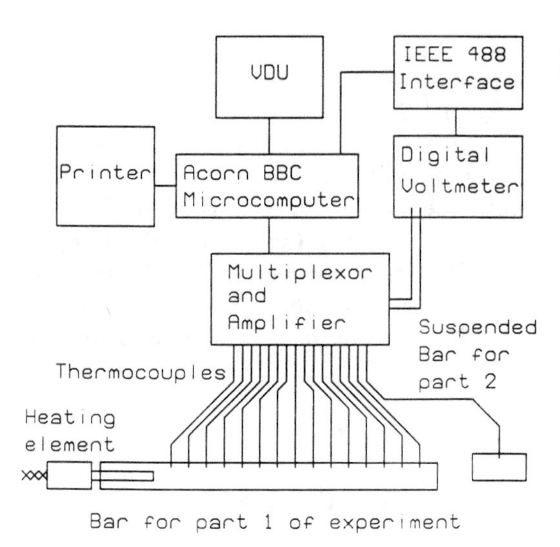

Fig. 4.6. The hardware connections

4.2.1 The Thermocouple as a Temperature Measuring Device

There is a wide variety of temperature measuring devices available, such as thermocouples, thermistors, mercury in glass thermometers, solid state sensors and resistance temperature detectors. The device which is most suitable for this particular application is the thermocouple. Thermocouples are inexpensive and accurate over the temperature range of this application; they also have reasonably good long term stability. Thermocouples possess a small sensing junction, and, because of their small size, they respond quickly to changes in temperature. This is especially important in the dynamic part of the Forbes' Bar experiment. It is for these reasons thermocouples were chosen as the temperature measuring devices.

Basically, a thermocouple consists of two junctions between dissimilar electrical conductors, usually either pure metals or alloys, in the form of thin wires. In use the two junctions are at different temperatures. A different electromotive force (e.m.f.) is generated at each junction by virtue of the fact that the number of free electrons in a metal depends on the composition of the metal and the temperature. The net e.m.f. is a measure of the temperature difference between the two junctions. The net e.m.f. can be measured if one of the wires is cut and the two free ends are connected to an instrument which is capable of measuring the e.m.f. developed. The two new junctions formed in this way should be at the same temperature so that any e.m.f. generated at these junctions produces no net effect.

When the temperature of the sensing junction differs from that of the other junction, the reference junction, the electromotive force generated is dependent upon the temperature difference between the two junctions. If the e.m.f.s which are generated at known temperatures are used to calibrate the thermocouple then the temperature of the sensing junction may be found.

Thus the thermocouple is a differential rather than an absolute measuring device and the temperature of the reference junction must be known if the temperature of the sensing junction is to be inferred from the overall output voltage.

The most common pairs of alloys used to form thermocouples are Copper/Constantin (T-type) and Chromel/Alumel (K-type), for which tables of thermocouple output voltages at specific temperatures exist, usually expressed with the reference junction maintained at zero degrees Celsius.

4.2.2 The AD595 Thermocouple Amplifier Integrated Circuit

To remove the necessity of using an ice bath to maintain the reference junction of all the thermocouples at zero degrees Celsius, a dedicated microcircuit has been developed by Analog Devices, which eliminates the need for an ice-bath.

The Analog Devices AD595 thermocouple amplifier integrated circuit relies on the fact that the temperature dependence of the sensitivity of silicon integrated circuit transistors is quite predictable and repeatable. Thus it is possible to produce a temperature-related voltage to compensate the reference junction of the thermocouple. The AD595 produces an internal voltage that is identical to that which would be produced by an external reference junction at 0° C. This reference junction is formed on the integrated circuit itself, thus requiring just the two leads from the sensing junction to be connected to this integrated circuit. The AD595 also houses an instrumentation amplifier. Combined with the ice-point compensator, a final output voltage of 10 millivolts per degree Celsius is produced directly from an input type K thermocouple voltage. A block diagram of the AD595 is shown in Fig. 4.7.

The AD595 behaves like two differential input amplifiers, with the summed output being used to control a high-gain amplifier. The output of the amplifier is fed back to the input via a differential inverting amplifier.

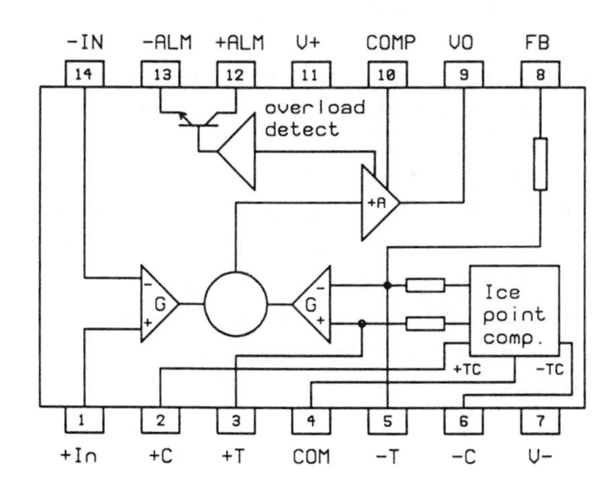

Fig. 4.7. The AD595 thermocouple amplifier

This feedback signal is driven to reduce the differential input signal to a small value. Provided that the gains of both differential amplifiers are identical, the feedback signal to the right hand differential amplifier will equal the thermocouple signal when the difference signal is zero.

The reference junction compensation voltage, which is proportional to the Celsius temperature of the AD595, adjusts the differential input so that the final amplifier output is adjusted, in turn, to restore the input to match the thermocouple voltage. Its effect on the final output is also ten millivolts per degree Celsius. Thus, the compensation voltage adds to the thermocouple voltage a voltage proportional to the difference between the temperature of the AD595 and zero degrees Celsius, thereby providing an ice-point reference.

The AD595 also provides an open circuit detector which can be used to light an LED should one of the thermocouple leads break or come loose. It is important to provide adequate high-frequency decoupling in the power supply leads of the AD595 to prevent the incidence of spurious signals in the output from the selected thermocouple.

4.2.3 Thermocouple Accuracy

Due to variations in the composition of the wires of commercially available thermocouples, the voltages they produce may only conform to within three or four degrees Celsius, even over relatively small temperature ranges. In measurement situations where temperature differences or temperature gradients are measured it is more important that the thermocouples match and track accurately. Making all the thermocouples needed for the experiment from the same reel of thermocouple wire ensures uniform composition and greatly improves the matching accuracy. By silver soldering the thermocouple junctions, using the same silver solder for each, uniformity is also enhanced and also a small junction size is achieved. By inserting the thermocouple into small holes drilled in the metal bar, the temperature of each individual isothermal can be measured due to the small junction size. Provided that the holes are drilled small enough for the thermocouples to fit snugly, thermal equilibrium between the bar isothermal and the couple will be established quickly, which also improves the accuracy of the measured temperature.

Thermocouples made from new wire tend to produce e.m.f.s which change gradually with time. This effect can be overcome by artificially ageing the wires by heating them to the maximum temperature which they will be measuring. This is achieved when the junctions are soldered at a high temperature. It is essential that temperature differences are prevented from existing at points of secondary couples. The thermocouple input connections to the AD595 must therefore be maintained at the same temperature.

4.2.4 Calibration of the Thermocouples

For the highest accuracy, the thermocouples should be three-point calibrated. This involves recording the voltage produced by each thermocouple at three known temperatures. The ice point and steam point of water provide two convenient calibration temperatures, but since at least three of the thermocouples will be reading temperatures in excess of one hundred degrees Celsius, a higher calibration temperature is also thought to be necessary. The hot end of the bar, after steady state has been achieved, provides a convenient third calibration point if its temperature can be measured with a calibrated thermometer.

By comparing the amplified output of the AD595 generated with each of the sixteen thermocouples it is possible to identify any of the thermocouples which exhibit a significantly different response to the others for any particular reason. In this way the thermocouples to be used in the experiment can be chosen as those which have similar response over the temperature range under consideration.

4.2.5 Thermocouple Selection Multiplexing Circuit

All sixteen thermocouples will be producing voltages which are temperature dependent. Since only one thermocouple measurement is possible at any one time, some form of selection circuit for one of the sixteen thermocouples is required. In addition, for full automation to be achieved this selection should be made under software control. The multiplexor is essentially an electronic single-pole, sixteen-way switch to provide the function shown in Fig. 4.8.

The HI 506-5 is a monolithic one of sixteen analogue CMOS multiplexor and it allows one input signal to be selected as the output signal depending on the state (logic 1 or 0) of four control lines. Two multiplexors are used so that the thermocouples can be individually selected. It is not possible to use a single multiplexor and a common lead for the second thermocouple lead since all the thermocouples will be in electrical contact with the steel bar at their junctions in order to measure the temperature at that point. (Good thermal contact is best obtained by a metal-to-metal contact which also forms good electrical contact).

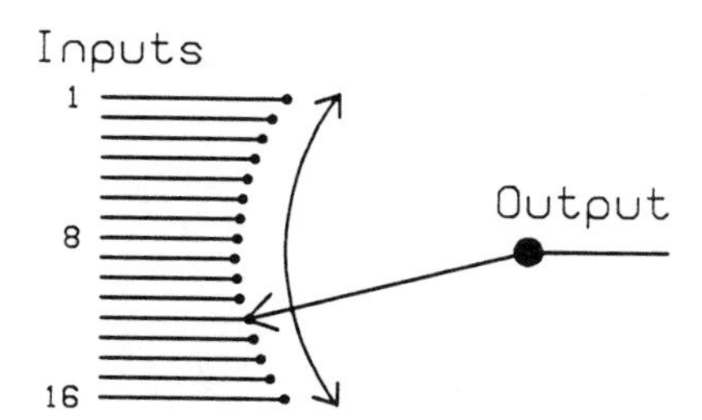

Fig. 4.8. The 16 way multiplexor function

4.2.6 Multiplexor Control

There are a number of ways in which the multiplexor may be controlled by a microcomputer using appropriate high level software and a hardware interface. The most widely available solution would be to use the parallel printer interface - but this solution is not advocated as it prevents printing at the same time. A better solution is to use a peripheral interface device such as the Intel 8255 or the Rockwell 6522. These devices are available either directly in some microcomputers, such as the BBC Acorn, or as plug-in interface boards for microcomputers such as the Apple or IBM PC. All that is necessary is direct connection of appropriate output pins of the 8255 or 6522 to the HI 506-5 and the output in software of the appropriate binary data to the registers of the device used. The data sheets for these devices provide all the necessary information. It is important here again to remember to provide adequate decoupling in the power supply leads to minimize the effects of high speed transients.

4.2.7 The IEEE-488 Bus Interface Unit

The Institute of Electrical and Electronic Engineers standard number 488 [4.3] has become the standard for interconnecting both programmable and non-programmable electronic apparatus to assemble complete instrumentation systems.

The IEEE-488 bus allows instruments to communicate with one another over a localised area, in this case a microcomputer and a digital multimeter (DVM). The computer constitutes the system manager which is known as the Controller. Both the computer and the voltmeter have the ability to *'listen'* and *'talk'*, The DVM would normally be the talker as it sends readings to the computer which is usually the listener. The computer, as system controller, issues the commands to tell the DVM whether to talk or listen by virtue of the fact that the DVM has a unique primary address which distinguishes it from any other instruments which may be connected to the bus.

The DVM also has secondary addressing capability which enables specific functions, such as the voltage range, to be selected by the computer remotely. More detailed information on the programming of the IEEE-488 interface bus is provided in the next section.

4.2.8 The Control and Measurement Software

The main tasks for the software are that it should (i) control the multiplexor selection circuitry, (ii) control the IEEE-488 bus interface and digital voltmeter, (iii) collect and display the thermocouple temperatures, (iv) calculate the thermal conductivity of the metal bar and (v) guide the experimenter through the experiment. The two measurement programs are listed in Appendix 4.A.

a) Control Commands The detailed method of sending control data to interfaced devices depends on the particular microcomputer and language used. In BBC Basic the command ?&FE60 = &1F sends the binary 00011111 to address FE60 (in hex). This and similar commands will be seen in the listing of the control program.

b) Control of the IEEE-488 Bus Interface and Digital Voltmeter The IEEE-488 bus interface used is provided with a library of procedures and functions on disc. Use has been made of a number of these routines so a brief explanation of each is given.

(i) PROCIEINIT (A%, S%) This is the initiation routine for the support software and has to be called before any other routines which access the bus. The argument A% is the device address of the bus interface unit, which is required by the other instruments on the bus for them to be able to communicate with the interface. In this case it is arbitrarily set to zero. The argument S% allows the interface unit to act either as system controller or as a passive device. As the unit is required to be system controlled, S% is set as TRUE. PROCIEINIT initialises the system variable, clears the bus and sends a remote enable (REN) signal to the voltmeter to enable it to be programmed by the interface unit.

(ii) PROCIEWRIT (D%, S%) This routine sets the bus system up ready to send data to any remote device. Effectively it sets the interface unit up as Talker and sets the remote device with the bus address D% to be sole Listener. The argument S% is an optional secondary address for the remote device. The digital voltmeter has a default bus address of 7, and since no secondary address is used, S% is set to −1.

(iii) PROCIEREAD (D%, S%) This routine performs the opposite function of PROCIEWRIT, and sets the bus system up ready to receive data from the voltmeter. Effectively it sets the voltmeter (D% = 7) up as Talker, sets the interface unit to be the sole listener and releases control ready for data transfers.

(iv) PROCIEPUTS (S$) This routine is used in conjunction with PROCIEWRIT to transmit data to remote devices. A string, S$ is sent over the bus with an end-or-identify (EOI) character terminating it. For this application, PROCIEPUTS is used to set the range of the voltmeter to range 2, to select the DC volts function and to set the display resolution to five and a half digits. This is achieved in the following two lines:

```
PROCIEWRIT (7,-1)
PROCIEPUTS ( "A1B0F0G0K0M0N0P0Q0R2S1T4W0Y0Z0X")
```

The commands are executed by sending an "X" character at the end.

(v) PROCIEGETA (A%, T%, L%) This routine is used in conjunction with PROCIEREAD to receive data from the voltmeter. The received data is stored in array A%. The incoming bytes are compared with T% to test for a terminating byte, and L% is the maximum number of bytes that may be received. This procedure is used to read all the voltmeter readings from the amplified thermocouple signals into the computer in the following routine:

```
PROCIEREAD (7,-1)
PROCIEGETA (IEBUF,13,14)
READING = VAL$ (IEBUF)
TEMPERATURE = READING*100
```

The received bytes are stored in the array IEBUF% and are terminated with the carriage return (ASCII code 10) character. The bytes read into IEBUF% are in a string which can be converted into a number using the VAL function.

A typical string of bytes sent over the bus might be:

" DCV1.042839"

which is converted into the number 1.042839, and since ten millivolts is equal to one degrees Celsius, multiplying this number by 100 will give the temperature as 104.2839 C.

(vi) PROCIECLR This is the only other procedure from the library that has been used. It is used to regain control once all data transfers have been completed. Essentially it unaddresses the talker and listeners and tidies the bus.

In order to explain how each of the software aims have been achieved each of the programs written to accompany the hardware will be discussed in turn. It should be noted that these programs load consecutively and the only reason why one combined program was not used was due to memory constraints. Full listings of the two measurement programs (in Acorn BBC Basic) are given in Appendix 4.A.

c) The PART1 Program This program has to be loaded in by the user. It includes a machine code routine for a screen dump which is used by it and by the PART2 program. The machine code is specific to the particular hardware used and for this reason it is not listed in the appendix with the rest of the program. The PART1 program is used throughout the first part of the experiment, in which the temperature distribution along the bar is measured to obtain a value for α in (4.4). The structure chart in Fig. 4.9 gives an overview of the PART1 program.

The program is made up of a number of procedures which are used to get the required data and to allow the experimenter to follow the changes in temperature distribution as the bar heats up to the steady state. The main procedures are described below.

PROCINITIALISE initialises the IEEE-488 bus, disables the front panel of the voltmeter, selects all the user port lines as outputs in the data direction register and dimensions the arrays used in the program.

PROCRECDATA1 calls PROCAXIS1 to draw, scale and label the axes for the graphs, takes readings of temperature from each of the fifteen thermocouples to a pre-programmed repeatability, plots these temperatures on the graph, connects all the ordinates with a straight line and, if required, sends a copy of the screen to the printer. A typical graph, obtained when

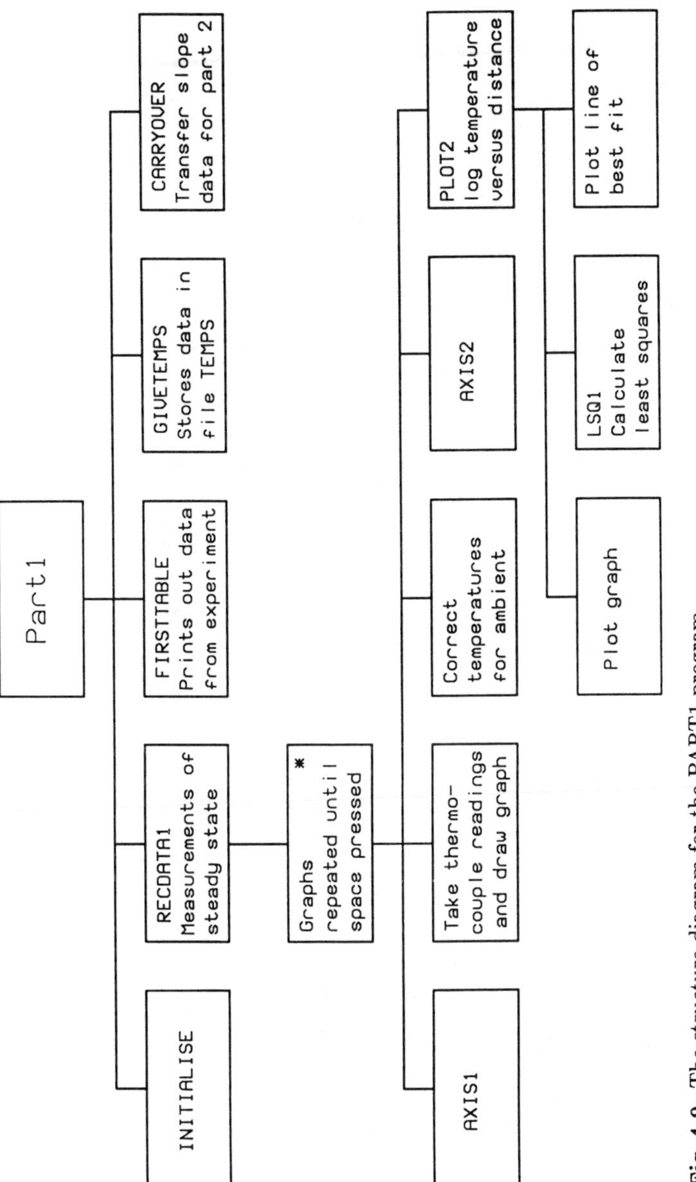

Fig. 4.9. The structure diagram for the PART1 program

the bar has reached the steady state, is shown in Fig. 4.10. After each set of readings of all fifteen thermocouples and the display of the graph of temperature against distance, a graph of the $\log_e$ of temperature against distance along the bar is plotted.

PROCAXIS2 draws, scales and labels the axes for the graph.

PROCPLOT2 plots the graph of $\log_e$ temperature against distance.

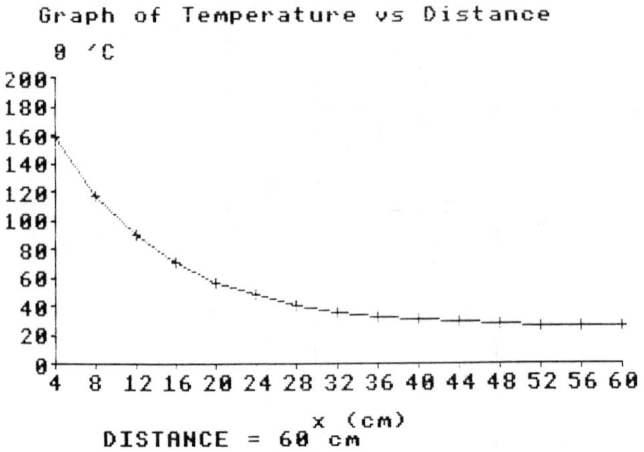

Graph of Temperature vs Distance

θ ′C

200
180
160
140
120
100
80
60
40
20
0

4 8 12 16 20 24 28 32 36 40 44 48 52 56 60

x (cm)

DISTANCE = 60 cm

Fig. 4.10. Typical output obtained during the first part of the experiment as the steady-state conditions are approached

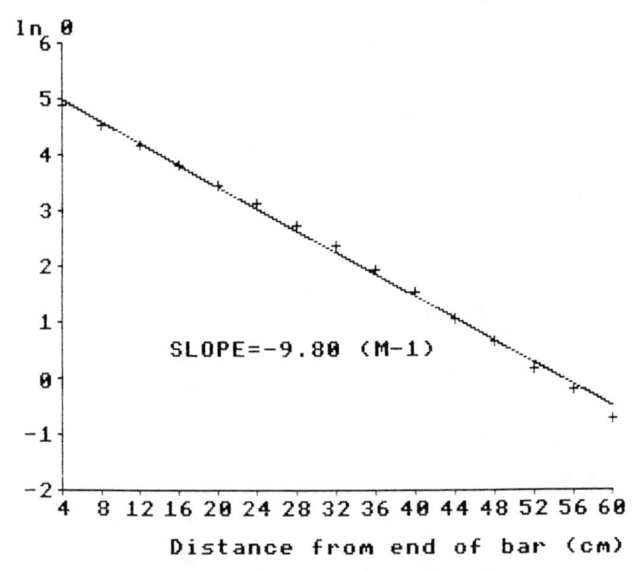

Fig. 4.11. Plot of $\log_e$ of temperature against distance from steady-state data to obtain α in (4.4)

PROCLSQ1 is called to calculate the line of best fit, which is then plotted and the slope is given on screen. A hard copy may then produced on the printer. Figure 4.11 shows a graph of $\log_e$ of temperature against distance along the bar. A line of best fit is calculated and drawn and a hard copy is produced, if required, on the printer.

PROCLSQ1 uses the method of least squares to calculate the line of best fit from the data.

From (4.4) it can be seen that this $\log_e$ graph will be a good straight line when the steady state is reached. The experimenter is given a short while to indicate that the data obtained is satisfactory for the first part of the experiment, after which the procedure repeats the process described above for another set of readings of the fifteen thermocouples. When a satisfactory set of data is indicated by the experimenter the program moves on to the next procedure to print the data obtained.

PROCFIRSTTABLE produces a table of the position and temperature of each thermocouple and the natural logarithm of its temperature and copies it to the printer.

PROCGIVETEMPS creates a file called "TEMPS" on the disc and stores the fifteen temperatures measured by the program for use in the PART2 program.

PROCCARRYOVER creates a file called "FROM1" on the disc and stores the value of α obtained for the slope of the graph for use in program PART2.

d) Program PART2 This program performs the second part of the experiment. The structure chart in Figs. 4.12 and 4.13 gives an overview of this program.

The temperature of the small bar is measured every ten seconds for two thousand seconds as it cools from a high temperature. A table of temperature as a function of time is sent to the printer, and graphs of temperature against time and log_e of temperature against time are plotted and sent to the printer. Typical graphs obtained for the cooling experiment are shown in Figs. 4.14 and 4.15.

Finally the program prompts for the physical constants needed and the thermal conductivity is calculated. The procedures are as follows:

PROCINITIALISE initialises the IEEE-488 bus, selects all the user port lines as outputs, chooses thermocouple 16 to be used for all readings and dimensions the arrays used in the program.

PROCPART2 instructs the user how to set up the second part of the experiment, reads in the temperature of the small bar every ten seconds and produces a table of the temperatures on the printer.

PROCPLOT3 plots a graph of temperature against time and copies it to the printer if required.

PROCPLOT4 plots a graph of $\log_e$ temperature against time and copies it to the printer if required.

PROCLSQ2 uses the method of least squares to calculate the equation of the line of best fit.

PROCGIVEADVICE prompts for various constants to be entered and calls PROCROUGH.

PROCROUGH calculates the thermal conductivity of the metal of the bar based on all the data collected.

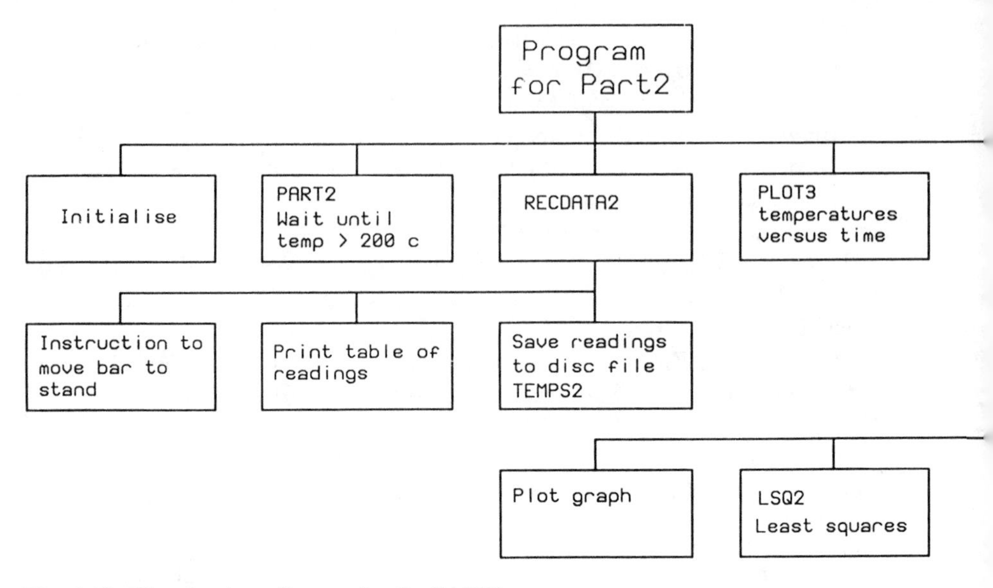

Fig. 4.12. The structure diagram for the PART2 program

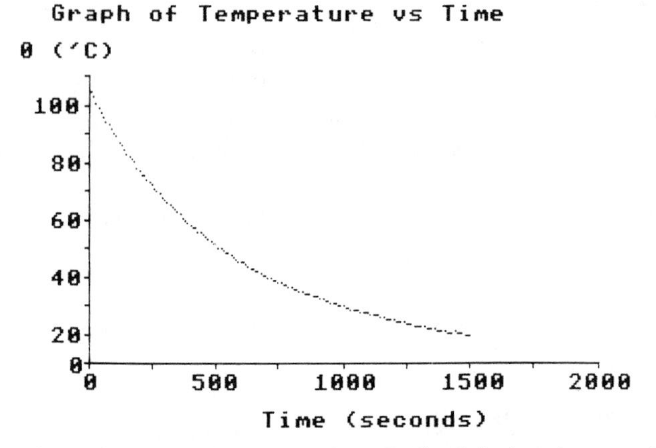

Fig. 4.14. The cooling curve data obtained during the second part of the experiment

e) Calculation of Thermal Conductivity To calculate a value for the thermal conductivity, the software prompts for the values of some physical constants. These are as follows:

(i) The width, a, and breadth, b, of the large bar.
(ii) The length, ℓ, of the small bar.
(iii) The density, ρ , of the metal used.
(iv) The specific heat, S, of the metal used.

For testing purposes, using mild steel, these values were taken to be:

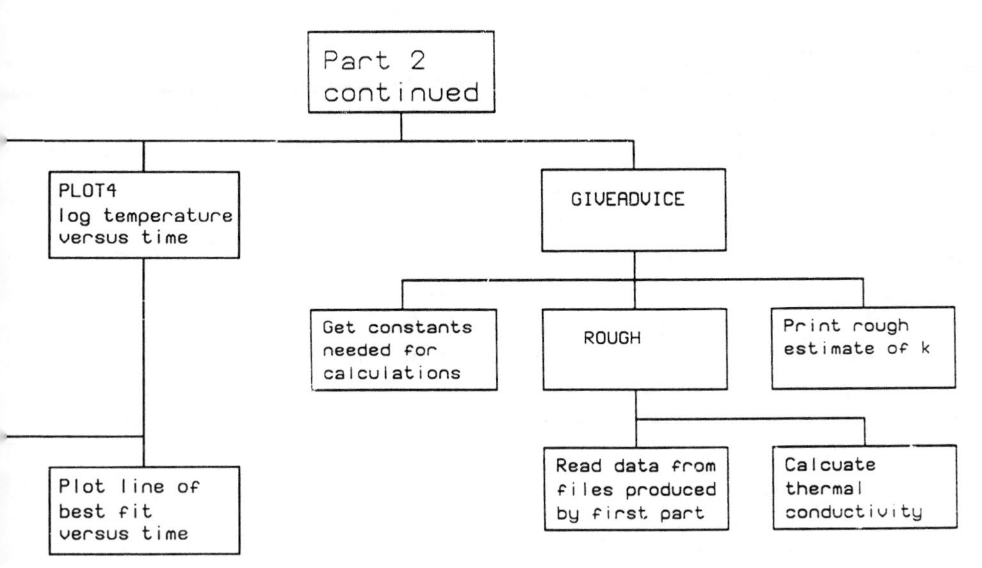

Fig. 4.13. Continuation of the structure diagram for the PART2 program

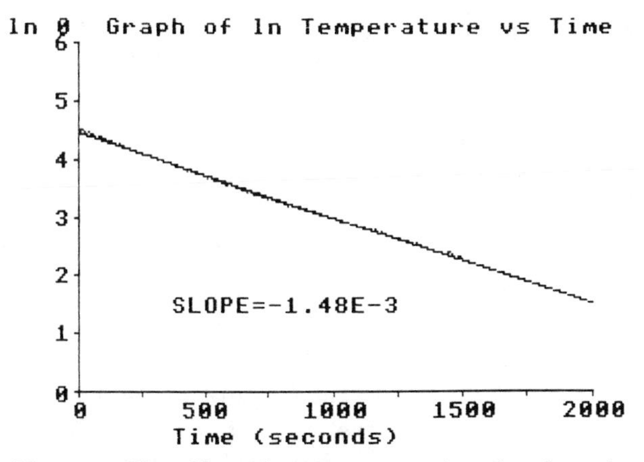

Fig. 4.15. Plot of $\log_e$ of temperature against time from the cooling curve data to obtain β in (4.5)

$$a = b = 0.0127\,\mathrm{m}$$
$$\ell = 0.043\,\mathrm{m}$$
$$\rho = 5800\,\mathrm{kgm}^{-3}$$
$$S = 420\,\mathrm{Jkg}^{-1}\mathrm{K}^{-1}$$

A typical value for the thermal conductivity using measured data and the equations given is 58.1 $\mathrm{Wm}^{-1}\mathrm{K}^{-1}$. The exact composition of the mild steel from which the bar was made is not known, but data books give values for the thermal conductivity ranging between 50 to 500 $\mathrm{Wm}^{-1}\mathrm{K}^{-1}$ for different compositions of polished and unpolished steels.

4.2.9 Discussion of the Experiment

There are several possible sources of error, which the experimentalist should investigate to explain any difference between the expected and observed values of K. From a design viewpoint, the cool end of the bar should ideally be at the same temperature as the surroundings. This is difficult to achieve in practice since heat should be allowed to flow across all of the bar to allow a significant temperature gradient to exist. The value used for room temperature has a significant effect on the calculated value of the thermal conductivity. Thermocouple sixteen can be used during the first part of the experiment to measure room temperature or an offset can be subtracted from thermocouple fifteen which can be shown to be close to room temperature. The measurement of thermal conductivity by this method depends upon the degree to which Newton's law of cooling is valid for the conditions during the experiment. It is advisable to arrange the apparatus so that it is not affected by draughts or variable environmental conditions. The thermocouple junctions have a finite size, which introduces a source of error since heat is conducted from the bar to the thermocouples and their position along the bar is not well defined.

The method used to calculate the thermal conductivity has been simplified from the method that Forbes used, in the way that the emissivity of the bar has been incorporated. Ideally, the two graphs of temperature against time and temperature against distance should be used to construct a third graph of $d\theta/dt$ from the cooling curve for each value of θ at known distances x from the temperature against distance curve. The area under this curve can then be used to find the total loss of heat per second between the two ends of the bar using the equation:

$$K\frac{(d\theta)}{(dx)} = S \times \text{ Area under graph} \quad .$$

This would involve some form of numerical integration for accurate results to be obtained, but could be incorporated into the existing program.

The program asks for the specific heat of the bar, S, to be input, but it would be possible to include the calculation of S in the program by guiding the experimentalist through the method of mixtures experiment. This could be added as a separate program which is run before the main Forbes' bar experiment.

A possible extension to the experiment would be to provide a selection of bars of different metals such as copper and aluminium, since the automation allows the experiment to be performed quite easily and in a relatively short amount of time. Comparisons could then be drawn between the properties of the different metals. Alternatively the experiment could be extended to involve the measurement of the heat capacity of a piece of the bar so that the all data about the material of the bar could be obtained by the experimenter.

4.3 The Computer Simulation

In both Searle's apparatus discussed in Sect. 4.1.1 and in Forbes' experiment discussed in Sect. 4.1.2 the measurements should only be taken when a steady state has been reached. In this simulation the intermediate temperature distributions are calculated and the time taken for the final "steady state" to be reached can be investigated. In theory it will take an infinite time for the true steady state to be reached. The simulation allows the accuracy of the measurements taken after various finite times to be studied, and thus actual experimental investigations can be enhanced by a better understanding of the approach to the "steady state".

The program uses an explicit finite difference approximation to solve the partial differential equations governing the heat flow. Consider a bar of length ℓ, cross sectional area A, density ρ, specific heat capacity S, thermal conductivity K, perimeter or circumference P and with heat transfer coefficient at the surface h. For the finite difference method the bar is considered to consist of a series of linked segments, each represented by a node at its centre. The temperature of the ith node after the nth time interval is T_i^n. The time step is taken to be δt and the nodes are evenly spaced a distance δx apart.

Consider the energy balance in node i within the material. We get rate of change of energy content = Σ rate of flow of heat in and out of segment.

$$\delta x A \rho S \frac{(T_i^{n+1} - T_i^n)}{\delta t} = \frac{KA}{\delta x}(T_{i-1}^n - T_i^n) + \frac{KA}{\delta x}(T_{i+1}^n - T_i^n) + h\delta x P(T_{ext} - T_i^n)$$

where T_{ext} is the external ambient temperature.

Re-arranging, we get

$$T_i^{n+1} = T_i^n + \frac{K\delta t}{\delta x^2 \rho S}(T_{i+1}^n + T_{i-1}^n - 2T_i^n) + \frac{\delta t h P}{\rho S A}(T_{ext} - T_i^n) \ . \qquad (4.7)$$

For stability we require

$$1 - \frac{2K\delta t}{\delta x^2 \rho S} - \frac{\delta t h P}{\rho S A} > 0$$

that is the coefficient of $T_i^n > 0$. The value of δt used is given by

$$\delta t = 0.9 / \left(\frac{2K}{\rho S \delta x^2} + \frac{hP}{\rho S A} \right) \ .$$

For the node at the heater end, it was assumed that the heater had a heat capacity equivalent to that of a length of the bar of $\delta x/2$. It was assumed that there is no heat loss out of the end of the heater away from the rod. The equation for energy balance is similar to (4.7) above with the addition of a term

$$\frac{VI\delta t}{\rho S A \delta x}$$

on the right hand side, where V is the voltage across and I is the current through the heater. The first bracketed term on the left-hand side becomes T_{i+1}^n - T_i^n since T_{i-1} no longer exists and the 2 factor is reduced to 1. The same stability condition applies.

For the node at the cold end (heat sink) water is assumed to flow through a coil with a temperature on entry of T_{in} and a flow rate of R litre per minute. The mass of the cooling coil was taken to be equivalent to a length of bar of $\delta x/2$. The energy balance equation then becomes

$$\rho SA\frac{\delta x}{\delta t}\left(T_i^{n+1}-T_i^n\right)=\frac{KA}{\delta x}\left(T_{i-1}^n-T_i^n\right)+\left(T_{in}-T_i^n\right)\frac{R4200}{60}$$

where 4200 JKg^{-1} is the specific heat capacity of water.

However, this energy balance equation possesses instabilities and so the final term was taken with T_i^{n+1} rather than T_i^n. The outflowing water temperature is taken to be that of the node, so that we get

$$\left(\frac{\rho SA\delta x}{\delta t}+70R\right)T_i^{n+1}=\frac{KA}{\delta x}T_{i-1}^n+\left(\frac{\rho SA\delta x}{\delta t}-\frac{KA}{\delta x}\right)T_i^n+T_{in}70R$$

and

$$\begin{aligned}T_i^{n+1}=&\left(KA/\delta x\left(\frac{\rho SA\delta x}{\delta t}+70R\right)\right)T_{i-1}^n\\&+\left(\left(\frac{\rho SA\delta x}{\delta t}-\frac{KA}{\delta x}\right)/\left(\frac{\rho SA\delta x}{\delta t}+70R\right)\right)T_i\\&+70T_{in}R/\left(\frac{\rho SA\delta x}{\delta t}+70R\right).\end{aligned}$$

The stability condition here is that the second term has a coefficient > 0, which is automatically satisfied if the previous condition is satisfied.

The program is listed in Appendix 4.B and is extensively commented and includes extensive explanatory text. The data inputs are all prompted and tested for acceptable range. The simulation can be done both in terms of Searle's experiment and Forbes's experiment. Putative values of thermal conductivity and instantaneous temperature distributions are produced throughout the simulations. Careful examination of the output data will lead to an enhanced understanding of the characteristics of these two experiments. Typical output from the program illustrating the approach to thermal equilibrium is shown in Figs. 4.16 and 4.17.

References

[4.1] G.F.C. Searle: Phil Mag **9** 125 (1905)
[4.2] A.G. Worthing and D. Halliday: Heat p169 Chapman and Hall New York (1948)
[4.3] IEEE Standard 488-1978 "Digital Interface for programmable instrumentation", The IEEE, Inc., 345 East 47th St., New York, NY, 1978

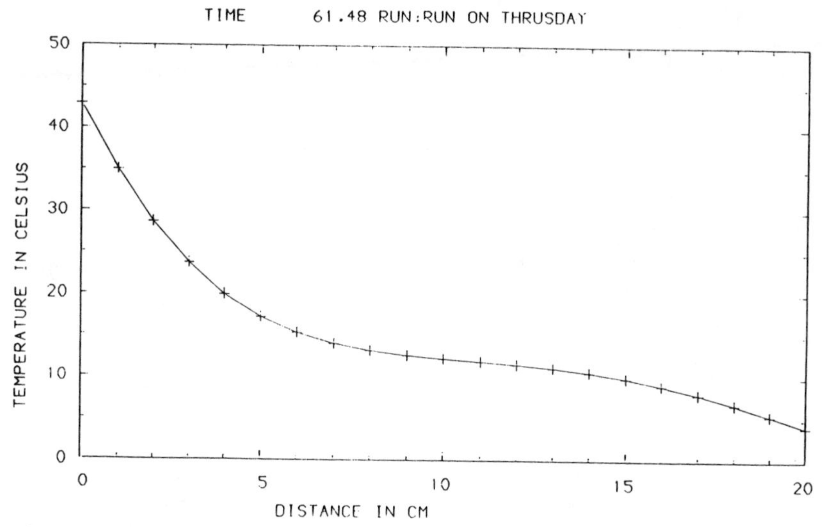

Fig. 4.16. Output from the simulation program as the Forbes bar is heated up

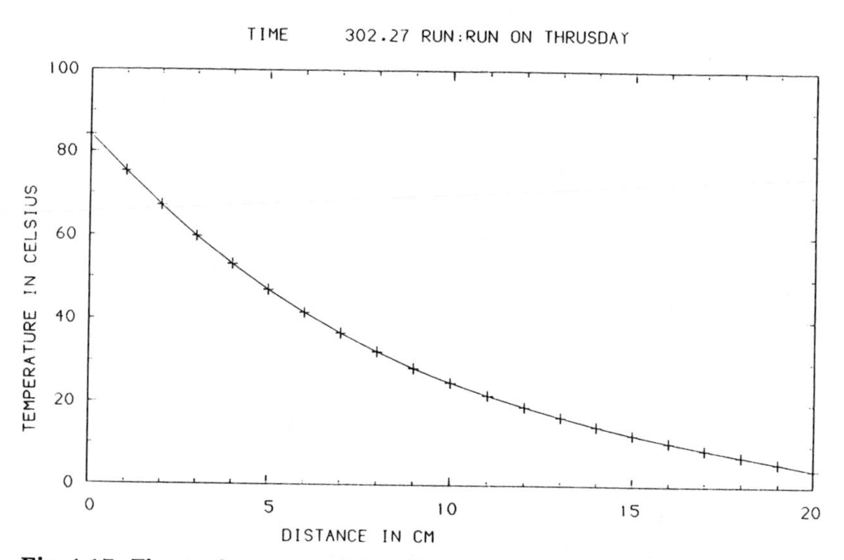

Fig. 4.17. The steady-state conditions illustrated by the simulation program

Appendix 4.A

Program for First Part of Experiment

```
  10 REM PART 1 OF FORBES BAR EXPERIMENT
  20 PROCINITIALISE
  30 PROCRECDATA1
  40 PROCFIRSTTABLE
  50 PROCGIVETEMPS
  60 REPEAT UNTIL FNgetkey=" "
  70 PROCCARRYOVER
  80 CHAIN "PART2" : REM Load and run PART2 program.
  90 END

5000 DEF PROCINITIALISE
5010 PROCIEINIT(0,TRUE)
5020 PROCIEWRIT(7,-1)
5030 PROCIEPUTS("A1B0F0G1K0M0N1P0Q0R2S1T4W0Y2Z0X")
5040 REM Set up DVM over IEEE 488 bus
5050 ?&FE62=&FF:?&FE60=30 : REM Write directly to
     memory to set up 6522 VIA.
5060 DIM TEMPONE(15),LTEMPONE(15),DIST(15)
5070 FACTOR=100:RDACC=0.00001:OFSET=0.5
5080 T%=13:L%=13 : REM Used for IEEE 488 bus message
     terminator, ASCII code for carriage return.
5090 FOR NUM=1TO15:DIST(NUM)=4*NUM:NEXT
5100 MODE 4 : REM 320 x 256 PIXEL (40 x 32 CHAR) 2-COLOUR
GRAPHICS MODE
5110 ENDPROC

7000 DEF PROCRECDATA1
7010 PROCIEREAD(7,-1):PROCIEGETA(IEBUF%,T%,L%)
7020 REPEAT
7030   PROCAXIS1
7050   FOR THERMOCOUPLE=1 TO 15
7060     ?&FE60=(15+THERMOCOUPLE) : REM select
         thermocouple by setting 6522 VIA
7065     PROCIEGETA(IEBUF%,T%,L%)
7070     N%=1
7090     NUMB2=VAL(FNIEGETS(T%,L%))
7092     REPEAT
7094        NUMB1=NUMB2
7100        NUMB2=VAL(FNIEGETS(T%,L%))
7110        N%=N%+1
7120        NUMB2=NUMB1*(N%-1)/N%+NUMB2/N%
```

```
7130       UNTIL ABS(NUMB2-NUMB1)<=RDACC
7140       TEMPONE(THERMOCOUPLE)=NUMB2*FACTOR
7150       PROClocate(5,23) : PRINT;" DISTANCE = ";
           DIST(THERMOCOUPLE);" cm  "
7155       PROClocate(5,25) : PRINT;" TEMPERATURE =           "
7160       @%=&20205 : REM Decimal number 5 digits wide to 2
           decimal places.
7165       PROClocate(20,25) : PRINT;TEMPONE(THERMOCOUPLE);" 'C  "
7170       @%=10 : REM 10 digit wide decimal or exponent
           format as required.
7180       X=(DIST(THERMOCOUPLE)*20)-80
7185       Y=TEMPONE(THERMOCOUPLE)*2.5
7200       MOVE X-8,Y : DRAW X+8,Y
7210       MOVE X,Y-8 : DRAW X,Y+8
7225       NEXT THERMOCOUPLE
7240     MOVE 0,TEMPONE(1)*2.5
7250     FOR NUM=2TO15:DRAW(DIST(NUM)*20)-80,TEMPONE(NUM)*2.5:NEXT
7260     IF FNcheckkey(500)=" " THEN 7400 : REM Possible entry of
         ambient not implemented here.
7400     AMBIENT=TEMPONE(15)-OFSET
7500     FOR I=1 TO 15
7510       TEMPONE(I)=TEMPONE(I)-AMBIENT
7520       IF TEMPONE(I)<OFSET THEN TEMPONE(I)=OFSET
7590       NEXT
7610     PROCAXIS2
7620     PROCPLOT2
7630     REM Printer dump of graph
7640     PROClocate(5,23) : PRINT;"PRESS SPACE BAR TO CONTINUE"
7650     PROCflush
7660     UNTIL FNcheckkey(1000)=" "
7900 ENDPROC

8000 DEF PROCAXIS1
8010 MODE 4 : REM 320 x 256 PIXEL (40 x 32 CHAR) 2-COLOUR
     GRAPHICS MODE
8020 PRINT TAB(2,0);"Graph of Temperature vs Distance"
8030 VDU29,100;400; : REM Move origin.
8040 MOVE 0,0 : DRAW 1120,0
8050 MOVE 0,0 : DRAW 0,500
8060 FOR X=0 TO 1120 STEP 80
8070   MOVE X,0 : DRAW X,-5 : NEXT X
8080 FOR Y=0 TO 500 STEP 50
8090   MOVE 0,Y : DRAW -5,Y : NEXT Y
8100 PROClocate(2,2) : PRINT;" 0 'C"
8110 PROClocate(18,22) : PRINT;" x (cm)"
```

```
8120 VDU5 : REM enable character printing at any pixel position.
8130 FOR NUM=4 TO 8 STEP 4
8140   MOVE -382+NUM*20,-20: PRINT NUM
8150   NEXT NUM
8160 FOR NUM=12 TO 60 STEP 4
8170   MOVE -363+NUM*20,-20 : PRINT NUM
8180   NEXT NUM
8190 MOVE -330,10
8200 FOR NUM=0 TO 200 STEP 20
8210   PRINT NUM : MOVE -330,60+NUM*2.5
8220   NEXT NUM
8230 VDU4  : REM Cancel VDU5 above. Characters only printed
     in character cells.
8240 ENDPROC

8500 DEF PROCFIRSTTABLE
8510 CLS : REM clear VDU screen.
8520 FOR NUM=1 TO 15
8530   LTEMPONE(NUM)=LN(TEMPONE(NUM))
8540   NEXT NUM
8545 VDU2 : REM Output to VDU screen and parallel printer port.
8550 PRINT : PRINT "   TABLE OF GRAPH DATA FOR PART 1"
8570 PRINT ''
8575 PRINT "-------------------------------------"
8580 PRINT "| x (cm)    |    0 ('C)    |    ln 0    |"
8590 PRINT "-------------------------------------"
8595 FOR NUM=1 TO 15
8596   @%=&00000002
8600   PRINT "|  ";DIST(NUM);
8610   IF DIST(NUM)>=10 THEN PRINT "       |"; ELSE PRINT "
       |";
8615   @%=&00020407 : REM Decimal number 7 digits wide to 2
       decimal places.
8620   PROClocate(14,NUM+7) : PRINT;INT(1000*TEMPONE(NUM))/1000;
8630   IF TEMPONE(NUM)>=100 THEN PRINT "  |"; ELSE PRINT "   |";
8640   PRINT "  ";INT(1000*LTEMPONE(NUM))/1000;"  |"
8650   NEXT NUM
8660 PRINT "-------------------------------------"
8665 VDU3 : REM Output to VDU screen only - printer output off.
8670 PRINT : PRINT " PRESS <SPACE> TO CONTINUE"
8680 IF FNgetkey<>" " THEN GOTO 8680
8690 ENDPROC

9500 DEF PROCGIVETEMPS
9510 B=OPENOUT"TEMPS" : REM Open a disc file called TEMPS.
```

```
 9520 FOR X=1 TO 15
 9530   PRINT#B,TEMPONE(X) : REM Write data to file assigned
        to B in OPENOUT line.
 9540   NEXT X
 9545 PRINT#B,AMBIENT
 9550 CLOSE#B
 9560 ENDPROC

10100 DEF PROCAXIS2
10110 CLS : REM Clear VDU screen.
10120 VDU29,100;380; : REM Move origin of graphics screen.
10130 MOVE 0,-200 : DRAW 1120,-200
10140 MOVE 0,-200 : DRAW 0,600
10150 FOR X=0 TO 1120 STEP 80
10160   MOVE X,-200 : DRAW X,-205 : NEXT X
10170 FOR Y=-200 TO 600 STEP 100
10180   MOVE 0,Y : DRAW -5,Y : NEXT Y
10190 PROClocate(0,0) : PRINT;"ln 0" : PROClocate(10,29) : PRINT;
      "Distance from end of bar (cm)"
10200 VDU5 : REM enable character printing at any pixel position.
10210 MOVE -12,-220: PRINT;4
10220 MOVE 68,-220: PRINT;8
10230 FOR NUM=12 TO 60 STEP 4
10240   MOVE NUM*20-110,-220 : PRINT; NUM
10250   NEXT NUM
10260 MOVE -45,10: PRINT "0"
10270 FOR NUM=1 TO 6
10280   MOVE -45,15+NUM*100 :PRINT ;NUM
10290   NEXT NUM
10300 MOVE-80,-85:PRINT;-1
10310 MOVE-80,-185:PRINT;-2
10320 VDU4 : REM enable character printing at any pixel position.
10430 FOR NUM=4 TO 8 STEP 4
10440   MOVE -382+NUM*20,-20: PRINT NUM
10450   NEXT NUM
10460 FOR NUM=12 TO 60 STEP 4
10470   MOVE -363+NUM*20,-20 : PRINT NUM
10480   NEXT NUM
10490 MOVE -330,10
10500 FOR NUM=0 TO 200 STEP 20
10510   PRINT NUM : MOVE -330,60+NUM*2.5
10520   NEXT NUM
10530 VDU4  : REM Cancel VDU5 above. Characters only printed in
      character cells
10530 ENDPROC
```

```
10540 DEF PROCPLOT2
10545 NOPTS%=15
10550 FOR NUM=15 TO 1 STEP -1
10555   LTEMPONE(NUM)=LN(TEMPONE(NUM))
10560   IF TEMPONE(NUM)<=OFSET THEN NOPTS%=NUM-1
10565   X=DIST(NUM)*20-80:Y=LTEMPONE(NUM)*100
10570   MOVE X,Y-8 :DRAW X,Y+8
10575   MOVE X-8,Y :DRAW X+8,Y
10580   NEXT NUM
10590 PROCLSQ1(NOPTS%)
10600 MOVE 0,A*100
10610 PLOT 69,1120,(100*SLOPE*(1120/20+4))+(A*100)
10620 SLOPE=SLOPE*100
10630 @%=&20206 : REM Decimal number 6 digit wide to 2 decimal
      places.
10640 PROClocate(10,18) : PRINT;"SLOPE=";SLOPE;" (M-1)"
10650 REM Printer Dump
10660 ENDPROC

10670 DEF PROCCARRYOVER
10680 B=OPENOUT"FROM1" : REM Open a disc file called FROM1.
10690 PRINT#B,SLOPE : REM write slope to file FROM1
10700 CLOSE#B
10710 ENDPROC

10720 DEF PROCLSQ1(NOPTS%) : REM least squares fit for slope.
10730 SUMX=0 : SUMXSQ=0
10740 SUMY=0 : SUMYSQ=0
10750 SUMXY=0
10760 FOR AA=1 TO NOPTS%
10770   SUMX=SUMX+DIST(AA)
10780   SUMY=SUMY+LTEMPONE(AA)
10790   SUMXSQ=SUMXSQ+DIST(AA)*DIST(AA)
10800   SUMYSQ=SUMYSQ+LTEMPONE(AA)*LTEMPONE(AA)
10810   SUMXY=SUMXY+DIST(AA)*LTEMPONE(AA)
10820   NEXT AA
10830 SXX=SUMXSQ-SUMX*SUMX/NOPTS%
10840 SYY=SUMYSQ-SUMY*SUMY/NOPTS%
10850 SXY=SUMXY-SUMX*SUMY/NOPTS%
10860 SLOPE=SXY/SXX
10870 A=(SUMY-SLOPE*SUMX)/NOPTS%
10880 R=SXY/SQR(SXX*SYY)
10890 ENDPROC
11600 DEF FNgetkey
11610 =GET$
```

```
11620 REM Wait for a single key press, then return that character.
11700 DEF FNcheckkey(T)
11710 =INKEY$(T)
11720 REM During T/100 seconds return character of next key
      pressed, or "" if none outstanding.

11900 DEF PROClocate(X,Y)
11905 REM Position character cursor at character coordinates X, Y.
11910 PRINT TAB(X,Y);
11920 ENDPROC

12100 DEF PROCflush
12110 *FX21,0
12115 REM Flushes the keyboard buffer of any outstanding
      keypresses.
12120 ENDPROC

30000 REM The special IEEE 488 Procedures are specific to
      interface used.  The main procedures are
      described in the text.
```

Program for Second Part of Experiment

```
   10 REM PART 2 READS IN TEMPERATURES EVERY 10 SECONDS
   20 PROCINITIALISE
   30 PROCPART2
   40 PROCRECDATA2
   40 PROCflush
   50 A$=FNgetkey
   60 PROCPLOT3
   70 REM Printer dump of graph
   80 PROCflush
   90 A$=FNgetkey
  100 PROCPLOT4
  110 REM Printer dump of graph
  120 PROCflush
  130 A$=FNgetkey
  140 PROCGIVEADVICE
  150 END

 1000 DEFPROCINITIALISE
 1001 B=("TEMPS")
 1002 FOR X=1 TO 16
 1003   INPUT#B,AMBIENT
 1004   NEXT X
```

```
1005 CLOSE#B
1010  PROCIEINIT(0,TRUE)
1020 ?&FE62=&FF : REM Write directly to memory to set up
     6522 VIA for output.
1030 DIM TEMPTWO(200)
1040 FACTOR=100:NOPTS%=200
1050 ?&FE60=31 : REM Select thermocouple 16 for cooling
     experiment using VIA.
1060 PROCIECLR
1070 PROCIEWRIT(7,-1)
1080 PROCIEPUTS("A1B0F0G1K0M0N0P0Q0R2S1T4W0Y0Z0X")
1090 REM Set DVM over IEEE 488 bus
1100 MODE 7 : REM Set VDU screen in 40 by 25 text mode.
1110 ENDPROC

1500 DEF PROCPART2
1510 PRINT "PART2 of Experiment"
1520  PRINT
1530 PRINT "Switch off the Forbes' bar heater and
     switch on the heater for
the small bar."
1540 PRINT "Place the small bar on the heater and
     wait for its temperature to reach 200'C The temperature
     of the bar will be displayed on screen."
1550 PRINT : PRINT :PRINT "      PRESS <SPACE> TO CONTINUE"
1560 PROCflush
1570 REPEAT UNTIL FNgetkey=" "
1575 CLS : REM Clear vdu screen.
1580 PRINT : PRINT : PRINT " PRESS <SPACE> WHEN THE
     TEMPERATURE HAS EXCEEDED 120'C"
1585 PROCflush
1590 PROCIEREAD(7,-1):PROCIEGETA(IEBUF%,13,14)
1600 REPEAT
1610   PROCIEGETA(IEBUF%,13,14)
1630   T16=VAL($IEBUF%)*FACTOR
1640   @%=&20205 : REM Print decimal number 5 digits wide
       to 2 decimal places.
1650   PROClocate(11,10) : PRINT ;"TEMP = ";T16;" 'C    "
1660   UNTIL FNcheckkey(200)=" "
1670  ENDPROC

2000 DEF PROCRECDATA2
2005 CLS : REM Clear VDU screen.
2010 PRINT : PRINT : PRINT "Now that the bar has attained a
     fairly  high temperature,the second part of the experiment
```

```
    can be performed." 2020 PRINT:PRINT "Using the tongs
    provided, remove the    small bar from the heater and place
    it  on the stand provided. The temperature  of the bar will
    be recorded every ten   seconds for two thousand seconds."
2030 PRINT : PRINT "The temperature readings will be sent tothe
    printer. Make sure that the printer is on-line."
2040 PRINT:PRINT " PRESS <SPACE> WHEN YOU HAVE PLACED"
2045 PRINT "  THE BAR ONTO THE STAND PROVIDED"
2050 PROCflush
2060 REPEAT UNTIL FNgetkey=" "
2065 VDU2 : REM Enable output to parallel printer and to VDU.
2070 CLS : REM Clear VDU screen.
2080 PRINT "TABLE OF GRAPH DATA FOR PART 2"
2085 PRINT
2090 PRINT "-------------------------"
2100 PRINT "|  TIME (s)  |   0 ('C)  |"
2110 PRINT "-------------------------"
2120 TIME=0
2140 FOR NUM=1 TO NOPTS%
2150   REPEAT UNTIL TIME>=1000*NUM
2160   PROCIEGETA(IEBUF%,13,14):PROCIEGETA(IEBUF%,13,14)
2170   TEMPTWO(NUM)=VAL($IEBUF%)*FACTOR
2180   @%=&00004 : REM Print numbers to 4 signicant digits.
2190   PRINT "|     ";NUM*10;
2200   IF NUM<=9 THEN PRINT "      |";
2210   IF NUM>9 AND NUM<100 THEN PRINT "     |";
2220   IF NUM>=100 THEN PRINT "    |";
2230   @%=&20205 : REM Print in decimal format, 5 digits wide
       and 2 places of decimal.
2240   PRINT TAB(16);TEMPTWO(NUM);TAB(25);"|"
2245   TEMPTWO(NUM)=TEMPTWO(NUM)-AMBIENT
2250   NEXT NUM
2260 PRINT "-------------------------"
2270 VDU3 : @%=10 : REM Turn output to printer off. Set free
    format for printing numbers
2280 B%=OPENOUT"TEMPS2" : REM Open disc file TEMPS2.
2290 FOR I%=1 TO NOPTS%
2300   PRINT#B%,TEMPTWO(I%) : REM Send data to file TEMPS2.
2310   NEXT I%
2320 CLOSE#B%
2330 PROCflush
2340 ENDPROC
2500 DEF PROCPLOT3
2502 B%=OPENIN"TEMPS2": REM Open disc file TEMPS2 for reading.
2504 FOR I%=1 TO NOPTS%
```

```
2506   INPUT#B%,TEMPTWO(I%) : REM Read data from TEMPS2.
2508   NEXT I%
2510 CLOSE#B%
2518 MODE 4 : REM 320 x 256 PIXEL (40 x 32 CHAR) 2-COLOUR
     GRAPHICS MODE
2520 PRINT "  Graph of Temperature vs Time"
2530 VDU29,140;400; : REM Move graphics origin.
2540 MOVE 0,0 : DRAW 1000,0
2550 MOVE 0,0 : DRAW 0,500
2560 FOR X=0 TO 1000 STEP 125
2570   MOVE X,0 : DRAW X,-5 : NEXT X
2580 FOR Y=0 TO 500 STEP 50
2590   MOVE 0,Y : DRAW -5,Y : NEXT Y
2600 PROClocate(0,2) : PRINT "0 ('C)"
2610 PROClocate(15,22) : PRINT "Time (seconds)"
2620 VDU5 : REM Enable character printing at any pixel position.
2630 NUM=0 : MOVE -10,-20 : PRINT; NUM
2640 NUM=500:MOVE 203,-20 : PRINT ;NUM
2660 FOR NUM=1000 TO 2000 STEP 500
2670   MOVE NUM/2-58,-20 : PRINT ;NUM
2680   NEXT NUM
2690 MOVE -40,10
2700 FOR NUM=0 TO 200 STEP 20
2710   PRINT ;NUM
2712   IF NUM<100 THEN MOVE -75,60+NUM*2.5
2714   IF NUM>=80 THEN MOVE -110,60+NUM*2.5
2720   NEXT NUM
2730 VDU4 : REM Printing of text characters at character
     cells only.
2740 FOR NUM=1 TO NOPTS%
2750   PROCplotpoint(NUM*5, TEMPTWO(NUM)*2.5)
2760   NEXT NUM
2770 ENDPROC

3000 DEF PROCPLOT4
3005 MODE 4 : REM 320 x 256 PIXEL (40 x 32 CHAR) 2-COLOUR
     GRAPHICS MODE
3010 VDU29,140;380; : REM Move graphics origin.
3020 PRINT "      Graph of ln Temperature vs Time"
3040 MOVE 0,0 : DRAW 1000,0
3050 MOVE 0,0 : DRAW 0,600
3060 FOR X=0 TO 1000 STEP 125
3070   MOVE X,0 : DRAW X,-5 : NEXT X
3080 FOR Y=0 TO 600 STEP 100
3090   MOVE 0,Y : DRAW -5,Y : NEXT Y
```

```
3100 PROClocate(10,22) : PRINT "Time (seconds)"
3110 PROClocate(0,0) : PRINT "ln 0"
3120 VDU5 : REM Enable character printing at any pixel position.
3130 NUM=0:MOVE -10,-20:PRINT ;NUM
3140 NUM=500:MOVE NUM/2-50,-20 : PRINT ;NUM
3160 FOR NUM=1000 TO 2000 STEP 500
3170   MOVE NUM/2-60,-20 : PRINT ;NUM
3180   NEXT NUM
3190 MOVE -45,10
3200 FOR NUM=0 TO 6
3210   PRINT ;NUM : MOVE -45,115+NUM*100
3220   NEXT NUM
3230 VDU4 : REM Printing of text characters at character
     cells only.
3240 FOR NUM=1 TO NOPTS%
3245   TEMPTWO(NUM)=LN(TEMPTWO(NUM))
3250   PROCplotpoint(NUM*5,TEMPTWO(NUM)*100)
3260   NEXT NUM
3270 PROCLSQ2
3275 VDU29,140;380; : REM Move graphics origin.
3280 FOR NUM=0 TO 1000
3290   PROCplotpoint(NUM,(SLOPE*NUM*200)+(A*100))
3300   NEXT NUM
3305 @%=&10309 : REM Print numbers in exponent format 9
     digits wide.
3310 PROClocate(10,15) : PRINT "SLOPE=";SLOPE
3320 ENDPROC

4010  DEF PROCLSQ2 : REM least squares fit for slope.
4020 SUMX=0 : SUMXSQ=0
4030 SUMY=0 : SUMYSQ=0
4040 SUMXY=0
4050 FOR AA=1 TO NOPTS%
4060   SUMX=SUMX+AA*10
4070   SUMY=SUMY+TEMPTWO(AA)
4080   NEXT AA
4090 FOR AA=1 TO NOPTS%
4100   SUMXSQ=SUMXSQ+(AA*10)^2
4110   SUMYSQ=SUMYSQ+TEMPTWO(AA)*TEMPTWO(AA)
4120   SUMXY=SUMXY+AA*10*TEMPTWO(AA)
4130   NEXT AA
4140 SXX=SUMXSQ-SUMX*SUMX/NOPTS%
4150 SYY=SUMYSQ-SUMY*SUMY/NOPTS%
4160 SXY=SUMXY-SUMX*SUMY/NOPTS%
4170 SLOPE=SXY/SXX
```

```
4180 A=(SUMY-SLOPE*SUMX)/NOPTS%
4190 R=SXY/SQR(SXX*SYY)
4200 ENDPROC

5000 DEF PROCGIVEADVICE
5010 MODE 7 : REM Put VDU in 40 by 25 text character mode.
5020 PRINT "To obtain the Thermal Conductivity it isalso
     necessary to measure various other physical constants :"
5030 PRINT
5040 PRINT "  (i) Lengths a and b (in m)"
5050 PRINT " (ii) The density, p"
5060 PRINT "(iii) The specific heat, s"
5070 PRINT " (iv) The length of the bar, l"
5080 PRINT
5090 PRINT "K can then be determined as outlined in the
     introduction"
5100 PRINT
5110 PROCROUGH
5140 PRINT
5150  @%=&20105 : REM Print number in decimal format
      5 digits wide with 1 decimal place.
5160 PRINT "A rough estimate of K based on this data is :
     ";ROUGH;" W/m/K"
5170 ENDPROC

5500 DEF PROCROUGH
5510 B=OPENIN "FROM1" : REM Open disc file FROM1 for reading.
5520 INPUT#B,SLOPE1 : Read data from FROM1 filew
5530 CLOSE #B
5540 k=-SLOPE/SLOPE1/SLOPE1*7900*420
5550 ROUGH=k/(1+2*.000161/.0508*.041)
5560 ENDPROC

11200 DEF PROCplotpoint(X,Y)
11210 PLOT 69,X,Y
11215 REM PLOT A SINGLE PIXEL AT SCREEN COORDINATES X, Y
11220 ENDPROC

11600 DEF FNgetkey
11610 =GET$
11620 REM Wait for a single key press, then return that
      character.

11700 DEF FNcheckkey(T)
11710 =INKEY$(T)
```

```
11720 REM During T/100 seconds return character of next key
      pressed, or "" if none outstanding.

11900 DEF PROClocate(X,Y)
11905 REM Put character cursor at character coordinates X, Y.
11910 PRINT TAB(X,Y);
11920 ENDPROC

12100 DEF PROCflush
12110 *FX21,0
12115 REM Flushes the keyboard buffer of any outstanding
      keypresses.
12120 ENDPROC

30000 REM The special IEEE 488 Procedures are specific to
      interface used.  The main procedures
      are described in the text.
```

Appendix 4.B

The FORTRAN Simulation Program

```
C     THIS IS A PROGRAM TO MODEL SEARLE'S BAR APPARATUS, USING A
C     FINITE DIFFERENCE APPROXIMATION FOR THE PARTIAL DIFFERENTIAL
C     EQUATION GOVERNING THE HEAT FLOW.  THE SCHEMA USED IS AN
C     EXPLICIT ONE.THE HEAT CAPACITY OF THE HEATER AND THAT OF THE
C     COOLING COIL ARE ASSUMED TO BE EQUAL TO THAT OF APPROXIMATELY
C     ONE HALF CENTIMETRE THE BAR.  HEAT IS SUPPLIED BY AN
C     ELECTRICAL HEATER OF A FIXED POWER OUTPUT, AND COOLING IS
C     ASSUMED TO BE DONE BY WATER PASSINGTHROUGH A COIL AT THE
C     OTHER END OF THE BAR.
C     THE BAR IS DIVIDED UP INTO SEGEMENTS OF APPROXIMATELY ONE
C     CENTIMETRE IN LENGTH.
      DIMENSION A(51),B(51),ITT(13),X(51),CHAR(5),JN(6)
      COMMON/A/DT,DX,TH1,ITH1,TH2,ITH2,NP,NIT
      COMMON/K/XN,FLO,TIN,VOLT,AMP
      COMMON/CALC/E1,E2,E3,E4,E5,E6,E7,E8,T1,T2
      DATA ITT(1),ITT(2),ITT(3),ITT(4)/0,8,2HDI,2HST/
      DATA JN(1),JN(2),JN(3),JN(4),JN(5)/2HOV,2HER,2H 1,2HO,,2HOO/
      DATA JN(6)/1HO/
      DATA ITT(5),ITT(6),ITT(7),ITT(8)/2HAN,2HCE,11,2HTE/
      DATA ITT(9),ITT(10),ITT(11),ITT(12)/2HMP,2HER,2HAT,2HUR/
      DATA ITT(13)/1HE/
```

```
C     OPEN OUTPUT FILE FOR PLOTTING INFORMATION
      OPEN(5,FILE='W5',STATUS='OLD')
C     OUTPUT BRIEF DESCRIPTION OF THE PROGRAM AND ASK IF MORE IS
C     WANTED.
  810 WRITE(*,800)
      READ(*,*) I
C     CHECK RESPONSE AND OUTPUT LONGER DESCRIPTION AS REQUIRED.
      IF (I.NE.1.AND.I.NE.0) GOTO 810
      IF (I.EQ.0) GOTO 820
      WRITE(*,10)
C     START OF GEOMETRIC DATA
  820 WRITE(*,20)
C     ASK FOR, READ IN, AND CHECK BOUNDS ON CROSS SECTIONAL AREA
   50 WRITE(*,30)
      READ(*,*) XN
      IF (XN.GE.1.0.AND.XN.LE.100.0) GOTO 360
      WRITE(*,60)
      GOTO 50
C     ASK FOR, READ IN, AND CHECK BOUNDS ON THE CIRCUMFERENCE.
  360 WRITE(*,370)
      READ(*,*) CIRC
      IF (CIRC/3.14158.GT.SQRT(XN/3.142)) GOTO 40
      WRITE(*,380)
      GOTO 360
C     ASK FOR, READ IN, AND CHECK BOUNDS ON THE LENGTH OF THE BAR.
   40 WRITE(*,70)
      READ(*,*) XL
      IF (XL.GE.10.0.AND.XL.LE.50.0) GOTO 80
      WRITE(*,60)
      GOTO 40
C     ASK FOR, READ IN, AND CHECK BOUNDS ON POSITION OF THE HOT
C     THERMOMETER.
   80 Z1=XL*0.1
      Z2=XL*0.7
      WRITE(*,100) Z1,Z2
      READ(*,*) XT1
      IF (XT1.GE.XL*0.1.AND.XT1.LE.XL*0.7) GOTO 110
      WRITE(*,60)
      GOTO 80
C     ASK FOR, READ IN, AND CHECK BOUNDS ON POSITION OF COLD
C     THERMOMETER.
  110 Z1=XL*0.9
      XTPLUS=XT1+0.1*XL
      WRITE(*,120) XTPLUS,Z1
      READ(*,*) XT2
```

```
      IF (XT2.GE.XTPLUS.AND.XT2.LE.XL*0.9) GOTO 130
      WRITE(*,60)
      GOTO 110
C     START OF THERMAL DATA.
  130 WRITE(*,140)
C     ASK FOR, READ IN, AND CHECK TEMPERATURE OF THE APPARATUS.
  170 WRITE(*,150)
      READ(*,*) TMP
      IF (TMP.GE.5.0.AND.TMP.LE.35.0) GOTO 160
      WRITE(*,60)
      GOTO 170
C     ASK FOR, READ IN, AND CHECK THE TEMPERATURE OF THE
C     INFLOWING WATER.
  160 WRITE(*,180)
      READ(*,*) TIN
      IF (TIN.GE.4.0.AND.TIN.LE.20.0) GOTO 700
      WRITE(*,60)
      GOTO 160
C     ASK FOR, READ IN, AND CHECK THE AMBIENT TEMPERATURE.
  700 WRITE(*,390)
      READ(*,*) TEXT
      IF (TEXT.GE.5.0.AND.TEXT.LE.35.0) GOTO 400
      WRITE(*,60)
      GOTO 700
C     ASK FOR, READ IN, AND CHECK THE DENSITY OF THE BAR.
  400 WRITE(*,220)
      READ(*,*) DNS
      IF (DNS.GE.4000.0.AND.DNS.LE.15000.0) GOTO 230
      WRITE(*,60)
      GOTO 400
C     ASK FOR, READ IN, AND CHECK THE SPECIFIC HEAT CAPACITY
C     OF THE BAR.
  230 WRITE(*,240)
      READ(*,*) SHC
      IF (SHC.GE.100.0.AND.SHC.LE.10000.0) GOTO 250
      WRITE(*,60)
      GOTO 230
C     ASK FOR, READ IN, AND CHECK THE THERMAL CONDUCTIVITY
C     OF THE BAR.
  250 WRITE(*,260)
      READ(*,*) XLAM
      IF (XLAM.GE.0.1.AND.XLAM.LE.1000.0) GOTO 270
      WRITE(*,60)
      GOTO 250
C     ASK FOR, READ IN, AND CHECK THE VOLTAGE ACROSS THE HEATER.
```

```
  270 WRITE(*,280)
      READ(*,*) VOLT
      IF (VOLT.GE.1.0.AND.VOLT.LE.30.0) GOTO 290
      WRITE(*,60)
      GOTO 270
C     ASK FOR, READ IN, AND CHECK THE CURRENT FLOWING THGOUGH
C     THE HEATER
  290 WRITE(*,300)
      READ(*,*) AMP
      IF (AMP.GE.0.2.AND.AMP.LE.8.0) GOTO 310
      WRITE(*,60)
      GOTO 290
C     ASK FOR, READ IN, AND CHECK THE RATE OF FLOW OF WATER
C     AT THE COLD END.
  310 WRITE(*,320)
      READ(*,*) FLO
      IF (FLO.GE.0.0.AND.FLO.LE.10.0) GOTO 330
      WRITE(*,60)
      GOTO 310
C     ASK FOR, READ IN, AND CHECK THE COEFFICIENT OF HEAT
C     TRANSFER BETWEEN THE SURROUNDS AND THE BAR.
  330 WRITE(*,340)
      READ(*,*) H
      IF (H.GE.0.0.AND.H.LE.15.0) GOTO 350
      WRITE(*,60)
      GOTO 330
  350 CONTINUE
C     CALCULATE THE NUMBER OF POINTS.
      NP=IFIX(XL)+1
C     CALCULATE THE DISTANCE BETWEEN ADJACENT NODES.
      DX=XL/(FLOAT(NP)-1.0)
C     INITIALIZE THE TEMPERATURE ARRAYS.
      DO 520 I=1,NP
      X(I)=DX*(FLOAT(I)-1.0)
      A(I)=TMP
  520 B(I)=TMP
C     SCALE THE VARIABLES TO S.I. UNITS
      CIRC=CIRC*0.01
      XN=XN*1.0E-04
      XL=XL*0.01
      DX=DX*0.01
C     CALCULATE THE TIME STEP TO BE USED.
      DT=0.9*DNS*SHC/(2.0*XLAM/(DX*DX)+H*CIRC/XN)
C     CALCULATE THE POSITIONS OF THE THERMOMETERS IN TERMS
C     OF THE NODES.
```

```
      TH1=XT1*0.01
      TH2=XT2*0.01
      ITH1=IFIX(TH1/DX)+1
      TH1=TH1-FLOAT(ITH1)*DX-DX
      ITH2=IFIX(TH2/DX)+1
      TH2=TH2-FLOAT(ITH2)*DX-DX
C     CALCULATE QUANTITIES TO USE IN THE ITERATION PROCESS.
      E1=VOLT*AMP*DT/(DNS*SHC*XN*DX)
      E2=XLAM*DT/(DNS*SHC*DX*DX)
      E3=H*CIRC*DT/(DNS*SHC*XN)
      E4=1.0-E2-E3
      E5=1.0-2*E2-E3
      E3=E3*TEXT
      E6=XLAM*XN/DX
      E7=TIN*70.0*FLO
      E8=DNS*SHC*XN*DX/DT
      EINT=E8+70.0*FLO
      E8=E8-E6
      E6=E6/EINT
      E7=E7/EINT
      E8=E8/EINT
C     END OF CALCULATED QUANTITIES
C     SET NUMBER OF ITERATIONS AND TIMES TO ZERO
      NIT=0
      ATIME=0.0
      BTIME=0.0
      TEND=0.0
C     SET VARIABLE FOR CHECKING IF PLOTTING DEVICE IS INITIALIZED
      ICB=0
C     SET THE NUMBER OF HARD COPIES PRODUCED TO ZERO
      ICC=0
C     ASK IF ONE WISHES TO CONTINUE WITH THE PROGRAM
  615 WRITE(*,901)
      READ(*,*) ICONT
      IF (ICONT.EQ.1) GOTO 610
      IF (ICONT.EQ.0) GOTO 635
      WRITE(*,540)
      GOTO 615
C     ASK FOR, READ IN, AND CHECK NEW TIME UP TO WHICH THE
C     PROGRAM WILL RUN.
  610 WRITE(*,480)
  430 READ(*,*) TNEW
      IF (TNEW.GE.TEND) GOTO 410
      WRITE(*,420) TEND
      GOTO 430
```

```
C     SEQUENCE FOR HALTING THE RUN OF THE PROGRAM
  635 IF (ICB.EQ.1) CALL DEVFIN
      IF (ICC.NE.0) WRITE(*,670) ICC
      WRITE(*,680)
      CALL COMI$('CONTINUE',8,6,ICODE)
      CALL ERRPR$(0,ICODE,'WRONG2',6,0,0)
      STOP
  410 TEND=TNEW
C     RESET CONTROL VARIABLE FOR OUTPUT OF HEADINGS FOR
C     THERMOMETERS.
      ICA=0
C     CHECK HOW OFTEN THE TEMPERATURES OF THE THERMOMETERS
C     WILL BE REQUIRED TO BE PRINTED.
  920 WRITE(*,900) DT
      READ(*,*) NOM
      IF (NOM.GE.1) GOTO 460
      WRITE(*,540)
      GOTO 920
C     CHECK IF WE ARE UP TO THE FINAL TIME
  460 IF (AMAX1(ATIME,BTIME).GE.TEND) GOTO 440
C     PERFORM AN ITERATION ON THE APPROPRIATE ARRAY.
      IF (ATIME.GE.BTIME) GOTO 805
      IF (ATIME.LT.BTIME) CALL IT(B,A,BTIME,ATIME,TIME,XK)
      GOTO 815
  805 CALL IT(A,B,ATIME,BTIME,TIME,XK)
  815 IF (MOD(NIT,NOM).NE.0) GOTO 460
      IF (ICA.EQ.0) WRITE(*,450)
      ICA=1
      IF (ABS(XK).LT.10000.) WRITE(*,940) TIME,T1,T2,XK
      IF (ABS(XK).GE.10000.) WRITE(*,942) TIME,T1,T2,(JN(I),I=1,6)
      GOTO 460
C     INITIALIZE A DEVICE IF NECESSARY
  440 IF (ICB.EQ.1) GOTO 470
      ICB=1
      CALL TTY
C     PRODUCE A PLOT AT THE TELETYPE/V.D.U.
  470 IF (ATIME.GE.BTIME) GOTO 500
      CALL FGPLT(X,B,NP,16,0,1,0,ITT)
      CALL NEWPAG
      WRITE(*,490) BTIME
      GOTO 510
  500 CALL FGPLT(X,A,NP,16,0,1,0,ITT)
      CALL NEWPAG
      WRITE(*,490) ATIME
C     CHECK IF FOUR HIGH QUALITY COPIES HAVE BEEN QUEUED AT THE
```

```
C     COMPUTING CENTRE.
  510 IF (ICC.LT.4) GOTO 660
      WRITE(*,650)
      GOTO 615
C     ASK IF A HIGH QUALITY COPY IS REQUIRED.
  660 WRITE(*,720) ICC
      READ(*,*) NCOPY
      IF (NCOPY.EQ.1.OR.NCOPY.EQ.0) GOTO 530
      WRITE(*,540)
      GOTO 660
  530 IF (NCOPY.EQ.0) GOTO 615
C     IF A HIGH QUALITY GRAPH HAS NOT BEEN PREVIOUSLY QUEUED,
C     THEN REQUEST A TITLE WHICH WILL APPEAR ON THE GRAPHS.
      IF (ICC.GE.1) GOTO 550
      WRITE(*,560)
      READ(*,580) (CHAR(J),J=1,5)
  550 ICC=ICC+1
C     OUTPUT THE TITLE, THE TIME, AND THE NUMBER OF POINTS
C     TO A PLOTTING FILE
      WRITE(5,580) (CHAR(J),J=1,5)
C     OUTPUT THE POSITIONS OF THE NODES AND THEIR TEMPERATURES
C     TO THE PLOTTING FILE.
      IF (ATIME.GE.BTIME) GOTO 620
      WRITE(5,590) BTIME,NP
      DO 630 I=1,NP
  630 WRITE(5,600) X(I),B(I)
      GOTO 615
  620 WRITE(5,590) ATIME,NP
      DO 640 I=1,NP
  640 WRITE(5,600) X(I),A(I)
      GOTO 615
  800 FORMAT(46HTHIS PROGRAM MODELS SEARLE'S BAR EXPT. DO YOU ,
     120HREQUIRE INFORMATION?,/,26HTYPE 0 FOR NO OR 1 FOR YES)
   10 FORMAT(46HTHIS PROGRAM TO MODELS THE TEMPERATURE PROFILE,
     116H IN SEARLE'S BAR,/,35HAS IT IS WARMING UP TO EQUILIBRIUM.,
     1/,6X,49HTHE PROGRAM WILL DRAW GRAPHS AT VARIOUS TIMES AS ,
     111HDIRECTED BY,/,32HTHE USER AND PRODUCE A HARD COPY,
     124H AT THE COMPUTING CENTRE,/,11HIF DESIRED.,
     1/,6X,51HHEATING IS ASSUMED TO BE BY A CONSTANT POWER DEVICE,
     14H AND,/,44HCOOLING BY WATER FLOWING AT A CONSTANT RATE ,
     119HWITH A STEADY INPUT,/,12HTEMPERATURE.,
     1/,6X,50HTHERMOMETER READINGS ARE PRINTED AT REGULAR TIMES.)
   20 FORMAT(/,/,14HGEOMETRIC DATA)
   30 FORMAT(45HWHAT IS THE CROSS SECTIONAL AREA OF THE BAR IN
     111H SQUARE, CM,/,36HTYPICAL VALUES ARE FROM 1.0 TO 100.0)
```

```
 60 FORMAT(22HPLEASE STICK TO BOUNDS)
 70 FORMAT(44HWHAT IS THE LENGTH OF THE BAR IN CENTIMETRES,/,
   131HTYPICAL VALUES ARE 10.0 TO 50.0)
100 FORMAT(46HHOW FAR, IN CM., FROM THE HOT END IS THE FIRST,
   112H THERMOMETER,/,22HACCEPTABLE BOUNDS ARE ,F4.1,5H AND ,
   2F4.1)
120 FORMAT(47HHOW FAR, IN CM., FROM THE HOT END IS THE SECOND,
   112H THERMOMETER,/,22HACCEPTABLE BOUNDS ARE ,F4.1,5H AND ,
   2F4.1)
140 FORMAT(/,/,12HTHERMAL DATA)
150 FORMAT(48HWHAT IS THE INITIAL TEMPERATURE OF THE APPARATUS,
   119H IN DEGREES CELSIUS,/,30HTYPICAL VALUES ARE 5.0 TO 40.0)
180 FORMAT(44HWHAT IS THE INTIIAL TEMPERATURE OF THE WATER,/,
   130HTYPICAL VALUES ARE 4.0 TO 20.0)
220 FORMAT(46HWHAT IS THE DENSITY OF THE BAR IN KG/METRE**3.,/,
   136HTYPICAL VALUES ARE 4000.0 TO 15000.0)
240 FORMAT(48HWHAT IS THE SPECIFIC HEAT CAPACITY OF THE BAR IN,
   15H J/KG,/,35HTYPICAL VALUES ARE 100.0 TO 10000.0)
260 FORMAT(46HWHAT IS THE THERMAL CONDUCTIVITY OF THE BAR ,
   18HIN W/M/K,/,32HTYPICAL VALUES ARE 0.1 TO 1000.0)
280 FORMAT(43HWHAT IS THE P.D. ACROSS THE HEATER IN VOLTS,/,
   130HTYPICAL VALUES ARE 1.0 TO 30.0)
300 FORMAT(46HWHAT IS THE CURRENT FLOWING THROUGH THE HEATER,
   18H IN AMPS,/,29HTYPICAL VALUES ARE 0.2 TO 8.0)
320 FORMAT(46HWHAT IS THE RATE OF FLOW OF WATER THROUGH THE ,
   1/,23HHEAT SINKIN LITERS/MIN.,/,
   230HTYPICAL VALUES ARE 0.0 TO 10.0)
340 FORMAT(48HWHAT IS THE COEFFICIENT OF HEAT TRANSFER BETWEEN,
   18H THE BAR,/,26HAND THE OUTSIDE IN W/K/M/M,9H TYPICAL ,
   122HVALUES ARE 0.0 TO 15.0,/,
   138HNATURAL FREE CONVECTION IN AIR IS 10.0)
480 FORMAT(45HAFTER WHAT TIME DO YOU WISH TO SEE A GRAPH OF,
   129H THE TEMPERATURE DISTRIBUTION)
420 FORMAT(35HPLEASE SPECIFY A TIME GREATER THAN ,F9.1)
450 FORMAT(22X,12HTEMPERATURES,13X,15HAPPARANT VALUES,/,3X,
   14HTIME,5X,13HTHERMOMETER 1,5X,13HTHERMOMETER 2,5X,
   213HOF K IN W/M/K)
490 FORMAT(20HTIME AFTER START IS ,F8.1,8H SECONDS)
720 FORMAT(50HDO YOU WISH A HARD COPY?  TYPE 1 FOR YES, 0 FOR NO,/,
   152HA MAXIMUM OF FOUR HARD COPY GRAPHS MAY BE PRODUCED. ,I1
   1,10H HAVE BEEN)
540 FORMAT(9HTRY AGAIN)
560 FORMAT(49HSPECIFY A TITLE FOR YOUR RUN.  YOU MAY USE UP TO ,
   113H20 CHARACTERS)
580 FORMAT(5A4)
```

```
  590 FORMAT(F10.2,I3)
  600 FORMAT(2E12.5)
  650 FORMAT(44HFOUR GRAPHS HAVE BEEN QUEUED AT THE COMPUTING
     129H CENTRE, NO MORE ARE ALLOWED.)
  670 FORMAT(I1,44H GRAPHS HAVE BEEN QUEUED AND MAY BE COLLECTED
     127H FROM, THE COMPUTING CENTRE)
  680 FORMAT(14HEND OF PROGRAM)
  370 FORMAT(42HWHAT IS THE CIRCUMFERENCE OF THE BAR IN CM,
     124H. NOT NECESSARILY ROUND.)
  380 FORMAT(34HTHIS CIRCUMFERENCE IS NOT POSSIBLE)
  390 FORMAT(46HWHAT IS THE AMBIENT TEMPERATURE.  THAT IS THE ,
     133HTEMPERATURE AROUND THE APPARATUS.
     1/,43HTYPICAL VALUES ARE FROM 5.0 TO 35.0 CELSIUS)
  940 FORMAT(F9.2,4X,F10.4,8X,F10.4,8X,F10.4)
  942 FORMAT(F9.2,4X,F10.4,8X,F10.4,8X,6A2)
  900 FORMAT(13HTIME STEP IS ,F10.4,24H THERMOMETER TEMPERATURES
     129H WILL, BE OUTPUT AT MULTIPLES,/,13HOF THIS TIME.,
     128H  WHAT MULTIPLE DO YOU WISH?)
  901 FORMAT(44HDO YOU WISH TO CONTINUE?  TYPE 1 FOR YES OR ,
     18H0 FOR NO)
      END
C
C
      SUBROUTINE IT(A,B,AT,BT,TIME,XK)
      DIMENSION A(51),B(51)
      COMMON/CALC/E1,E2,E3,E4,E5,E6,E7,E8,T1,T2
      COMMON/K/XN,FLO,TIN,VOLT,AMP
      COMMON/A/DT,DX,TH1,ITH1,TH2,ITH2,NP,NIT
C     CALCULATE THE NEW TIME.
      BT=AT+DT
      TIME=BT
C     CALCULATE THE TEMPERATURE AT THE HEATER END.
      B(1)=E1+E2*A(2)+E3+E4*A(1)
      NP1=NP-1
C     CALCULATE THE TEMPERATURE FOR THE MAIN BODY OF THE BAR.
      DO 10 I=2,NP1
   10 B(I)=E3+E2*(A(I-1)+A(I+1))+E5*A(I)
C     CALCULATE THE TEMPERATURE FOR THE COLD END OF THE BAR.
      B(NP)=E6*A(NP1)+E8*A(NP)+E7
C     CALCULATE THE READINGS ON THE TWO THERMOMETERS
      T1=(B(ITH1+1)*TH1+B(ITH1)*(DX-TH1))/DX
      T2=(B(ITH2+1)*TH2+B(ITH2)*(DX-TH2))/DX
C     CALCULATE THE HEAT FLOW
      QDOT=VOLT*AMP+(B(NP)-TIN)*70.0*FLO
      QDOT=QDOT*0.5
```

```
C     CALCULATE THE TEMPERATURE GRADIENT BETWEEN THE TWO
C     THERMOMETERS.
      DTHETA=(T2-T1)/(FLOAT(ITH1)*DX+TH1-FLOAT(ITH2)*DX-TH2)
      IF (DTHETA.EQ.0.0) GOTO 20
      XK=QDOT/(XN*DTHETA)
      NIT=NIT+1
      RETURN
C     IF THERE IS A ZERO TEMPERATURE GRADIENT THEN ASSIGN AN
C     ARBITARY VALUE TO THE THERMAL CONDUCTIVITY BECAUSE
C     IT WILL NOT BE CALCULABLE
   20 XK=1.0E+20
      NIT=NIT+1
      RETURN
      END
```

Part III

Solid State Physics

5. Experiments with High-T_c Superconductivity

M. Ottenberg and H.M. Staudenmaier

The discovery of the new high-temperature superconductors in 1986 by Bednorz and Müller [5.1] has led to a remarkable growth in the amount of research and the number of scientists working in this exciting field. It is therefore important that students also become acquainted with these new phenomena and one of the best methods is to introduce such high-T_c superconductors into an undergraduate or an advanced laboratory course. Particularly the development of materials that show superconductivity at temperatures above the boiling point of liquid nitrogen — i.e. $YBa_2Cu_3O_7$ — allowed the demonstration of superconductivity in laboratory experiments that could be simple in design and low in cost. In such experiments the high-T_c material may be simply immersed into a dewar filled with liquid nitrogen and then the most interesting part — the phase transition to superconductivity — will occur within a few seconds. The rapidity of this phase transition may have caused problems in earlier times, but the application of modern on-line computing techniques allows us to collect and record a multitude of experimental data even during the short transition period. In this article we describe a computerized high-T_c experiment using the conventional four point probe technique to measure the zero resistivity property, and with help of a tunnel diode oscillator method the exclusion of the magnetic field near T_c will be demonstrated.

5.1 Experimental Setup

5.1.1 The Apparatus

As superconducting material a crescent-shaped $YBa_2Cu_3O_7$ piece of 35mm in length and 2.5mm thickness was used (Fig. 5.1). The preparation of the sample was done by a usual solid-state sinter reaction under oxygen atmosphere at 950°C[1]. The sample was contacted with four point contacts (Indium pressed) and together with a Fe-Constantan-thermocouple the sample was placed into a quartz tube of 20mm diameter and 50cm in length.

[1] We are grateful to the Kristall- u. Materiallabor d. Universität Karlsruhe for preparing the high-T_c material and the thermocouple for us.

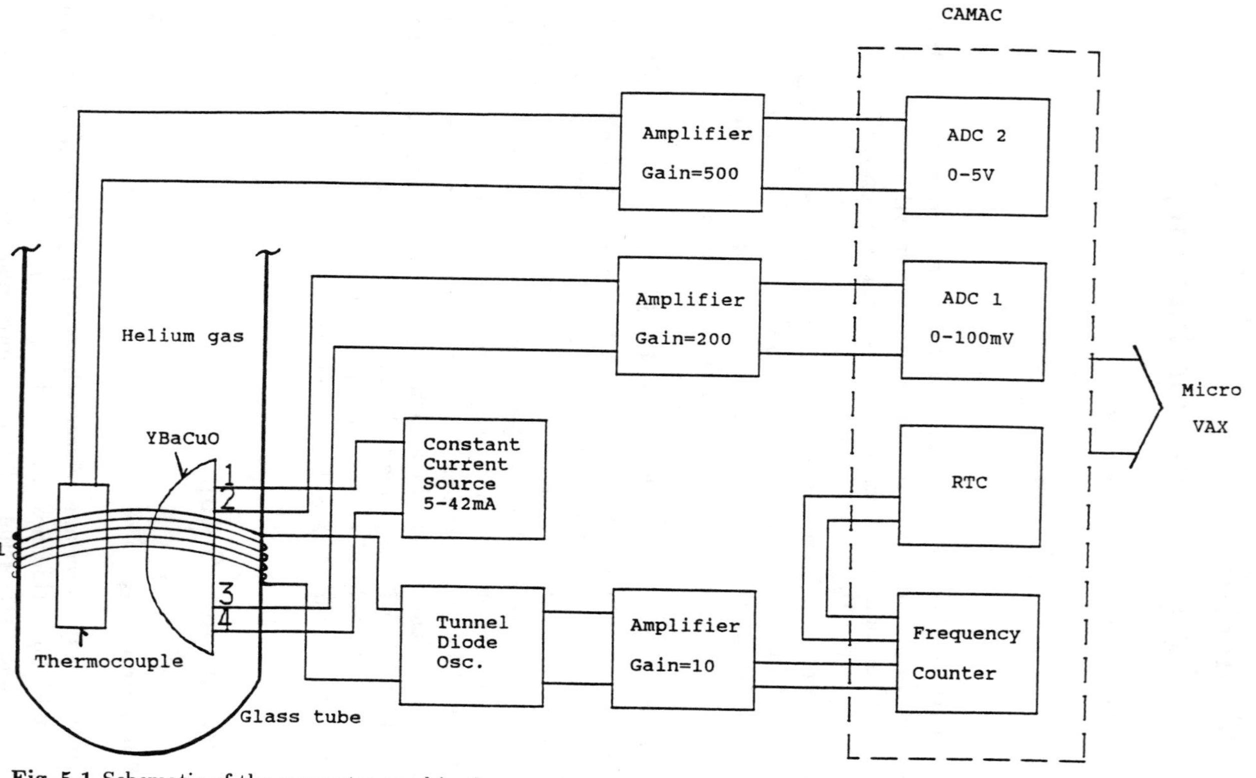

Fig. 5.1 Schematic of the apparatus used in the experiment

The tube was evacuated and filled with helium gas, mainly to protect the $YBa_2Cu_3O_7$ sample from oxygen. In addition changes of the He-pressure allowed us to increase or decrease the cooling rate during the experiment. As the main component of the tunnel diode oscillator (TDO), a coil (20mm diameter with windings of 0.1mmØ-Cu wire) is slipped on that end of the quartz tube containing the $YBa_2Cu_3O_7$ sample. Within this article we will not enter into the physics of superconductivity or details of the $YBa_2Cu_3O_7$ material. We mainly refer the reader to various review articles or monographs, e.g. [5.2] and [5.3].

5.1.2 Electronics

a) Electronics for Resistance Measurements *Constant Current Source:* A diagram is given in Appendix 5.A, Fig. 5.A1, the constant current was adjustable within the range from 5 to 42mA. With help of switch S it was possible to decouple the source from the high-T_c sample and then the voltage drop between the inside contacts (No. 2 and 3, Fig. 5.1) should be exactly zero. With this procedure a very precise offset balance could be achieved at the *instrumentation amplifier* connected to the inside contacts. A constant current $I_c \approx 30\text{mA}$ led to a voltage drop of $\approx 500\mu\text{V}$. To measure this voltage with the AD-converter ($ADC1$ — Fig. 5.1) an amplification of 200 is needed. Further requirements to the instrumentation amplifier were low noise performance, good drift stability, pin programmable gains and high gain accuracy. The chip AD624-Precision Instrumentation Amplifier from Analog Devices fulfilled these criteria. A circuit diagram is given in Appendix 5.A, Fig. 5.A2. To measure the thermocouple voltage a *second instrumentation amplifier* of the same type was used with an amplification factor of 500. The output of this amplifier was digitized with an 11-bit AD-converter ($ADC2$) which gives a temperature resolution of $\approx 0.2\text{K}$ in the whole region from 77K to 300K. To calculate absolute temperatures from our Fe-Constantan-thermocouple voltages we used a quadratic polynomial relation, whereas the coefficients were fitted to a calibration data set. The data set contained thermovoltages given in a table by the manufacturer [5.4] and in addition the boiling point of liquid nitrogen. More details on thermocouples as measuring devices may be found in Chap. 4.

b) Electronics for TDO-Measurements The TDO was first used by Glover and Wolf [5.5] to measure the paramagnetic susceptibility. The resonant frequency of a normal LC-oscillator $\nu_{\text{LC}} \sim \frac{1}{\sqrt{L \cdot C}}$ depends on the inductance of the coil and if material is inserted into the coil, the frequency will change. The change in coil inductance will therefore result in a frequency change that can be easily measured with a frequency counter. The application of this method to superconductivity was given by Fox et al. [5.6]. The jump of coil inductance at the critical temperature T_c is due to the exclusion of the magnetic field from inside the superconducting material by the induction of superconducting screening currents at the sample surface. (Though it looks rather similar this induction effect should not be interchanged with the Meissner effect occurring at dc conditions.) In our experiment we have used a more simple version of a TDO and the circuit diagram is given in Appendix 5.A, Fig. 5.A3. The TDO is selfresonant and with the help of the potentiometer P the resonant frequency can be changed slightly from 650 to 950kHz, then the circuit oscillates with a pulse height of ≈ 0.1V giving a rather rectangular signal.

5.1.3 Computer, Interface and Software

As shown in Fig. 5.1 we use a Micro-Vax and a CAMAC interface to control the experiment, but these rather expensive parts are not necessary at all; a simple personal computer with interface should be sufficient. We used the rather large scale equipment because it was already available and the high-T_c experiment is part of a special course called Computer Assisted Instruction Laboratory (CAIL), that has been organized in Karlsruhe for many years (see [5.7]). One advantage of the CAMAC interface is its modular structure, since process peripherals such as AD or DA converters, may simply be plugged into the CAMAC crate. To control the experiment under consideration we used four modules:

- ADC_1 AD-converter to measure voltage drops (9 bit), $\approx 100\mu$sec conversion time.
- ADC_2 AD-converter to measure temperature by digitizing thermocouple voltages (11 bit), $\approx 30\mu$sec conversion time.
- FC Frequency counter (100MHz), gated with RTC.
- RTC Real time clock, provides gate signals of programmable length for the frequency counter, FC.

The CAMAC interface is read out with the data taking program shown in Fig. 5.2. The main part of this program was written in FORTRAN 77 and two assembler subroutines are used to control the direct input/output with the interface. Each read-out of the interface delivers measured information, which is stored in the main memory of the computer. At the end of a measurement these data are written as file on a hard disk, typical lengths of such files are 100–200kB, depending on the selected data taking rate and cooling time of the experiment. Thereafter the data are analyzed off-line using graphics software and fit programs available on our computer. As special software used in this experiment a helpful spline-fit program SPLFIT should be mentioned and details will be given in Appendix 5.B.

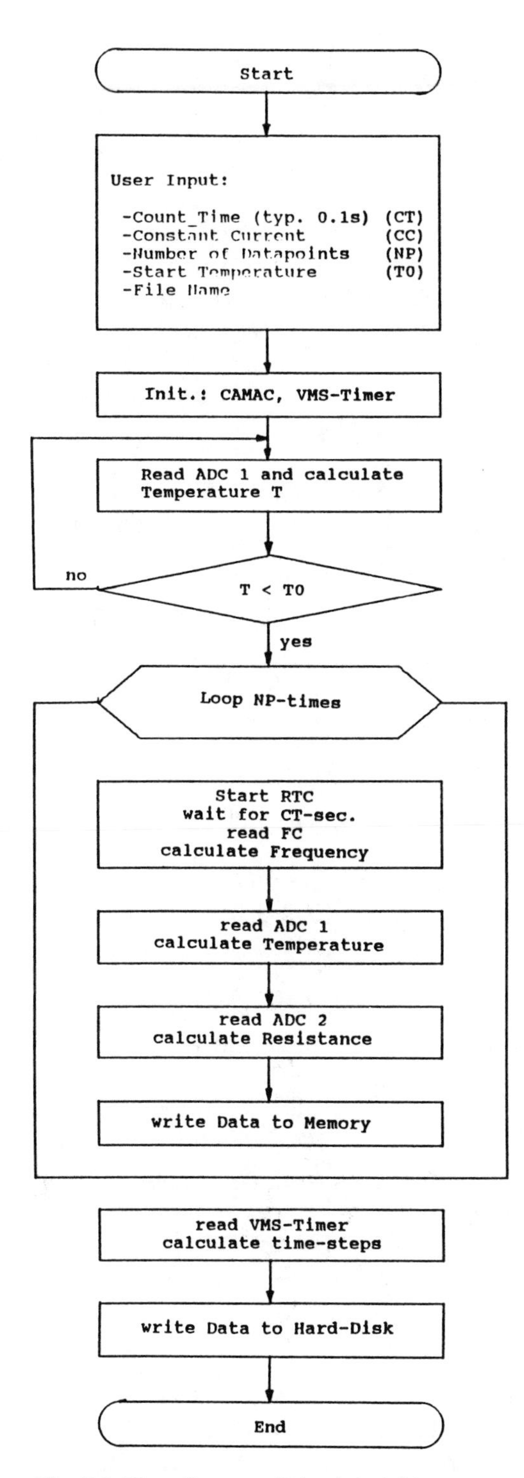

Fig. 5.2 Flow diagram of the data taking program

5.2 Measurements

5.2.1 Resistance Measurement

The fact that the electric resistance vanishes near the critical temperature T_c is one of the unique characteristics of the superconducting state, and it is the aim of this section to measure and demonstrate this feature of superconductivity. The $YBa_2Cu_3O_7$ sample has to be contacted with four leads, (Fig. 5.1) but the application of this four point resistance measuring technique requires several precautions to get reliable results. (The reader is referred to the textbook Kresin-Wolf [5.3].) The constant current is supplied to outside current leads and the voltage drop across the inside leads is read by the ADC_1 with a resolution of 0.2mV. At the beginning of the measurement the constant current has to be set e.g. $I_c = 10$mA and an offset balance of the instrumentation amplifier should be made. Then the data-taking program (Fig. 5.2) is started. The quartz tube containing the $YBa_2Cu_3O_7$ sample is then slowly immersed into the nitrogen dewar. The phase transition itself performs within 1 or 2 seconds and it is important to collect as much data as possible within this short period. A typical measurement starting from room temperature to $\approx$ 77K takes 4min and $\approx$ 2000 data points will have been collected within this time interval. As soon as the sample has reached the temperature of the liquid nitrogen ($\approx$ 77K) the measuring process is stopped and the data are stored on hard disk.

The quartz tube may then be raised above the liquid nitrogen and warmed up in ambient temperature. If desired, data taking can be continued during the warm-up process until room temperature is reached. This method of rather rapid resistance measurement was called by Pechan and Horvat [5.8] "quasi-equilibrium" technique (QE), because no precautions (e.g. no cryostat or temperature control) have been applied to assure thermal equilibrium between sample and thermocouple. But just this simplicity of the QE-technique makes the experiment inexpensive and well suited for an advanced undergraduate laboratory; nevertheless the results show that the transition temperature T_c can be measured quite accurately and reproducibly. The measured resistivity ρ — see Fig. 5.3 — has a rather linear dependence on temperature above $\approx$ 100K. This behaviour well known from metallic conductors is surprisingly valid also for $YBa_2Cu_3O_7$ ceramics. This behaviour is discussed by Halbritter et al. [5.9] on the basis of Matthiessen's rule:

$$\rho(T) = \rho_0 + \alpha T \qquad (5.1)$$
$$(100\text{K} < T < 600\text{K}) \,.$$

ρ_0 is a temperature independent term related only to impurity scattering. The slope $\alpha = d\rho/dT$ may be determined form our data using a fit program (e.g. MINUIT [5.10]) and the results are (the errors given by the MINUIT program are purely statistical):

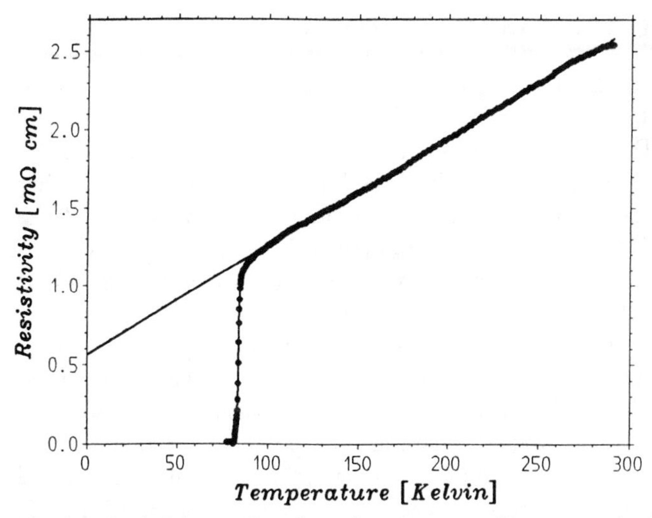

Fig. 5.3 Resistivity as function of temperature (I_c = 25mA). Solid line: fit with $\rho = \rho_0 + \alpha T$

$$\rho_0 = 0.57 \pm (1 \cdot 10^{-3})\text{m}\Omega \cdot \text{cm} \quad \alpha = 6.8 \pm (0.7 \cdot 10^{-2})\mu\Omega \cdot \text{cm/K} \quad (5.2)$$
$$(100\text{K} \leq T \leq 290\text{K}) .$$

These parameters agree well with those given by [5.9]. As mentioned at the beginning of this section, the data taking rate in our experiment was rather high and Fig. 5.4 shows the resistance measurement near the critical temperature with full resolution. Below T_c an interesting structure near $T \approx 86$K can be seen, and we will discuss this effect in more detail in Sect. 5.3.1 together with a determination of T_c.

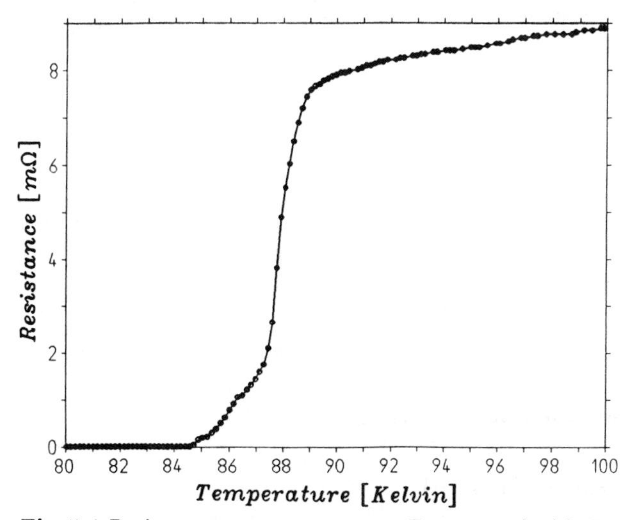

Fig. 5.4 Resistance measurement near T_c measured with I_c = 25mA

5.2.2 Tunnel Diode Oscillator Measurement

As described in Sect. 5.1.2b the exclusion of the magnetic field will be shown in this experiment by using the TDO method. At the beginning of the measurement the TDO frequency is set to the desired value, e.g. 800kHz, and then the quartz tube is immersed into the liqud nitrogen dewar. During the cooling process the frequency of the TDO is measured with help of the frequency counter FC (see Fig. 5.1) — e.g. every 0.5 second — and read by the computer. Normally the resistance measurements (see Sect. 5.2.1) and the TDO measurements were done in parallel, i.e. all four CAMAC modules were read more or less at the same time by the data taking program shown in Fig. 5.2. The TDO frequency is plotted versus temperature measured with the thermocouple. Within a wide range of temperature (240K to 95K) the frequency increases only slightly (see Fig. 5.5). This behaviour is mainly caused by the change of resistance of the coil immersed totally into liquid nitrogen. Near $T_c \approx 88$K however a large frequency jump of ≈ 10kHz can be seen, indicating that the ac field is rapidly excluded from the sample (i.e. from the area enclosed by the coil) except for a small layer near the surface of the sample (see Sect. 5.1.2b). If the temperature is reduced further the ac field is more and more excluded also from this surface layer and this results in an additional small rise of the TDO frequency on cooling to 80K. (see Fig. 5.5 and Fig. 5.7a). TDO measurements could be repeated at different frequencies and then the absolute values of the frequency jumps at T_c may be used to calculate interesting parameters of the superconducting material; details are given in [5.6]. The TDO measurement near T_c indicates also an interesting structure as seen already in Fig. 5.4 at the resistance measurements, therefore both results will be analyzed together in Sect. 5.3.1.

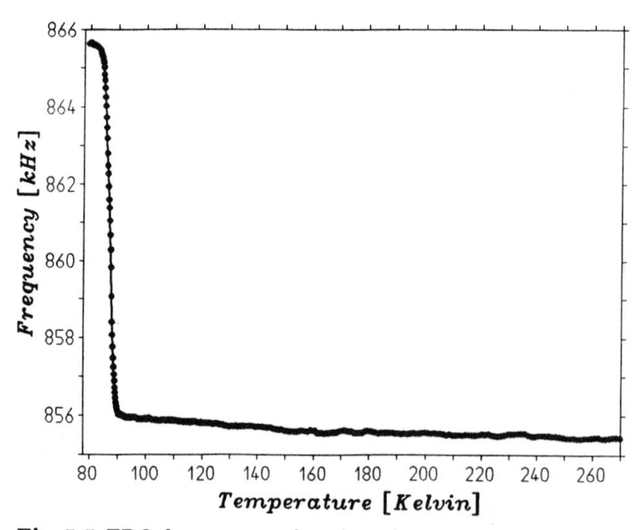

Fig. 5.5 TDO frequency as function of temperature

5.3 Results

5.3.1 Detailed Analysis of the Resistance and TDO Measurements

During the experiment many and rather different physical quantities have been recorded by the computer. Therefore an off-line analysis of the data into very distinct directions may be made and several examples will be given within this section. At first we investigate the resistance measurement in the region of the critical temperature (see Fig. 5.4). To extract details, the data of measurements with four different constant currents $I_c = 5$, 10, 30 and 42.5mA were fitted using our program SPLFIT (see Appendix 5.B) and in Fig. 5.6a–d the first derivative of this fit $\partial R/\partial T$ is plotted versus the temperature. The phase transition at T_c is taken as dominant peak in Fig. 5.6a–d. From these figures the critical temperature of our $YBa_2Cu_3O_7$ sample can be determined to be $T_c = 87.8 \pm 1.5$K. The value 1.5K is the average of the FWHM (full width half maximum) of the peaks shown in Fig. 5.6 and it can be taken as width of the sc-transition or as a measure of the inhomogenity of the sample.

As indicated in Sect. 5.2, our measurements show an interesting structure below T_c near 85...86K and this becomes particularly evident as a secondary peak or shoulder in the plots of the first derivatives (Fig. 5.6a–d). From our measurements we can not conclude if the significance of the observed second peak[2] is depending on the current I_c. A similar two-peak structure in $YBa_2Cu_3O_7$ samples has recently been reported by Pureur et al. [5.11] and Goldschmidt [5.12]. Both have measured the resistive transition of $YBa_2Cu_3O_7$ ceramics and they conclude that the transition to the zero resistance state is achieved in two steps. They use an inhomogenous granular-like model of the high-T_c oxydes to interpret their data [5.11]: "The higher temperature peak should then correspond to the superconducting transition occurring inside the grains, while the current dependent (and sometimes not well defined) low temperature peak would indicate the establishment of long-range order through the weak links." It should be pointed out that both experiments referenced used the resistance measurement technique and there are several disadvantages that make such measurements difficult, e.g. heating effects by currents, large contact resistance, cracks, parasitic thermal voltages, etc. Certainly it would be of great importance to prove that the two-peak structure may be found also in other measurement techniques of superconductivity. As mentioned in Sect.2.2 in our experiment, the resistance and TDO measurements are done simultaneously and the TDO method had the advantage of being a contact-free way of measurement. We have therefore plotted the data of simultaneous measurements

[2] It should be mentioned that the shape and height of the second peak is slightly dependent on the weight parameter w used in the spline-fit program SPLFIT.

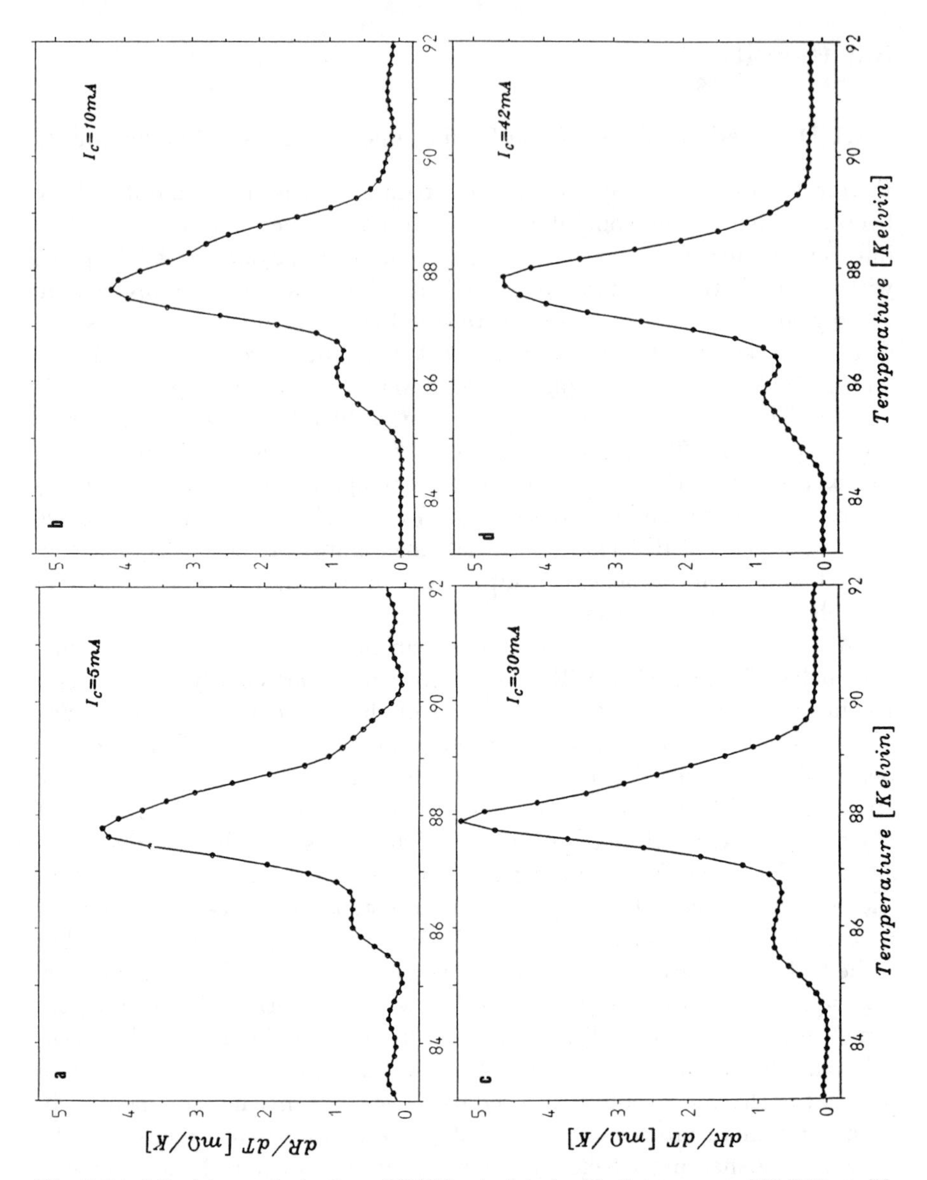

Fig. 5.6a-d Resistance derivatives $\partial R/\partial T$ calculated with the program SPLFIT at different constant currents I_c. ○ = base points connected by straight lines

Fig. 5.7a–c. (**a**) Resistance measurement (○) and TDO frequency measurement (△) both as function of temperature. (**b**) Resistance derivative of Fig. 5.7a. (**c**) Frequency derivative of Fig. 5.7a

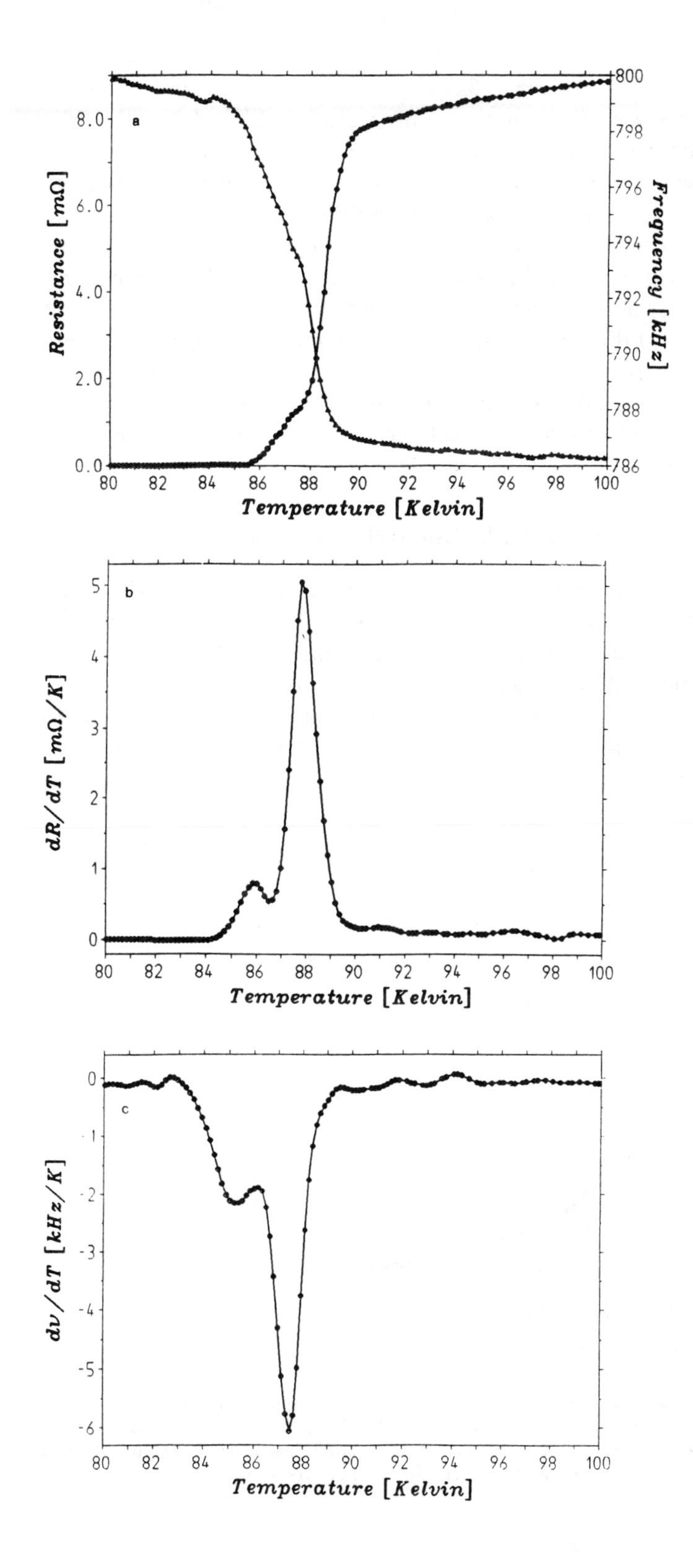

a
Resistance [mΩ]
Frequency [kHz]
Temperature [Kelvin]
b
dR/dT [mΩ/K]
Temperature [Kelvin]
c
dν/dT [kHz/K]
Temperature [Kelvin]

made at $I_c = 25$mA and $\nu_{TDO} = 786$kHz into the same diagram (Fig. 5.7a). Both measurements were fitted with the program SPLFIT (Appendix 5.B) and the first derivative of the resistance data (Fig. 5.7b) show the two-peak structure discussed in this section. The frequency jump of the TDO measurement leads to a dip in the corresponding derivative $\partial\nu/\partial T$ shown in Fig. 5.7c. It can clearly be seen that the main peak in Fig. 5.7b and the large dip in Fig. 5.7c occur at the same temperature $T = T_c$. The new result, however, is a two-dip structure in the contact free TDO-measurement data, with the second dip appearing at the same temperature as the second peak in Fig. 5.7b.[3] Up to our maximum frequency (950kHz) the shape and significance of the second dip remains the same in the corresponding derivative plots. This second dip can be explained, as the structure in Fig. 5.6, assuming granularity of our high-T_c sample.

5.3.2 Thermodynamic and Calorimentric Results

a) Cooling Law Another interesting field for off-line data analysis is the detailed study of the cooling procedure and several wanted and unwanted effects can be worked out. According to Newton's law of cooling an exponential decay of temperature as function of time is expected and a first glance at our data — see Fig. 5.8 — seems to confirm this. Attempts, however, to fit these data with a single exponential law (with help of the fit program MINUIT [5.10]) gave very bad χ^2 values and demonstrated that the cool-

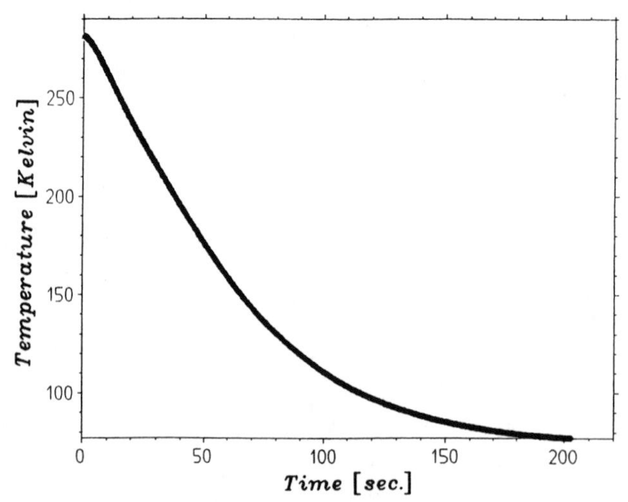

Fig. 5.8 Temperature as function of time

---

[3] For control purposes we have repeated both measurements separately i.e. the resistance measurement without frequency and the TDO measurement with current $I_c = 0$, but the characteristics found in Fig. 5.7 remain essentially unchanged. The same holds for measurements with reduced filling gas pressures.

ing period within our experiment seemed to be more complicated. Several mechanisms normally take part in the cooling process, e.g. heat conduction by the quartz tube and lead-in wires, heat transport by the He-filling gas, cooling by radiation. The sum of these effects mainly determines the shape of the cooling curve and in principle one could try to disentangle these contributions by fitting the data with sums of different functions. We found that in the region below 150K our data could satisfactorily be fitted with a single exponential:

$$T(t) = a + b\mathrm{e}^{-ct} \tag{5.3}$$

and this will be used in the next section.

b) Cooling Rate and Filling Gas Pressure As mentioned, various effects contribute to our cooling rate and the main part is the heat transport by the helium gas. The heat transport carried out by heat conduction and convection should therefore depend on the gas pressure in the quartz tube. A clear reduction of the helium pressure would lead to reduced heat transport from the $YBa_2Cu_3O_7$ sample to the surrounding liquid nitrogen. To demonstrate this we have made resistance and TDO measurements at six different helium pressures and the resulting cooling curves are shown in Fig. 5.9. We assume that the heat transport is on the whole proportional to the parameter c in the exponential law of (5.3). One can get these parameters using the fit-procedure described in Sect. 5.3.2a ($T < 150$K). The parameters c with errors, as given by the MINUIT-program [5.10] are plotted as function of the helium pressure in Fig. 5.10. The heat conductivity of gases can be calculated with the kinetic theory of gases and it remains fairly constant at moderate pressures. At low pressures a rapid drop is expected, because then the mean free path of the atoms between two collisions is of the order

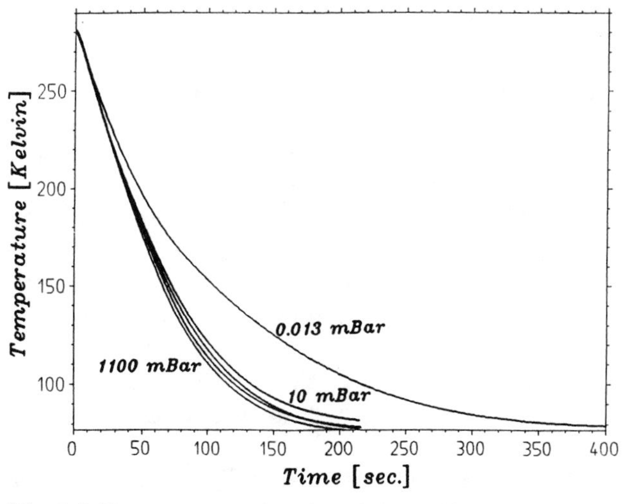

Fig. 5.9 Temperature as function of time with He-pressure as parameter

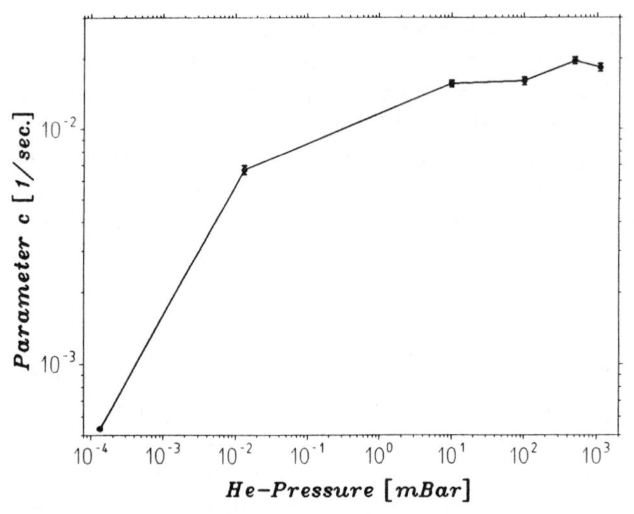

Fig. 5.10 Slope parameters c of cooling curves

of the dimensions of the apparatus (see [5.13]). The same behaviour can be observed in our results shown in Fig. 5.10 where the rapid drop starts at helium pressures of $\approx 10^{-2}$mbar.

5.3.3 Experience Within the Laboratory Course

The experiment has been in use now for more than six months and we still use the original $YBa_2Cu_3O_7$ sample. Within this period no changes of the measured characteristics have been observed and the method to store the sample in a helium-filled tube seems to be well suited for our purpose. It was mentioned already that our high-T_c-experiment is part of a special laboratory course called CAIL [5.7], that was started to familiarize students with real-time applications of computers in physics experiments. At the beginning of the course participants learn to program the computer. Additional topics are e.g. plot and program libraries, input/output operations with the interface and some more details of the computer system. With this basic knowledge the students should be able to write their own data taking programs for the high-T_c experiment and to perform selected topics of the off-line analysis. It should, however, cause no problems to integrate the experiment into a normal graduate laboratory course. Participants of such courses are in general not familiar with the computer or interface used and therefore prepared software packages for data taking and analysis methods should be available to the students. This allows the organizer of the course to adapt the exercises to the participant's level of knowledge and the time available for the experiments.

The authors thank Prof. Dr. H. Wühl and Dr. G. Müller-Vogt for helpful discussions and comments.

References

5.1. G. Bednorz and K. A.Müller, Z. Phys. **B64**, 189 (1986)
5.2. Physics Today, (special issue on superconductivity), Vol. 44, No. 6, June 1991
5.3. V. Z. Kresin and S. A. Wolf, Fundamentals of Superconductivity, Plenum Press New York and London, 1990, ISBN 0-306-43474-1
5.4. H. W. Fricke (Ed.), "Temperaturmessungen mit Philips Miniatur-Mantel-Thermoelementen, Philips Elektronik Industrie, D-3500 Kassel
5.5. R. B. Clover and W. P. Wolf, Rev.Sc.Instr., **41**,617 (1970)
5.6. J. N. Fox, F. A. Rustad and R. W. Smith, Am.J.Phys., **56**, 980 (1988)
J. N. Fox and J. U. Trefuy, Am.J.Phys., **43**, 622 (1975)
5.7. F. Kaiser and H. M. Staudenmaier, Comp. Phys. Comm. **15**, 335 (1978)
H. M. Staudenmaier, Eur. J. Phys. **3**, 144 (1982)
H. M. Staudenmaier in E. F. Redish, J. S. Risley (Ed's), Conf. Proceedings Computers in Physics Instruction, Addison-Wesley 1990
5.8. Pechan and Horvat, Am. J. Phys. **58**, 642 (1990)
5.9. J. Halbritter, M. R. Dietrich, H. Küpfer, B. Runtsch and H. Wühl, Z. Phys. **B71**, 411 (1988)
5.10. F. James and M. Roos, MINUIT, D506 CERN-Program Library (Public Domain Software)
5.11. P. Pureur, J. Schaf, M. A. Gusmao and J. V. Kunzler, Physica **C 176**, 257 (1991)
5.12. D. Goldschmidt, PR **B 39**, 9139 (1989) and PR **B 39**, 2372 (1989)
5.13. K. Tödheide et al. in Landolt-Börnstein II/5b, 39 (Springer 1968)
5.14. P. Jacob and S. Jancar in Vieweg Programmothek 4 (Ed. H. Kohler), Vieweg Braunschweig 1985, ISBN 3-528-04396-2

Appendix 5.A: Electric Circuit Diagrams

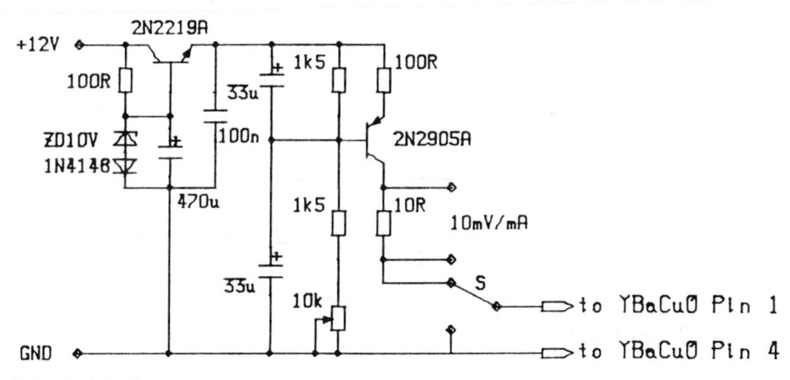

Fig. 5.A1 Constant current source

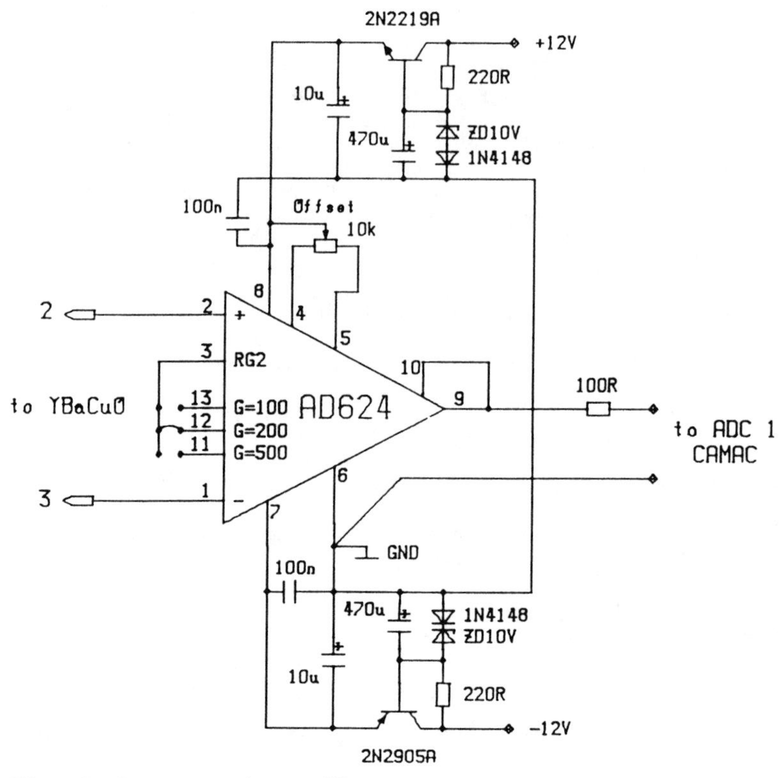

Fig. 5.A2 Instrumentation amplifier

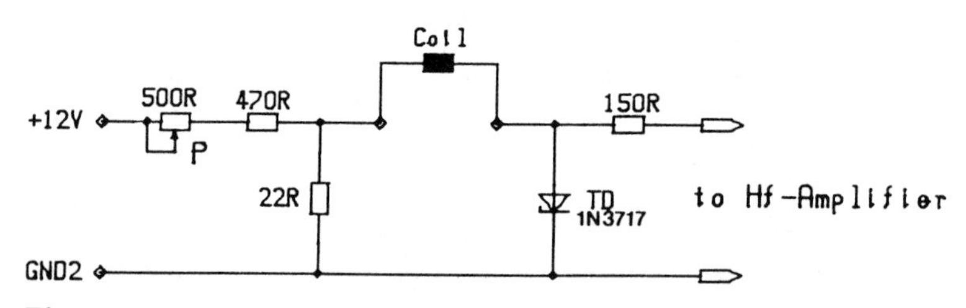

Fig. 5.A3 Tunnel-diode oscillator

Appendix 5.B: Spline Fit Program SPLFIT

The listed FORTRAN program uses an algorithm described by P. Jacob and S. Jancar [5.14], details and examples concering the influence of the weight parameter W may be found there. The arguments of the FORTRAN subroutine SPLFIT are as follows:

NMAX: Number of data or base points
X: Input vector containing the measured x-base points
Y: Input vector containing the measured y-values
W: Weight parameter (e.g. W= 500), influences the flexibility of the fit and the accuracy of the derivatives
A: Output vector with spline functions at x-base points
B: Output vector with first derivative of spline functions
C2: Output vector with second derivative of spline functions

```
      SUBROUTINE SPLFIT(NMAX,X,Y,W,A,B,C2)

      PARAMETER (NN2=5000, E=1.E-3, MX=1000)
      REAL X(0:NMAX),Y(0:NMAX),W(0:NMAX)
      REAL A(0:NMAX),B(0:NMAX),C(0:NN2),C2(0:NMAX),D(0:NN2)
      REAL H(0:NN2),M(0:NN2,0:4),E,C1,D1,DF
      INTEGER I,N,MX,K,J
C
C
      DO 1000 I=0,NMAX-1,1
         H(I)=X(I+1)-X(I)
         IF (H(I).LT.0.) THEN
            PRINT*,' *** ERROR:  X-VALUES NOT MONOTONE ***'
            WRITE(*,*) X(I),X(I+1),I
         END IF
 1000 END DO

C
C*********************** CALCULATION MATRICE ***********************
C
      DO 1300 I=1,NMAX-1,1
         D1=2.*(H(I-1)+H(I))+6./W(I-1)/H(I-1)/H(I-1)+
     &        6./W(I)*(1./H(I-1)+1./H(I))*(1./H(I-1)+
     &        1./H(I))+6./W(I+1)/H(I)/H(I)
         M(I,0)=0.
         M(I,1)=0.
         M(I,2)=0.
         M(I,3)=0.
         IF (I.EQ.1) GOTO 1100
         M(I,0)=6./W(I-1)/H(I-2)/H(I-1)/D1
         M(I,1)=(H(I-1)-6./W(I-1)/H(I-1)*(1./H(I-2)+1./H(I-1))-
     &          6./W(I)/H(I-1)*(1./H(I-1)+1./H(I)))/D1
 1100    CONTINUE
         IF (I.EQ.NMAX-1) GOTO 1200
         M(I,2)=(H(I)-6./W(I)/H(I)*(1./H(I-1)+1./H(I))-
     &          6./W(I+1)/H(I)*(1./H(I)+1./H(I+1)))/D1
         M(I,3)=6./W(I+1)/H(I)/H(I+1)/D1
 1200    CONTINUE
         M(I,4)=3.*((Y(I+1)-Y(I))/H(I)-(Y(I)-Y(I-1))/H(I-1))/D1
 1300 END DO
C
C************************* INITIAL VALUES *************************
C
      DO 1400 I=0,NMAX,1
         C(I)=0.
 1400 END DO
C
C*********************** GAUSS-SEIDEL-ITERATION ********************
C
      DO 1900 K=0,MX,1
         DF=0.
         DO 1800 I=1,NMAX-1,1
            C1=M(I,4)
            IF (I.EQ.1) GOTO 1500
            C1=C1-M(I,0)*C(I-2)-M(I,1)*C(I-1)
```

```
 1500          CONTINUE
               IF (I.EQ.NMAX-1) GOTO 1600
               C1=C1-M(I,2)*C(I+1)-M(I,3)*C(I+2)
 1600          CONTINUE
               IF(ABS(C1-C(I)).LE.DF) GOTO 1700
               DF=ABS(C1-C(I))
 1700          C(I)=C1
 1800       END DO


C           WRITE(*,*)DF
            IF (DF.LE.E) GOTO 2000
 1900     END DO
          PRINT*,' *** ERROR: GIVEN ACCURANCY NOT ACHIEVED ***'
C
C************************* COEFFIZIENTS **************************
C
 2000     A(0)=Y(0)+2./W(0)/H(0)*(C(0)-C(1))
          A(NMAX)=Y(NMAX)-2./W(NMAX)/H(NMAX-1)*(C(NMAX-1)-C(NMAX))
          DO 2100 I=1,NMAX-1,1
            A(I)=Y(I)-2./W(I)*(C(I-1)/H(I-1)-C(I)/H(I-1)-C(I)/H(I)+
     &          C(I+1)/H(I))
 2100     END DO
          DO 2200 I=0,NMAX-1,1
            B(I)=(A(I+1)-A(I))/H(I)-H(I)/3.*(C(I+1)+2.*C(I))
            D(I)=(C(I+1)-C(I))/3./H(I)
            C2(I)=2*C(I)
 2200     END DO

          RETURN
          END
```

Fig. 5.B1 Spline Fit Program SPLFIT

6. Computer Control of Low Temperature Specific Heat Measurement

G. Keeler

The measurement of specific heat is, in principle, simple and straightforward, requiring the measurement of the temperature rise of a body of known mass when a measured quantity of heat is supplied to the body. The main practical difficulty is in avoiding or compensating for the heat losses to (or from) the surroundings.

Most of the heat losses can be avoided by insulation, but some radiation loss, in particular, cannot be avoided, so it is important to correct for the heat losses.

The value of the specific heat at low temperature is of particular interest for comparison with theoretical predictions. Other than at very low temperatures, this does not impose any major extra problems, except for the initial cooling of the specimen and a greater requirement for some degree of radiation shielding. At very low temperatures, difficulties increase in obtaining and maintaining the low temperature, which requires the use of liquid helium, and more effective shielding is required. More fundamentally, it is more difficult to make the measurements at very low temperature because the specific heats of both specimen and surroundings become extremely small, so that only very small amounts of heat are needed, and these must be delivered rapidly because the surroundings may also change temperature quickly.

The present experiment uses a modified form of the apparatus introduced by Nernst and Eucken in 1910 [6.1]. A metal specimen in the form of a cylinder has a heater coil and a separate copper resistance thermometer wound onto it, and it is suspended in a vacuum with radiation shielding to minimize heat losses. A microcomputer is used to make the measurements of temperature and heat supplied to the specimen. The microcomputer also controls the experiment by switching the heater on at an appropriate setting for the amount of heat required, monitoring the temperature until a sufficient rise has occurred, and timing the period of heating needed for that temperature rise.

By cooling the specimen and its surroundings to low temperature initially, the microcomputer can carry out a whole series of heating measurements as the specimen is warmed up, and obtain specific heat data over a wide range of temperature. The microcomputer program has been designed to give the user control over the range of measurements to be made and the temperature increment for each reading; the program makes a calculation of the correction needed for heat losses.

This system has been used in our Department for several years in the final year undergraduate laboratory, with specimens of either copper, erbium or dysprosium. The copper illustrates the nature of the Debye specific heat curve, while erbium and dysprosium have substantial specific heat anomalies caused by magnetic transitions.

6.1 Basic Physics

6.1.1 Specific Heat

If the molar energy density of a system (in our case a metallic solid) is U, then the specific heat is defined as

$$C = \frac{dU}{dT} . \tag{6.1}$$

Theoretically, it is usual to calculate the *constant volume* specific heat, C_V

$$C_V = \left(\frac{\partial U}{\partial T}\right)_V , \tag{6.2}$$

since this then excludes any work done in expanding the solid.

Experimentally, especially for solids, the easiest quantity to measure is the constant pressure specific heat, C_P.

In order to make a comparison with theoretical predictions, C_P must be converted to C_V, which can be done using the thermodynamic relationship

$$C_P - C_V = K \, \alpha_V{}^2 \, T \tag{6.3}$$

where K is the bulk modulus, and α_V is the volume expansion coefficient, defined as

$$\alpha_V = \frac{1}{V}\left(\frac{\partial V}{\partial T}\right)_P . \tag{6.4}$$

According to classical theory of equipartition of energy, the molar energy of a solid is made up of $\frac{1}{2}kT$ per degree of freedom. The atoms in a solid can be looked at as three separate simple harmonic oscillators, for the three orthogonal directions of vibration, and since each direction has both kinetic energy and potential energy, due to the atom being in a lattice, there are six degrees of freedom per atom. Since a mole contains N atoms, where N is Avagadro's number, the molar energy density is

$$U = 3NkT . \tag{6.5}$$

Thus

$$C_V = \left(\frac{\partial U}{\partial T}\right)_V = 3Nk = 3R \tag{6.6}$$

which is a constant value independent of temperature.

Experimentally, this is found to be correct at high temperature, but at low temperature the specific heat is found to decrease more and more.

6.1.2 Low Temperature Specific Heat

To explain the low temperature form of the specific heat it is necessary to adopt a quantum mechanical approach. A full treatment of lattice specific heat is given in many texts, including Kittel [6.2].

A very simple model was developed by Einstein in 1907 [6.3], which is really only applicable to gases, but which gave a qualitative explanation of the low temperature decrease for any material. He treated a system of N atoms as a set of $3N$ one-dimensional independent simple harmonic oscillators and applied a simple statistical treatment.

The energy of a single simple harmonic oscillator is quantized in units of $\hbar\omega$, and is given by

$$E_n = \left(n + \frac{1}{2}\right)\hbar\omega\ . \tag{6.7}$$

According to the Boltzmann distribution law the probability of a system with discrete energy levels having energy E_n is proportional to

$$\exp\left[\frac{-E_n}{kT}\right]\ . \tag{6.8}$$

Since the system must be in one of its states, the probability of the system being in state E_n must be given by

$$P_n = \frac{\exp\left[\frac{-E_n}{kT}\right]}{\sum_{n=0}^{\infty}\exp\left[\frac{-E_n}{kT}\right]}\ . \tag{6.9}$$

In the case of a simple harmonic oscillator this probability becomes

$$P_n = \frac{\exp\left[\frac{-(n+\frac{1}{2})\hbar\omega}{kT}\right]}{\sum_{n=0}^{\infty}\exp\left[\frac{-(n+\frac{1}{2})\hbar\omega}{kT}\right]}\ . \tag{6.10}$$

For a system of simple harmonic oscillators, as postulated by Einstein, the average energy is given by summing the possible energies multiplied by the probability of the oscillators having that energy:

$$\langle E\rangle = \sum P_n\,E_n\ . \tag{6.11}$$

Substituting for E_n and P_n, and multiplying the expression for P_n top and bottom by $\exp\left[\frac{\frac{1}{2}\hbar\omega}{kT}\right]$, this becomes

$$\begin{aligned}\langle E\rangle &= \frac{\sum_{n=0}^{\infty}\left(n+\frac{1}{2}\right)\hbar\omega\exp\left[\frac{-n\hbar\omega}{kT}\right]}{\sum_{n=0}^{\infty}\exp\left[\frac{-n\hbar\omega}{kT}\right]} \\ &= \frac{1}{2}\hbar\omega + \frac{\sum_{n=0}^{\infty}n\hbar\omega\exp\left[\frac{-n\hbar\omega}{kT}\right]}{\sum_{n=0}^{\infty}\exp\left[\frac{-n\hbar\omega}{kT}\right]}\,. \end{aligned} \tag{6.12}$$

To simplify this expression we can temporarily make the substitution

$$x = \frac{\hbar\omega}{kT} \qquad \text{and define} \qquad S = \sum_{n=0}^{\infty} e^{-nx} \tag{6.13}$$

where S is a standard sum which has the value $1/(1-e^{-x})$.

Further, by differentiating S we obtain

$$\frac{\partial S}{\partial x} = \sum_{n=0}^{\infty}(-n)\,e^{-nx} = \frac{\partial}{\partial x}\left(\frac{1}{1-e^{-x}}\right) = \frac{-e^{-x}}{(1-e^{-x})^2}\,. \tag{6.14}$$

The top and bottom sums in the expression for $\langle E\rangle$ are S and $-\hbar\omega\,\frac{\partial S}{\partial x}$ respectively, so

$$\begin{aligned}\langle E\rangle &= \frac{1}{2}\hbar\omega + \hbar\omega\frac{e^{-x}/\left(1-e^{-x}\right)^2}{1/\left(1-e^{-x}\right)} \\ &= \frac{1}{2}\hbar\omega + \hbar\omega\frac{e^{-x}}{1-e^{-x}}\,. \end{aligned} \tag{6.15}$$

Multiplying top and bottom by e^{-x} and substituting back for x, this finally becomes

$$\langle E\rangle = \frac{1}{2}\hbar\omega + \frac{\hbar\omega}{\exp\left[\frac{\hbar\omega}{kT}\right]-1}\,. \tag{6.16}$$

For Einstein's system of $3N$ independent oscillators, the molar energy density is simply $U = 3N\langle E\rangle$. The specific heat can therefore be calculated by differentiating the above expression for $\langle E\rangle$:

$$\begin{aligned}C_V &= 3N\hbar\omega\frac{\partial}{\partial T}\left(\frac{1}{\exp\left[\frac{\hbar\omega}{kT}\right]-1}\right) \\ &= 3N\hbar\omega\frac{\frac{\hbar\omega}{kT^2}\exp\left[\frac{\hbar\omega}{kT}\right]}{\left(\exp\left[\frac{\hbar\omega}{kT}\right]-1\right)^2} \\ &= 3Nk\left(\frac{\hbar\omega}{kT}\right)^2\frac{\exp\left[\frac{\hbar\omega}{kT}\right]}{\left(\exp\left[\frac{\hbar\omega}{kT}\right]-1\right)^2}\,. \end{aligned} \tag{6.17}$$

At high temperature (or more accurately, where $kT \gg \hbar\omega$) this reduces to the classical value $3Nk$, but at lower temperatures it falls off more and more rapidly, giving a good qualitative explanation of the observed effect.

6.1.3 The Debye Model for the Specific Heat

Although Einstein's model is quite good for gases, it has one major defficiency for solids, in that the simple harmonic oscillators are, in fact, strongly coupled, and have a continuous spectrum of vibrational frequencies. Moreover, in a realistic model the spectrum is a complex one. The calculation of C_V still follows most of the process outlined above, but the total energy involves an integration over the frequency spectrum.

A satisfactory simplification, first introduced by Debye [6.4], is to treat the vibrational spectrum as that for an elastic continuum. There are then three vibrational modes, one longitudinal and two transverse, which correspond to longitudinal and transverse sound waves. If the sound wave velocities are $c_{\rm L}$ and $c_{\rm T}$ respectively, the vibrational frequencies are inversely proportional to the wavelengths λ of the sound waves, as

$$\omega_{\rm L} = \frac{2\pi c_{\rm L}}{\lambda} \quad \text{and} \quad \omega_{\rm T} = \frac{2\pi c_{\rm T}}{\lambda} . \tag{6.18}$$

The integration over this spectrum in 3D is still too complex a calculation to deal with here, but it is treated in many solid state physics texts. It eventually results in the expression for specific heat

$$C_V = \frac{kV}{2\pi^2}\left(\frac{1}{c_{\rm L}^3} + \frac{2}{c_{\rm T}^3}\right) \int_0^{\omega_{\rm D}} \left(\frac{\hbar\omega^2}{kT}\right)^2 \frac{\exp\left[\frac{\hbar\omega}{kT}\right]}{\left(\exp\left[\frac{\hbar\omega}{kT}\right] - 1\right)^2}\, d\omega . \tag{6.19}$$

V is the molar volume and $\omega_{\rm D}$ is a cut-off frequency, required because the model has no natural frequency limit. It can be evaluated by the requirement that the total number of vibration modes is $3N$. It is usual again to introduce for $\frac{\hbar\omega}{kT}$ the parameter x, and to express the cut-off frequency in terms of an equivalent temperature, the *Debye Temperature*, $\Theta_{\rm D} = \frac{\hbar\omega_{\rm D}}{k}$. The specific heat then becomes

$$C_V = 9Nk\left(\frac{T}{\Theta_{\rm D}}\right)^3 \int_0^{\Theta_{\rm D}/T} x^4 \frac{e^x}{(e^x - 1)^2}\, dx \tag{6.20}$$

where

$$\Theta_{\rm D}^3 = \frac{18\, N\, \pi^2\, \hbar^3}{k^3\, V\left(\frac{1}{c_{\rm L}^3} + \frac{2}{c_{\rm T}^3}\right)} . \tag{6.21}$$

This specific heat has the same qualitative features as the Einstein model, tending to $3Nk$ at high temperature and rolling off at low temperature, but it behaves quite differently at very low temperature, and is a much better approximation to the behaviour of real solids. Over the temperature range of the present experiment, however, both Einstein and Debye models give a satisfactory approximation to the observed results.

6.1.4 Specific Heat Anomalies

For most solids the lattice energy is the main contributor to the internal energy of the solid, and hence its specific heat, although for metals at low temperature the electronic energy becomes paramount. However, some materials have other internal energy contributions. These contributions frequently occur over a small temperature interval, and so make a large contribution to the specific heat over the small temperature region where the internal energy change is occurring, resulting in *specific heat anomalies*. A good example of this type of mechanism is the internal energy resulting from a disordering process, such as when a magnetic material is undergoing a phase change. For instance, dysprosium has a small specific heat anomaly at the Curie temperature, where it changes from ferromagnetic to antiferromagnetic, and a very large anomaly around the Néel temperature, where it changes from antiferromagnetic to paramagnetic, a major disordering process.

6.2 Experimental Setup

6.2.1 Specimen

The specimen is machined into a cylinder with typical dimensions of about 2 cm diameter and 2 cm length, having a mass of approximately 100 g. The mass of the specimen (including the copper thermometer) is determined once before the apparatus is assembled, and then provided for students as one of the parameters of the system.

A heater coil and a copper resistance thermometer are wound onto the specimen. The whole specimen is first covered with an insulating layer of tissue cemented down with a solvent-based adhesive.

The heater consists of a few turns of constantan resistance wire, that has only a small variation of resistance with temperature. A length with about 10 ohms resistance is used, and it is covered with another layer of adhesive coated tissue to provide thermal contact to the specimen. The ends are each soldered to pairs of copper leads to provide separate measurement of current and voltage; it is important that these ends are cemented to the specimen so that all of the heater wire is thermally anchored, otherwise the power dissipation in the free section of wire can cause local heating sufficient to burn out the wire.

The copper thermometer consists of several hundred turns of about 50 SWG (25 micron diameter) copper wire in order to produce a substantial resistance, typically about 5000 ohms. The winding is carried out on a lathe. Again it is important that the ends, which are soldered to more substantial connection leads, are cemented to the specimen, this time mainly to provide support for the fragile thermometer wire.

The specimen is supported by nylon threads for thermal insulation. Conduction down the heater and thermometer connection leads can form a substantial form of heat loss (or more likely, heat input, since the specimen is usually colder than room temperature) so the leads are made of the finest wire that is practical, and are long lengths of wire coiled up in the form of a spring to increase the thermal path.

6.2.2 Apparatus

Figure 6.1 shows a diagram of the apparatus. The specimen is suspended at the bottom of a double-walled glass dewar. The dewar has silvered walls for radiation shielding, and is sealed at the top with a plate that provides access for electrical leads, gas handling system and an inlet port for a liquid helium transfer tube.

The dewar is surrounded by an open outer dewar. For low temperature work the outer dewar is filled with liquid nitrogen, which acts as a 77 K radiation shield for the inner dewar. For higher temperature work the liquid nitrogen shield can act as a substantial radiation absorber and the liquid nitrogen is, therefore, removed. If available, solid carbon dioxide is used to provide an intermediate temperature shield.

The gas handling system can be used to evacuate the inner chamber and the dewar walls, thus providing thermal insulation for the specimen. A combination of rotary pump and diffusion pump are used to obtain a vacuum of around 10^{-5} torr, adequate for insulation purposes. The gas handling system will also allow the introduction of a few cm pressure of helium exchange gas into the inner chamber and dewar walls. This is used to cool down the

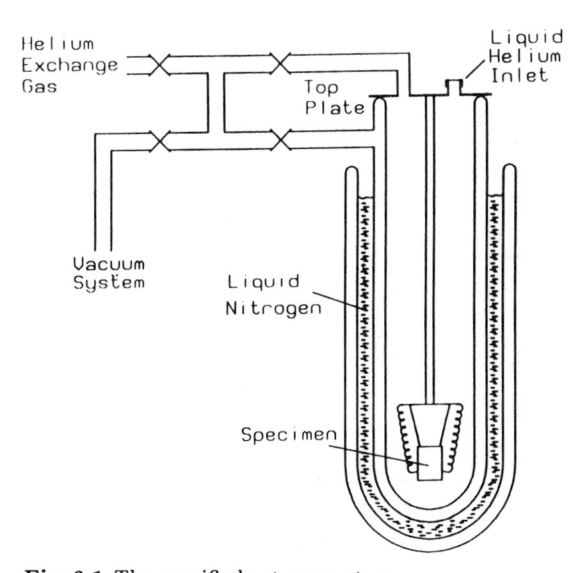

Fig. 6.1 The specific heat apparatus

specimen to the same temperature as the outer dewar by forming a good thermal link.

To cool the specimen further still, down to liquid helium temperature, the inner dewar walls can be evacuated and liquid helium introduced through the inlet port in the top plate. Once the specimen has reached liquid helium temperature, before a substantial amount of liquid has collected, the transfer is stopped and the inner chamber evacuated to thermally isolate the specimen.

At present all the gas handling is done manually; it does not seem practical or necessary to use computer control for this aspect of the work.

6.2.3 Electronics

Figure 6.2 shows a block diagram of the arrangement of the electronics. There are three circuits:

a) Thermometer Circuit The thermometer is read as a four terminal device, with separate connections for current and voltage measurement. Since the thermometer is to be calibrated, there is no need for the current to be measured. It is supplied by an accurate constant current supply, a circuit diagram of which is shown in Appendix 6.A. It has an internal adjustment that gives a small range of current, and it is usually set to 10 μA.

The voltage across the thermometer must be measured by the microcomputer. Since it is too small to measure directly, it is first amplified by a chopper stabilized operational amplifier in the non-inverting configuration. The circuit diagram of the amplifier is shown in Appendix 6.A. The main amplifier chip is a 7650 and it incorporates a clamp circuit to prevent input overloads; this is implemented automatically by connecting pin 9 to the in-

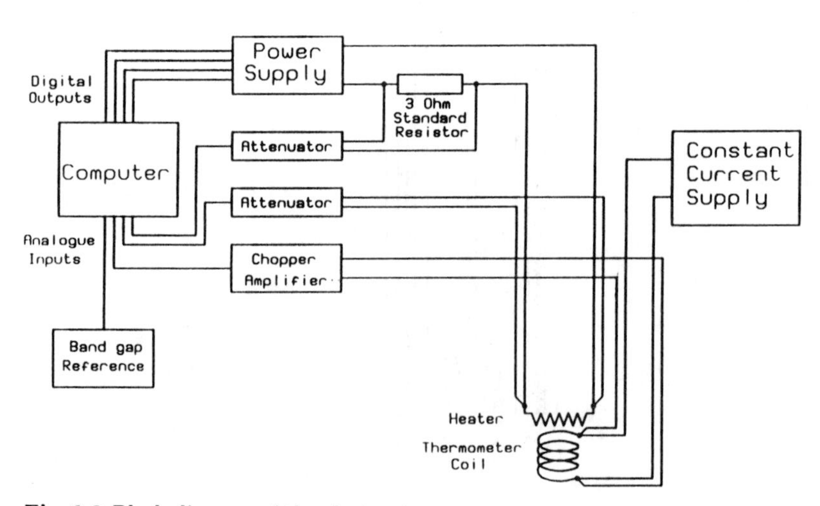

Fig. 6.2 Block diagram of the electronics

verting input. The gain is controlled by the external resistors R_1 and R_2, and is given by

$$A_{\mathrm{VCL}} = \frac{R_1 + R_2}{R_1} \ . \tag{6.22}$$

In the present circuit this value is set at 10.

b) Heater Circuit This is very simple, consisting of another four terminal connection to the heater on the specimen. The current is supplied from the computer-controlled power supply; the voltage across the heater is measured by one of the digital inputs of the microcomputer via a simple isolating attenuator. It would be possible to compute the heater power simply from the voltage, knowing the heater coil resistance and assuming it to be constant, but it is just as easy, and more accurate, to pass the heater current through a standard 3 ohm resistor in series, and monitor the voltage across the standard resistor, again via an attenuator, with another of the analogue inputs to the microcomputer. The isolating feature of the attenuators is important, since they feed into the A/D converters that have a common ground.

c) Power Supply Circuit The power supply is a standard laboratory stabilized power supply (manufactured by Kingshill), modified to give four set output levels under computer control.

The computer control circuit is shown in Fig. 6.3. Four single bit outputs from the computer are buffered by opto-isolators, and used to switch different loads across the power supply control circuit. The normal operation of the power supply is via two variable resistors, $RV1$ and $RV2$, giving respectively fine and coarse control of the voltage output. The voltage output is proportional to the combined resistance of $RV1$ and $RV2$.

The original circuit with $RV1$ and $RV2$ is broken, and replaced by a transistor switch, $T1$. When the computer is not controlling the power sup-

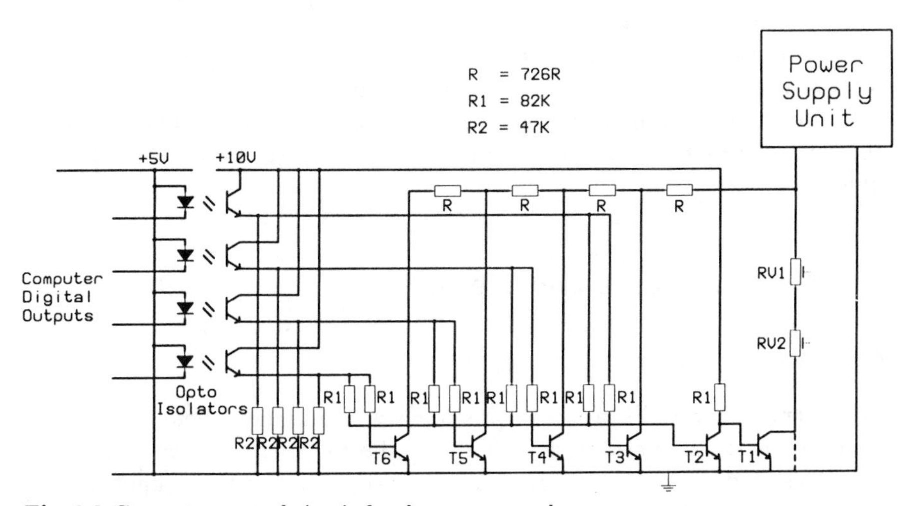

Fig. 6.3 Computer control circuit for the power supply

ply, $T1$ is switched on and the power supply functions normally. (The normal state is in practice to have $RV1$ and $RV2$ turned down so that the power output is zero.) When any one of the computer output lines is switched on, $T1$ is switched off, breaking the normal connection. Instead, one of the transistors $T3$ to $T6$ is switched on. These each connect into the power supply one of the points on the ladder of four resistors, R, that are in parallel with $RV1$ and $RV2$. The four resistors are equal, and chosen so that the output of the power supply is 2.5 V, 5 V, 7.5 V or 10 V.

6.2.4 Microcomputer Control

The microcomputer originally used in the experiment was an Acorn BBC model B microcomputer. This microcomputer is ideally suited for controlling an experiment because it is supplied with all the requirements needed: a sophisticated version of Basic (BBC Basic) that is adequate for control purposes; an inbuilt timing capability; four 12-bit analogue-to-digital inputs; and various digital inputs and outputs, including the complete 8-bit Port A of a 6522 VIA that can have each line programmed either for input or output. The A to D inputs are designed for voltages in the range 0 to 1.8 V, and the digital lines are standard TTL levels.

The system has since been adapted to run also on an IBM compatible PC microcomputer. The PC does not have any built-in I/O facilities, but a number of commercial plug-in interface boards are available that provide digital or analogue I/O, or both. The board used in this experiment is the Salford Analogue and Digital I/O Board, designed in this Department and now being marketed by E & L Instruments Ltd [6.5]. It has 24 digital I/O lines (the three 8-bit Ports of an 8255 Peripheral Interface Adapter) and 4 analogue inputs to a 12-bit ADC (a MAX182 analogue-to-digital Converter), as well as three timer/counters and analogue outputs that are not used in this particular application.

The present experiment uses four of the digital I/O lines as outputs, to control the heater power supply: three analogue inputs are needed to measure the thermometer voltage and the heater voltage and current, and the fourth is used to calibrate the readings by measuring the voltage of a band gap reference.

6.3 Measurements and Results

6.3.1 Measurement Principles

a) Thermometer Calibration A preliminary requirement before the main measurements can be taken is to calibrate the copper resistance thermometer. This is done by taking readings at either two or three known temperatures, and using a linear calibration either through the two points or determined by a least squares fit to the three points. This is adequate for temperatures above 77 K. For lower temperatures a more sophisticated fit would be necessary, along the lines used to calibrate standard platinum resistance thermometers.

The two obvious calibration points, always used, are ice point and liquid nitrogen. Some students use room temperature as the third, relying on the fact that at the start of the experiment the apparatus is in good equilibrium and the room temperature is determinable with sufficient accuracy. These assumptions are probably valid, but room temperature is too close to ice point to make a good least squares fit. The ideal third point is the sublimation point of solid carbon dioxide, if it is available, and carbon dioxide can also form a useful intermediate radiation shield temperature. However, solid carbon dioxide is not often available, and the present experiment has frequently had to rely on two calibration points.

For fairly pure copper between 77 K and 300 K the linear fit assumed in this experiment should be a good approximation [6.6].

b) Specific Heat Measurement In principle all that is needed for specific heat determination is measurement of an amount of energy delivered, and the corresponding temperature rise in a specimen of known mass.

In practice, heat losses (or heat gains) are a major complicating factor. For the purposes of all following discussions, these will be referred to as heat losses, but in practice there are times when the specimen is colder than its surroundings and so the isolated specimen will be warming up, rather than cooling down.

The three sources of potential heat loss are conduction through the support wires and electrical leads, convection and conduction through the surrounding gas, and radiation. With the design of apparatus used, the first two are negligible compared with the effects of radiation, which are unavoidable without elaborate techniques such as surrounding the specimen with a temperature controlled shield maintained at the same temperature as the specimen.

The radiation itself can be broadly subdivided into two contributions – that from the surrounding inner dewar walls, and that from the top plate above (in practice this can include the upper part of the dewar wall, above any external coolant). The dewar walls can easily be maintained at either liquid nitrogen temperature or room temperature, but the top plate will always be at room temperature. When the specimen is at low temperature, this will be a significant source of heating, even though it subtends a relatively small solid angle at the specimen. It can be reduced somewhat by radiation baffles, but not eliminated easily since the baffles themselves will be heated by the top plate radiation.

The main complication of the measurement process, therefore, is the correction for heat losses. Two straightforward methods can be envisaged.

Measurement of Heating and Cooling Gradients If the rate of heat loss of the specimen with the heater off is L (where L is treated as positive when there is a heat loss to the surroundings), then the rate of cooling of the specimen with the heater off will be

$$\left(\frac{dT}{dt}\right)_{\mathrm{C}} = \left(\frac{dT}{dQ}\right)\left(\frac{dQ}{dt}\right) = -\frac{1}{mS}L \tag{6.23}$$

where the specific heat of the specimen is given by

$$S = \frac{1}{m}\frac{dQ}{dT} \,. \tag{6.24}$$

When the heater is on, the rate of heating will be

$$\left(\frac{dT}{dt}\right)_{\mathrm{H}} = \left(\frac{dT}{dQ}\right)\left(\frac{dQ_{\mathrm{total}}}{dt}\right) = \frac{dT}{dQ}\left(\frac{dQ_{\mathrm{heater}}}{dt} - L\right) \tag{6.25}$$

and again eliminating the term $\frac{dT}{dQ}$

$$\left(\frac{dT}{dt}\right)_{\mathrm{H}} = \frac{1}{m\,S}\left(\frac{dQ_{\mathrm{heater}}}{dt} - L\right) . \tag{6.26}$$

The term L can now be eliminated from the two expressions for the rate of heating:

$$\left(\frac{dT}{dt}\right)_{\mathrm{H}} = \frac{1}{m\,S}\left(\frac{dQ_{\mathrm{heater}}}{dt}\right) + \left(\frac{dT}{dt}\right)_{\mathrm{C}} . \tag{6.27}$$

The first and last terms represent the heating and cooling temperature gradients respectively ($(\frac{dT}{dt})_{\mathrm{C}}$ will be negative if heat losses predominate). Thus measurements of the heating and cooling temperature gradients, for a given heater power $\left(\frac{dT}{dQ}\right)$, will enable the heat loss to be eliminated and the specific heat determined.

Correction of the Temperature Rise During Heating If the expression above for $(\frac{dT}{dt})_{\mathrm{H}}$ is integrated for the period of the heating burst, this will give

$$\Delta T_{\mathrm{H}} = \frac{1}{m\,S}\Delta Q_{\mathrm{heater}} + \Delta T_{\mathrm{C}} \tag{6.28}$$

or

$$S = \frac{1}{m}\frac{\Delta Q_{\mathrm{heater}}}{\Delta T_{\mathrm{H}} - \Delta T_{\mathrm{C}}} \,. \tag{6.29}$$

The quantity $-\Delta T_{\mathrm{C}}$ represents the amount by which the specimen temperature would have fallen if the heater had not been switched on. By extrapolating cooling curves measured before and after the heating burst, $-\Delta T_{\mathrm{C}}$ can be calculated, or more simply the vertical difference between the two cooling curves represents the total temperature excursion, $(\Delta T_{\mathrm{H}} - \Delta T_{\mathrm{C}})$.

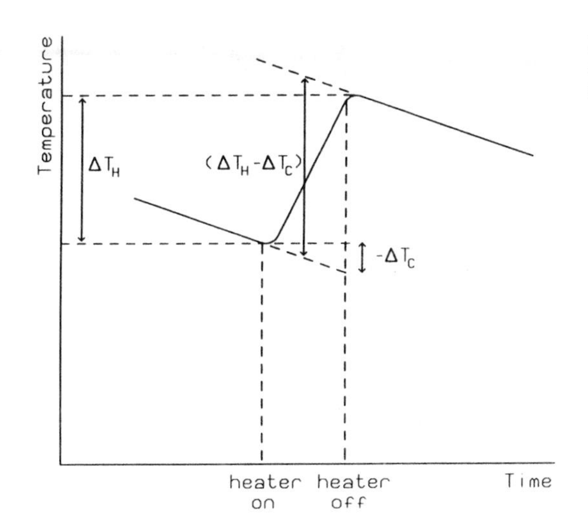

Fig. 6.4 The extrapolation process to calculate the temperature excursion

The extrapolation process is shown in Fig. 6.4.

The interesting feature of this technique is that the measurement of $(\Delta T_\mathrm{H} - \Delta T_\mathrm{C})$ is easier, and more accurate, than either ΔT_H or ΔT_C separately. Time lags between the heater switching on and the heat diffusing through the solid will have the effect of 'rounding off' the points where the heater is switched on and off, as shown in the figure, but this does not affect $(\Delta T_\mathrm{H} - \Delta T_\mathrm{C})$. It also becomes apparent that measurements during heating are less important than those of the cooling curves before and after (provided that the actual time of switching the heater on and off are known).

In an extreme case, the cooling 'curves' would indeed be curved rather than straight. The most pronounced result of this would be that the cooling curves on either side of the heating burst would have different slopes, as shown in Fig. 6.5, and the value of $(\Delta T_\mathrm{H} - \Delta T_\mathrm{C})$ would then depend on where the vertical line between the two curves is drawn. Ideally, the line should be drawn at the mid point of the *actual* heating section (as indicated by the broken line on the figure). In practice the difference between the slopes is always small, so drawing the line at the midpoint of the heating burst should be reasonably accurate.

This second technique has the advantage over the method using heating and cooling gradients that it is not necessary to determine the heating gradient in what may be a short heating burst without a well-defined slope, and by using the cooling gradients on either side of the heating it is more accurate, as explained above. It is, therefore, the preferred method of making the specific heat measurements. It has the added advantage that the rate of heating does not have to be constant, and this feature is useful so that the rate of heating can be varied by the microcomputer to find the optimum rate.

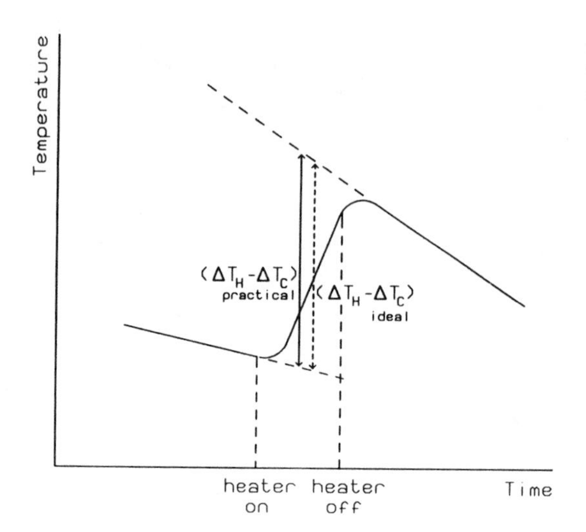

Fig. 6.5 The effect of different gradients of the two cooling curves

c) Temperature Variation of Specific Heat Since the essence of this experiment is to measure how the specific heat varies with temperature, the measurements should ideally be done at spot temperatures. However, the accuracy of the measurements clearly improve as the temperature increment $(\Delta T_{\mathrm{H}} - \Delta T_{\mathrm{C}})$ becomes larger. Thus a compromise is necessary in the temperature increment over which the specific heat is measured, and the results will represent the average specific heat over this temperature interval. The effect is particularly important in the measurement of the specific heat anomalies in erbium or dysprosium, and here the compromise could be different to that for copper where the specific heat only changes slowly with temperature.

In the computer-controlled experiment, the temperature increment over which the measurement is made is one of the most important parameters under the control of the user.

6.3.2 Using the Computer Program

The program to control the specific heat experiment was originally written in BBC Basic for the Acorn BBC microcomputer. It has now been converted to run on a PC, and is available both in Turbo Pascal and Microsoft QuickBasic. Appendix 6.B gives a listing of the Turbo Pascal version of the Program.

The first process that must be performed on any program run is the calibration of the thermometer. Because calibration is a long and tedious process, there is an option to enter calibration values determined on a previous occasion. Otherwise, the program performs a calibration from either two or three fixed point measurements of the thermometer when the specimen is held at the fixed point temperature. The program waits for the student to enter a calibration temperature, which should be that of the fixed point. The program then makes and averages a whole series of rapid measurements of the thermometer voltage, to try and reduce noise and achieve an accurate calibration. When the two or three fixed points have been determined to the student's satisfaction, a linear calibration is made, either through the two points or fitted to three points. The linear fit parameters are reported on screen, and the student needs to record these if he wishes to use them again on a later occasion.

At this point an option is offered to save all the subsequent measurements to disc, so that they can be stored for later use if desired.

The main menu of the program is now displayed, along with a dynamic reading of the thermometer temperature and the heater voltage and current. There are four possible options:

1. Manual temperature set
2. Automatic temperature set
3. Manually set heating curve
4. Automatically set heating curve

The four options relate to the different ways in which the experiment can be operated. The method used will either be specified by the laboratory instructor, or left to the student's discretion.

The first two options are provided to allow the operator to warm up the specimen between sets of measurements or to reach a suitable temperature to start measurements. These options may not always be used, if the student decides to make successive specific heat measurements at consecutive temperature intervals.

The automatic temperature setting option asks the user for the desired temperature, and then warms up the specimen until the temperature is reached. To try and avoid overshooting, it uses a technique where full heater power is used up to 20 degrees below the target temperature, then the heater power is reduced progressively.

The manual temperature setting option is provided as an alternative if the automatic setting is found to be unsatisfactory. The program offers the four heater settings, which are selected by entering the appropriate number, with an option to end the heating process. Thus the user simply varies the heater voltage at will, and a dynamic display shows the current temperature. In this method the heating process can be terminated at any time.

Menu options 3 and 4 carry out the main specific heat determination. In each case, the program performs an initial cooling curve, followed by a heating curve, and then a second cooling curve, based on the measurement technique described in Sect. 6.3.1b. While the measurements are in progress the program displays a temperature/time graph on the screen. Measurements are made and displayed every 3 seconds, using the internal timing facility of the microcomputer.

The difference between the manual and automatic methods is that, in the former case, the user selects the heater power, and decides by pressing a key when to start and stop the heating to the specimen (judging when to do so with the aid of the graphical display of the temperature). In both cases a second cooling curve is finally carried out, this time for up to three minutes to allow for any time lag in the heating cycle, but the user can end this cooling curve earlier if it seems from the graphical display that a satisfactory cooling curve has been obtained. Deciding when to end the second cooling curve is always a difficult judgement that must be made 'manually', because there is a considerable delay before the burst of heat has diffused through the specimen.

In the automatic case, a temperature increment for the heating cycle is asked for, and then the program controls the heater directly. (Note that the temperature increment will be the quantity denoted in Sect. 6.3.1 as $\Delta T_{\rm H}$, not $(\Delta T_{\rm H} - \Delta T_{\rm C})$). The first cooling curve is performed for one minute. The heater is then switched on until the chosen temperature rise has occurred.

As with the automatic temperature setting, the program attempts to select the appropriate heater power. In this case, it starts with the lowest setting, and monitors the rate of heating. If it is too slow, the heater voltage is increased progressively. The heat input is calculated for each 3 second measurement and the total heat supplied is accumulated, so that any increase in the heater power during the heating period, as usually occurs in the automatic method, is taken into account automatically.

Figure 6.6 shows an actual screen display at the end of a cooling/heating/cooling cycle. The figure shows the extent of the delay that typically occurs before the heat diffusion process is completed. After each cooling/heating/cooling cycle, both options 3 and 4 go on to carry out the calculation of the specific heat, using the method of correcting the heating temperature from the cooling curve on either side.

A simple least squares fit is made to a set of 21 readings (i.e. readings over a 1 minute time interval) for each cooling curve. For the first cooling curve this is the set of readings immediately before heating starts: in the manual method this is done by discarding all readings previous to the last 20 on the initial cooling curve. For the second cooling curve the last 20 readings before the user stops the experiment are used. Thus the user must allow the second cooling to run for at least a minute after the temperature has settled following the heating burst.

The two straight line fits are then extrapolated to the midpoint of the heating cycle and the temperature difference between the two lines at that point is used as the effective rise in temperature due to the heating. Figure 6.7 shows the same data as Fig. 6.6 with the two fitted lines and the estimated temperature difference.

If the user has chosen it, the number of readings and the array of temperature readings, plus the calculated results, are finally written to disc.

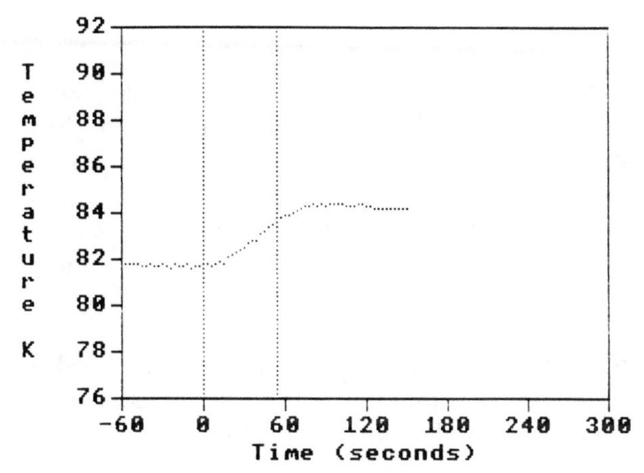

Run complete: heating time = 54.00sec.

Heater Volts = 3.548: Amps = 0.3405
Power = 1.2083 Watts

Press any key to process these results

Fig. 6.6 Screen display of a typical measurement cycle

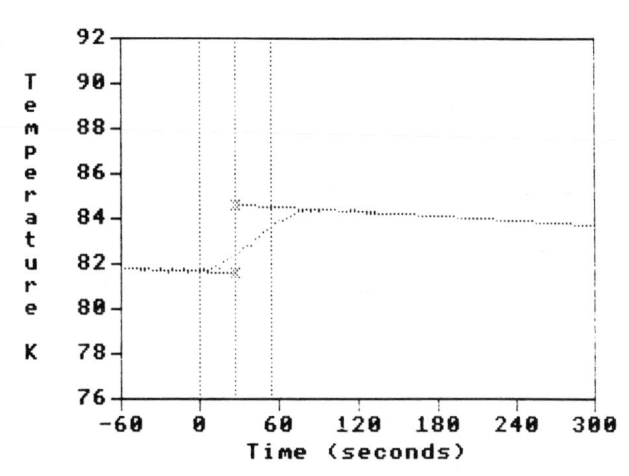

Total heat input = 65.246J
Temperature rise = 3.025K
Mean temperature = 83.137K

Press any key to review these results

Fig. 6.7 The fitting process to determine the temperature difference ($\Delta T_\mathrm{H} - \Delta T_\mathrm{C}$)

6.3.3 Typical Results

Figure 6.6 shows a typical screen display for a cooling/heating/cooling cycle where the heat losses are reasonably small, as is the case most of the time. It can be seen that the points show little noise and can be fitted well by straight lines in the 'cooling' regions. Figure 6.8 shows the effect in a region where the unheated specimen is cooling rapidly. Figure 6.9 shows a similar display in a region where the unheated specimen is warming up rapidly. Figure 6.10 shows a recording where the time lag for the heat diffusion is particularly severe.

The mean temperature of the specific heat determination can be taken as the midpoint of the effective temperature rise, and a specific heat curve plotted against temperature. The sort of result obtained for dysprosium is

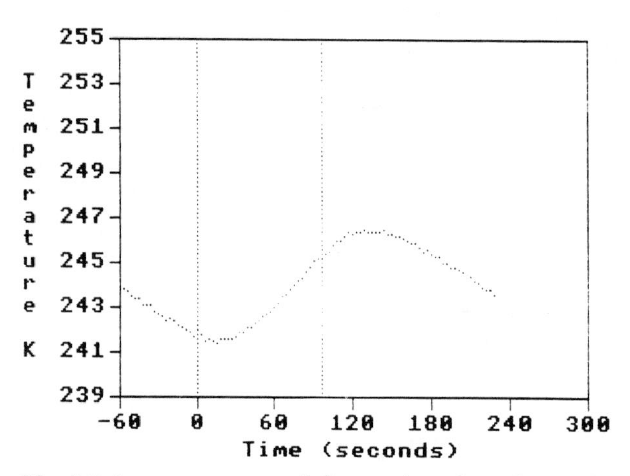

Fig. 6.8 A measurement cycle in a region where the specimen is cooling rapidly

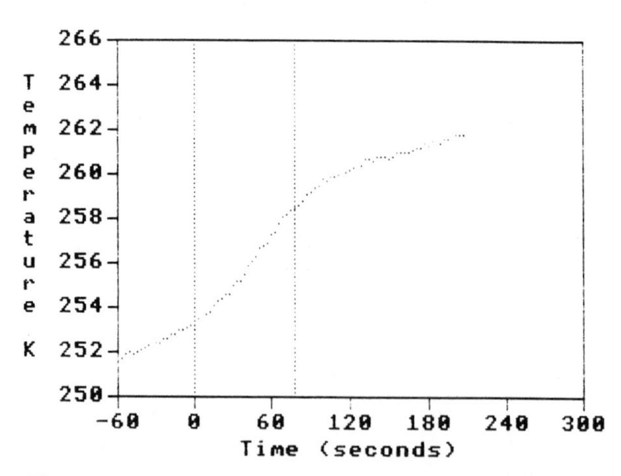

Fig. 6.9 A measurement cycle in a region where the specimen is warming rapidly

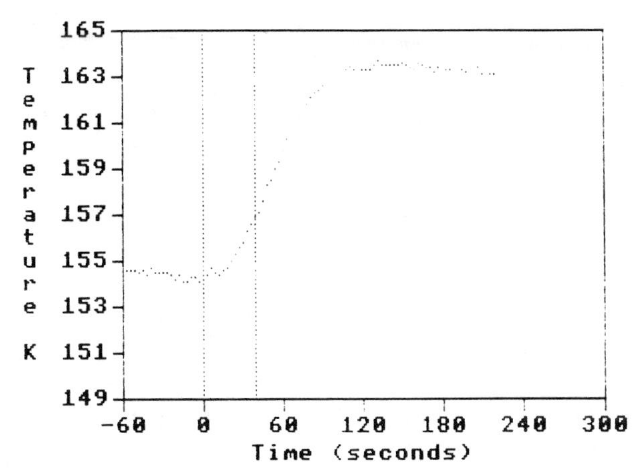

Fig. 6.10 A measurement cycle with a severe time lag in the heating

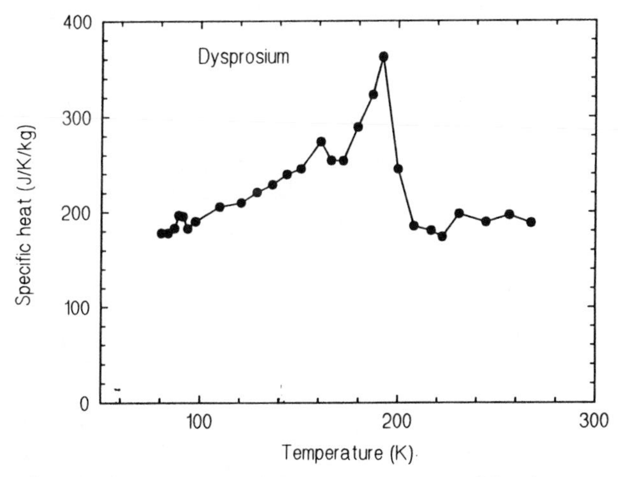

Fig. 6.11 Specific heat of dysprosium measured by the apparatus

shown in Fig. 6.11. The large anomaly that occurs at the transition from the antiferromagnetic to paramagnetic state is dramatically illustrated. There is another, very much smaller, anomaly at around 85 K at the transition from the ferromagnetic to antiferromagnetic state. There is a suspicion of this anomaly but it is not clearly detected.

6.4 Discussion

Microcomputer control of this experiment is not an essential requirement except for work at very low temperatures; indeed the experiment is based on a manual version that we still use in the laboratory for those students that prefer it. However, the microcomputer version does have a number of advantages.

Firstly, it shows how the same experiment can be performed with or without the aid of a microcomputer, and it is a good example of microcomputer control, where the computer is used to run and control the experiment, not just as a passive data recording and analysis device.

Secondly, the student has a much better feel for exactly what is going on in the experiment as it progresses. There are two main reasons for this: i) the microcomputer is able to make rapid calculations to convert readings of the thermometer voltage directly to temperature, so the student knows what the true temperature of the specimen is during the experiment; and ii) the microcomputer is able to display the results graphically as they are recorded, so that the student gains an immediate feel for the rate of heating or cooling, how smooth the data are and when time delays in the heat spreading through the specimen have settled down.

Thirdly, readings can be recorded much more rapidly than by hand, so each specific heat determination can be made more rapidly, enabling the whole experiment to be completed more quickly, or more determinations to be made.

Fourthly, the microcomputer results should be more accurate because the timings can be made much more precisely, and without the human bias that occurs when meter readings are recorded periodically, nor the chance of human error in recordings.

Finally, the microcomputer measurements are essential for very low temperature work because the readings need to be taken rapidly, and heating periods are too short for accurate measurements, and in particular timings, to be made manually.

The program for the experiment has been written in such a way that the experiment can be performed at different levels; for example, the manual and automatic control over the sequence of cooling and heating curves allow for more or less student control, and this can be specified to the student by the instructor.

If the execution of the experiment is not precisely specified by the instructor, there are a number of operational features which require intelligent decisions by the student for best results (in the main these are not features of the microcomputer control, but they are nevertheless of interest).

It often fails to occur to the student that he or she can heat the specimen up between the heating curves that are recorded in detail for specific heat determinations. One result of this is that the student might decide to make ten determinations, calculate that the temperature range is about 100 K to

300 K, so choose 20 K temperature increments, which are much too large. (The need to select the temperature increment in the automatic mode of operation of the microcomputer experiment does have the effect of bringing this aspect of the operation to the student's attention.)

An alternative result of the student making consecutive readings is that he or she chooses sensible temperature increments (say 3 to 5 degrees), but does not cover a wide enough range of temperatures.

Students seldom think carefully about when is the appropriate time to remove the liquid nitrogen radiation shielding. This gives the instructor an opportunity to pick up the mistake and emphasize the unexpected effect of the T^4 power law of radiation. In fact a simple calculation can be made to work out when to remove the shielding. Neglecting the radiation from the top of the dewar and assuming the same proportionality constant for radiation from specimen to shield and shield to specimen, the heat loss of the specimen is proportional to

$$T_S^4 - T_R^4 \tag{6.30}$$

where T_S and T_R are the temperatures of specimen and radiation shield respectively. The ideal point to remove the liquid nitrogen shielding is when the heat loss with liquid nitrogen shielding is the same as the heat gain with room temperature shielding. Taking liquid nitrogen temperature and room temperature to be 77 K and 300 K respectively, this gives

$$T^4 - 77^4 = 300^4 - T^4 \tag{6.31}$$

$$2T^4 = 300^4 - 77^4 \,. \tag{6.32}$$

This yields a cross-over point of 252 K – rather higher than the midpoint of 190 K that students tend to use. Indeed, it makes essentially no difference whether the shield is at 77 K or 4 K compared with the effect of room temperature radiation.

Finally, it is very important to carry out a preliminary experiment to work out roughly what heater power is needed to give a sensible heating rate, how long the heater will need to be on for, and how much the specific heat changes between 77 K and 300 K. In the case of the erbium and dysprosium specimens, it should be possible to get a rough idea of where the main specific heat anomalies occur. A convenient time to make this preliminary experiment is after cooling the specimen down for the low temperature calibration of the thermometer.

References

6.1. G K White: *Experimental Techniques in Low Temperature Physics* (Oxford University Press, 1959)
6.2. C Kittel: *Introduction to Solid State Physics* (Wiley, New York 1953)
6.3. A Einstein: Ann. d. Physik **22**, 180 (1907)
6.4. P Debye: Ann. d. Physik **39**, 789 (1912)

6.5. The Salford Analogue and Digital I/O Board, supplied by E & L Instruments Ltd, Rackery Lane, Llay, Wrexham, Clwyd, UK.
6.6. T M Dauphinee and H Preston-Thomas: Rev. Sci. Instrum. **25**, 884 (1954)

Appendix 6.A – Circuit Diagrams

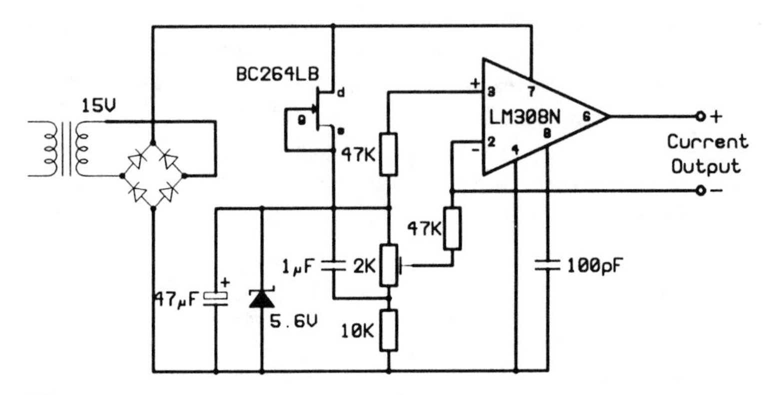

Fig. 6.A1 Constant current supply for the thermometer

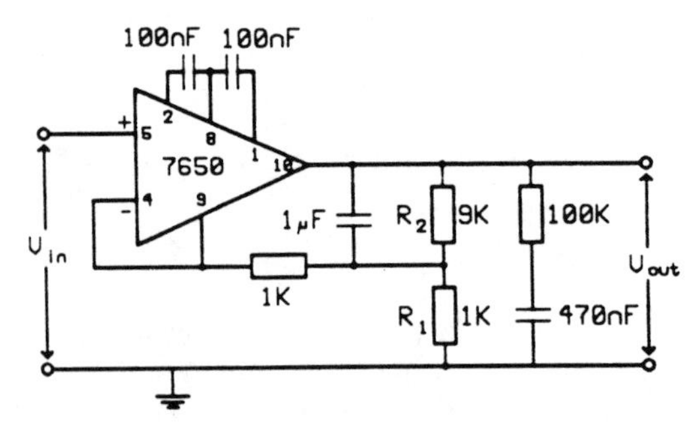

Fig. 6.A2 Thermometer voltage amplifier

Appendix 6.B – Program Listing

```
PROGRAM spht;
USES Graph, CRT, DOS;

CONST tick = 4; chsize = 8; Vref = 1.243; RES = 3; AV = 100;
      Vatn = 5.0574; Iatn = 1.4797;
      conv = 798; LSB = 798; MSB = 799;
      PortC = 786; PCR = 787;
      MODEFLAG = 128;
```

```
      PAMODE0 = 0; PBMODE0 = 0;
      PAIN   = 16; PBIN   = 2;
      PCLOUT = 0; PCUIN = 8;

      {tick - Length of ticks etc
       chsize - Size of characters
       Vref - Reference Voltage
       RES - Standard resistor
       AV - no of ADC readings to be averaged
       Vatn - Voltage attenuator
       Iatn - Current attenuator
       conv - address to write to to start conversion on ADC
       LSB, MSB - addresses to read result from ADC
       PortC - address of Port C on 8255 PPI
       PCR - Program Control Register on 8255 PPI
       MODEFLAG, PAMODE0 etc - bits to write to program PCR}

TYPE calarray = ARRAY [0..3, 0..1] OF REAL;
     temparray = ARRAY [0..120] OF REAL;

VAR GrDriver, GrMode, Errorcode : INTEGER;
    X0, Y0, Xscale, Yscale, Xrange, Yrange : REAL;
    X, X1, Y1, X2, Y2, slope, constant, graphslope : REAL;
    M, C, Sx, Sy, Sxx, Syy, Sxy, chk, volts : REAL;
    tim, vol, amp, pow, heat, starttime, stoptime : REAL;
    temp1, temp2, reqtemp, diff, starttemp, tempinc : REAL;
    T, V, I, oldT, oldV, oldI, inittemp : REAL;
    choice, mmchoice, noofpoints, pointno : INTEGER;
    n, q, itemp, v10, dash, pwr, VSS : INTEGER;
    mid, cool, coolend, heatpts, stoppts, rdgs, hpower : INTEGER;
    yn, dummy, key : CHAR;
    filename : STRING;
    save, repeating, OK : BOOLEAN;
    cal : calarray;
    temp : temparray;

{***************************************************************
  **************************************************************
  SYSTEM DEPENDENT ROUTINES.
  ALL THE ROUTINES THAT ACCESS THE INTERFACE BOARD ARE
  COLLECTED IN THIS FIRST SET OF PROCEDURES AND FUNCTIONS
  **************************************************************
  **************************************************************}
```

```
PROCEDURE initioboard;
BEGIN
   PORT[PCR] := MODEFLAG+PAMODE0+PBMODE0+PAIN+PBIN+PCLOUT+PCUIN;
   {PROGRAM THE 8255 PROGRAMMABLE PERIPHERAL INTERFACE, BY
    WRITING A CONTROL WORD TO THE CONTROL REGISTER AT ADDRESS
    PCR. THE LOWER HALF OF PORT C (BITS 0 TO 3) IS PROGRAMMED FOR
    OUTPUT. THE REMAINING UNUSED LINES ARE PROGRAMMED FOR INPUT.}
END;

FUNCTION readADC (ch:INTEGER):REAL;
VAR KY: CHAR; TP:REAL; CODE : INTEGER;
BEGIN
   {READ THE ANALOGUE INPUT VOLTAGE ON CHANNEL CH}
   Port[CONV] := ch - 1;
   {START READING ON THE ADC BY WRITING CHANNEL NO TO BITS D0 AND
    D1. THE PSEUDO-ARRAY Port[] IS USED TO ADDRESS THE I/O LINES.
    WRITING A CHANNEL NUMBER TO ADDRESS 'CONV' OF THE MAX182
    FORCES THE START OF A READING ON THAT CHANNEL (WHERE THE
    CHANNEL NUMBER IS IN THE RANGE 0 TO 3)}

   REPEAT UNTIL Port[MSB] < 128;
   {READ THE STATUS ON BIT 7 OF MSB.
    BIT 7 IS ZERO WHEN THE READING HAS ENDED.}

   readADC := ((Port[MSB] mod 16) * 256 + Port[LSB])/819.2;
   {THE 12+BIT READING IS AVAILABLE OVER THE 8+BIT DATA BUS BY
    READING FROM TWO ADDRESSES. ADDRESS 'LSB' CONTAINS THE LEAST
    SIGNIFICANT BYTE. THE LOWER 4 BITS ON ADDRESS 'MSB' CONTAIN
    THE HIGH 4 BITS OF THE 12+BIT READING. THE RESULTING READING
    IS IN THE RANGE 0 TO 4095. DIVIDING BY 819.2 CONVERTS TO A
    VOLTAGE IN THE RANGE 0 TO 5V.}
END;

PROCEDURE setpower (P:INTEGER);
VAR byte, i : INTEGER;
BEGIN
   IF (P > 0) AND (P < 5) THEN BEGIN
      byte := 1;
      FOR i := 2 TO P DO byte := byte * 2;
      PORT[PortC] := 15 - byte END
   ELSE PORT[PortC] := 15;
   {SWITCHES OFF BIT P OF THE DIGITAL OUTPUT LINES ON PORT C OF
    8255. THESE 4 LINES ARE USED TO CONTROL THE 4 POWER LEVELS
    SET UP ON THE POWER SUPPLY. A LOW IS USED TO SWITCH ON A
    POWER LEVEL.}
END;
```

```
{*****************************************************************
  *****************************************************************
  LANGUAGE DEPENDENT ROUTINES.
  ALL THE ROUTINES THAT CALL NON-STANDARD TURBO-PASCAL ROUTINES
  ARE COLLECTED IN THIS NEXT SET OF PROCEDURES AND FUNCTIONS
  *****************************************************************
  *****************************************************************}

PROCEDURE initialize;
BEGIN
   TextMode(0); {SELECT 40 COLUMN B/W TEXT MODE}
   GrDriver := cga; GrMode := 3;
   InitGraph(GrDriver, GrMode, '\TP\BGI');
   SetGraphMode(3);
   {SELECT 320 x 20 PIXEL (40 x 25 CHAR) CGA GRAPHICS MODE}
   initioboard;
   {INITIALIZE THE I/O BOARD}
END;

PROCEDURE Graphics;
BEGIN
   SetGraphMode(3);
   {THIS INSTRUCTION SWITCHES TO THE GRAPHICS SCREEN.
    IF IN TEXT MODE, THE SCREEN IS CLEARED.}
END;

PROCEDURE clrtext;
BEGIN
   RestoreCRTMode;
   {THIS INSTRUCTION SWITCHES BACK TO TEXT MODE.}
   CLRSCR; {CLEAR THE TEXT SCREEN.}
END;

PROCEDURE flush;
VAR dummy : CHAR;
BEGIN
   WHILE KeyPressed DO dummy := ReadKey;
   {FLUSHES THE KEYBOARD BUFFER OF ANY OUTSTANDING KEYPRESSES}
END;

FUNCTION time:REAL;
VAR hour, min, sec, centisec : WORD;
BEGIN
   GetTime(hour, min, sec, centisec);
   time := centisec/100 + sec + 60 * min + 3600 * hour;
   {RETURNS A TIME IN SECONDS}
END;
```

```
PROCEDURE setscales1(VAR X0,Y0,Xscale,Yscale,Xrange,Yrange:REAL);
BEGIN
   {SCALING FACTORS FOR THE GRAPHS OF TEMP. VS VOLTAGE}
   X0 := 52; Y0 := 52; {ORIGIN OF GRAPHS}
   Xscale := 128; Yscale := 0.4;
   {SCALE FACTOR FOR ONE DIVISION ON AXES}
   Xrange := Xscale * 2; Yrange := Yscale * 350;
   {FULL SCALE OF AXES}
END;

PROCEDURE setscales2(VAR X0,Y0,Xscale,Yscale,Xrange,Yrange:REAL);
BEGIN
   {SCALING FACTORS FOR THE GRAPHS OF TEMP. VS TIME}
   X0 := 52; Y0 := 64; Xscale := 2; Yscale := 8;
   Xrange := Xscale * 120; Yrange := Yscale * 16;
END;

PROCEDURE drawline (X1, Y1, X2, Y2:REAL);
BEGIN
   LINE(ROUND(X1), ROUND(200 - Y1), ROUND(X2), ROUND(200 - Y2));
   {DRAW A LINE FROM SCREEN COORDINATES X1, Y1 TO X2, Y2
    WHERE Y IS MEASURED FROM THE   B O T T O M   OF THE SCREEN}
END;

PROCEDURE plotpoint (X, Y:REAL);
BEGIN
   PutPixel (ROUND(X), ROUND(200 - Y), 1);
   {PLOT A SINGLE PIXEL AT SCREEN COORDINATES X, Y}
END;

PROCEDURE ruboutpoint (X, Y:REAL);
BEGIN
   PutPixel (ROUND(X), ROUND(200 - Y), 0);
   {PLOT A SINGLE POINT IN BACKGROUND COLOUR}
END;

PROCEDURE cursorpsn (X, Y:REAL);
BEGIN
   IF (X > 0) AND (X < 41) AND (Y > 0) AND (Y < 26) THEN
      GotoXY(ROUND(X), ROUND(Y));
   {POSITION CHARACTER CURSOR AT CHARACTER COORDINATES X, Y ON
    TEXT SCREEN}
END;
```

```
PROCEDURE cursorto (X, Y:REAL);
BEGIN
   MoveTo(ROUND(X), 200 - ROUND(Y));
   {POSITION CHARACTER CURSOR TO PRINT CHARACTER ON GRAPHICS
    SCREEN WITH TOP LEFT CORNER AT GRAPHIC COORDINATES X, Y}
END;

FUNCTION getkey: CHAR;
VAR ch: CHAR;
BEGIN
   getkey := ReadKey;
   {WAIT FOR A SINGLE KEY PRESS, THEN RETURN THAT CHARACTER}
END;

FUNCTION checkkey: CHAR;
VAR ch: CHAR;
BEGIN
   IF KeyPressed THEN
      checkkey := ReadKey
   ELSE
      checkkey := CHR(0);
   {RETURN CHARACTER OF NEXT KEY PRESSED, OR CHR(0) IF NONE
    OUTSTANDING}
END;

PROCEDURE printtext(txt : STRING);
VAR xp, yp, ch : INTEGER;
BEGIN
   xp := GetX; yp := GetY;
   SetViewPort(xp, yp, xp + chsize * LENGTH(txt) - 1,
               yp + chsize - 1, TRUE);
   ClearViewPort;
   SetViewPort(0, 0, GetMaxX, GetMaxY, TRUE);
   MoveTo(xp, yp); OutText(txt);
   {OUTPUT STRING txt ON GRAPHICS SCREEN AT CURRENT (GRAPHICS)
    CURSOR POSITION. OutText PRINTS CHARACTERS ON A TRANSPARENT
    BACKGROUND. THEREFORE, TO OBLITERATE ANY OLD CHARACTERS,
    A VIEWPORT IS SET UP THE SIZE OF THE TEXT TO BE PRINTED, AND
    THAT SECTION OF THE SCREEN IS CLEARED WITH ClearViewPort.}
END;

PROCEDURE formprint(num : REAL; wdth, dpts : INTEGER);
VAR s : STRING;
BEGIN
   STR(num:wdth:dpts, s); printtext(s);
```

```
   {CONVERT A NUMBER TO A STRING WITH FIELDWIDTH wdth, dpts
    DECIMAL POINTS, SO THAT IT CAN BE PRINTED TO THE GRAPHICS
    SCREEN WITH THE printtext PROCEDURE. EQUIVALENT TO
    write(num:wdth:dpts);}
END;

PROCEDURE newline;
BEGIN
   MoveTo(0, GetY + chsize);
   {MOVE TO THE NEXT LINE ON THE GRAPHICS SCREEN. EQUIVALENT TO
    writeln;}
END;

PROCEDURE cleos (L: INTEGER);
BEGIN
   SetViewPort(0, L, GetMaxX, GetMaxY, TRUE);
   ClearViewPort;
   SetViewPort(0, 0, GetMaxX, GetMaxY, TRUE);
   {CLEAR THE GRAPHICS SCREEN FROM LINE L TO THE BOTTOM}
END;

PROCEDURE writedata (filename : STRING; coolend : INTEGER;
                     temp1, temp2, heat : REAL; temp : temparray);
VAR n : INTEGER; f : TEXT;
BEGIN
   ASSIGN (f, filename); REWRITE (f);
   WRITELN(f, coolend:4); {WRITE A NUMBER TO THE FILE}
   FOR n := 1 TO coolend DO BEGIN
   WRITELN(f, temp[n]:7:2); END;
   WRITELN(f, temp1:7:2, temp2:7:2, heat:7:2);
   CLOSE(f);
END;

{************************************************************
  ************************************************************
  END OF LANGUAGE AND SYSTEM DEPENDENT ROUTINES
  ************************************************************
  ************************************************************}

FUNCTION intlen(num : INTEGER) : INTEGER;
VAR len : INTEGER;
BEGIN
   IF num<0 THEN len := 2 ELSE len := 1;
   WHILE ABS(num)>10 DO BEGIN
      len := len + 1; num := num DIV 10 END;
   intlen := len;
END;
```

```
FUNCTION eval(s : STRING) : INTEGER;
VAR num, code : INTEGER;
BEGIN
   IF (s >= '0') AND (s <= '9') THEN BEGIN
      VAL(s, num, code);
      IF code <> 0 then BEGIN
         clrtext; WRITELN('ERROR IN VAL'); END;
      eval := num; END
   ELSE
      eval := -1;
END;

PROCEDURE spc(n : INTEGER);
VAR i : INTEGER;
BEGIN
   FOR i := 1 TO n DO WRITE(' ');
END;

PROCEDURE verticalprint (X, Y : REAL; P : STRING);
VAR CHNUM : INTEGER;
BEGIN
   FOR CHNUM := 1 TO LENGTH(P) DO BEGIN
      cursorto(X, Y - (CHNUM - 1) * chsize);
      printtext(P[CHNUM]);
   END;
END;

PROCEDURE waitnext (T : REAL);
VAR TM : REAL;
BEGIN
   TM := T * TRUNC(time / T);
   REPEAT  UNTIL time > (TM + T);
END;

PROCEDURE printtemp (T : REAL);
BEGIN
   cursorto(0, 2 * chsize); printtext('Temperature = ');
   formprint(T, 7, 2); printtext('K    ');
END;

PROCEDURE drawcross (X, Y : REAL);
BEGIN
   drawline(X - tick, Y - tick, X + tick, Y + tick);
   drawline(X - tick, Y + tick, X + tick, Y - tick);
END;
```

```
PROCEDURE graphcont;
VAR A : CHAR;
BEGIN
   printtext('   Please press any key to continue: ');
   A := getkey;
END;

PROCEDURE cont;
VAR A : CHAR;
BEGIN
   WRITE('   Please press any key to continue: ');
   A := getkey;
END;

PROCEDURE headings;
BEGIN
   cursorpsn(11, 2); WRITE('Temperature = ');
   cursorpsn(3, 5); WRITE('Htr Volts =          Htr Amps =');
END;

PROCEDURE prvals (T, V, I, oldT, oldV, oldI : REAL);
BEGIN
   IF oldT <> T THEN BEGIN
      cursorpsn(25, 2); WRITE(T:7:1, '    '); oldT := T;
   END;
   IF oldV <> V THEN BEGIN
      cursorpsn(15, 5); WRITE(V:5:2, ' '); oldV := V;
   END;
   IF oldI <> I THEN BEGIN
      cursorpsn(34, 5); WRITE(I:5:3, ' '); oldI := I;
   END;
END;

PROCEDURE readall (M, C : REAL; VAR T, V, I : REAL);
VAR RDG : INTEGER; cal : REAL;
BEGIN
   {READ A/D VALUES}
   T := 0; V := 0; I := 0;
   FOR RDG := 1 TO AV DO BEGIN
      cal := Vref / readADC(1);
      T := T + M * readADC(2) * cal + C;
      V := V + readADC(3) * cal * Vatn;
      I := I + readADC(4) * cal * Iatn / RES;
   END;
   T := T / AV; V := V / AV; I := I / AV;
END;
```

```
PROCEDURE linreg (first, last : INTEGER; temp : temparray;
                  VAR slope, constant : REAL);
VAR Sx, Sy, Sxx, Syy, Sxy, SSx, SSy, SPxy : REAL;
    q, npts : INTEGER;
BEGIN
   {LIN. REG. ON COOLING CURVES}
   Sx := 0; Sy := 0; Sxx := 0; Syy := 0; Sxy := 0;
   npts := last - first + 1;
   FOR q := first TO last DO BEGIN
      Sx := Sx + q; Sy := Sy + temp[q];
      Sxx := Sxx + q * q; Syy := Syy + temp[q] * temp[q];
      Sxy := Sxy + q * temp[q];
   END;
   SSx := Sxx - Sx * Sx / npts; SSy := Syy - Sy * Sy / npts;
   SPxy := Sxy - Sx * Sy / npts; slope := SPxy / SSx;
   constant := Sy / npts - slope * Sx / npts;
END;

PROCEDURE calreading(pointno : INTEGER; VAR cal : calarray);
VAR which : STRING;
    E : CHAR; RDG : INTEGER;
BEGIN
   {SET POINT TEMP. READINGS}
   IF pointno = 1 THEN which := 'first'
      ELSE IF pointno = 2 THEN which := 'second'
      ELSE which := 'third';
   cursorpsn(1, 5); spc(38); WRITELN; WRITELN;
   WRITELN('Press "E" to enter the ', which, ' point ');
   WRITELN; spc(38);
   REPEAT
      cal[pointno, 0] := 0;
      FOR RDG := 1 TO AV DO
         cal[pointno, 0] := cal[pointno, 0]
                              + readADC(2) / readADC(1);
      cal[pointno, 0] := Vref * cal[pointno, 0] / AV;
      cursorpsn(1, 2);
      WRITELN('Thermometer voltage = ',cal[pointno,0]:7:3, ' V');
      E := checkkey;
   UNTIL (E = 'E') OR (E = 'e');
   cursorpsn(1, 9); WRITE('Temperature (K) = ');
   READLN(cal[pointno, 1]);
   cursorpsn(1, 12 + 2 * pointno); WRITE(pointno:1, ': ');
   WRITE(cal[pointno, 0]:8:3, ' Volts at ');
   WRITELN(cal[pointno, 1]:7:2, ' K      ');
END;
```

```
PROCEDURE hgraph(repeating : BOOLEAN);
VAR rdg, tm, temp2, n, dash : INTEGER;
{** NOTE - ALL OTHER VARIABLES IN THIS PROCEDURE ARE GLOBAL **}
BEGIN
   {HEATING GRAPH}
   IF NOT repeating THEN BEGIN
      clrtext;
      cursorpsn(9, 10); WRITELN('Setting Temperature scale:');
      cursorpsn(9, 12); WRITELN('****** Please  wait ******');
      inittemp := 0;
      FOR rdg := 1 TO 10 DO BEGIN
         readall(M, C, T, V, I); inittemp := inittemp + T;
      END;
      inittemp := inittemp/10; VSS := 2*TRUNC((inittemp - 5)/2);
   END;
   setscales2(X0,Y0,Xscale,Yscale,Xrange,Yrange);
   Graphics; drawline(X0, Y0, X0 + Xrange, Y0);
   drawline(X0 + Xrange, Y0, X0 + Xrange, Y0 + Yrange);
   drawline(X0 + Xrange, Y0 + Yrange, X0, Y0 + Yrange);
   drawline(X0, Y0 + Yrange, X0, Y0);
   FOR temp2 := 0 TO 8 DO BEGIN
      drawline(X0, Y0 + 2 * temp2 * Yscale,
               X0 - tick, Y0 + 2 * temp2 * Yscale);
      cursorto(X0 - tick - 7 * chsize / 2,
               Y0 + 2 * temp2 * Yscale + chsize / 2);
      formprint((VSS + 2 * temp2), 3, 0);
   END;
   FOR n := -1 TO 5 DO BEGIN
      tm := n * 60;
      drawline(X0 + (tm + 60) * Xscale / 3, Y0,
               X0 + (tm + 60) * Xscale / 3, Y0 - tick);
      cursorto(X0 + (tm + 60) * Xscale / 3
               - chsize * (3 - intlen(tm) / 2) + 1,
               Y0 - 3 * tick / 2);
      formprint(tm, 3, 0);
   END;
   FOR dash := 0 TO (ROUND(Yrange) DIV tick) - 1 DO
      drawline(X0 + 60 * Xscale / 3, Y0 + dash * tick,
               X0 + 60 * Xscale / 3, Y0 + dash * tick + tick/2);
   verticalprint(0, Y0 + Yrange / 2 + chsize * 13 / 2,
                 'Temperature K');
   cursorto(X0 + Xrange / 2 - 14 * chsize / 2,
            Y0 - 3 * tick / 2 - 3 * chsize / 2);
   printtext('Time (seconds)');
END;
```

```
PROCEDURE calc; {TEMP.RISE CALC'S}
VAR n, dash : INTEGER;
{** NOTE - ALL OTHER VARIABLES IN THIS PROCEDURE ARE GLOBAL **}
BEGIN
   cleos(21 * chsize); cursorto(0, 4 * chsize);
       printtext('Run complete: heating time =');
   formprint(tim, 6, 2); printtext('sec. ');
   newline; printtext('Heater Volts ='); formprint(vol, 6, 3);
   printtext(': Amps ='); formprint(amp, 6, 4); newline;
   printtext('Power ='); formprint(pow, 7, 4);
   printtext(' Watts ');
   newline; printtext('Press any key to process these results');
   flush; dummy := getkey;
   heat := pow * tim;
   mid := (ROUND(tim) DIV 2) DIV 3 + 20;
   cool := (ROUND(tim) + 61) DIV 3 + 20;
   FOR dash := 0 TO (ROUND(Yrange) DIV tick) - 1 DO
      drawline(X0 + mid * Xscale, Y0 + dash * tick,
               X0 + mid * Xscale, Y0 + dash * tick + tick / 2);
   linreg(0, 20, temp, slope, constant);
   temp1 := slope * mid + constant;
   X1 := X0; Y1 := Y0 + (constant - VSS) * Yscale;
   X2 := X0 + mid * Xscale; Y2 := Y0 + (temp1 - VSS) * Yscale;
   graphslope := (Y2 - Y1) / (X2 - X1);
   FOR n := (ROUND(X1) DIV tick) TO (ROUND(X2) DIV tick) - 1 DO
   BEGIN
      X := n *tick;
      drawline(X, Y1 + graphslope * (X - X1), X + tick / 2,
               Y1 + graphslope * (X - X1 + tick / 2));
   END;
   drawcross(X2, Y2);
   linreg(coolend - 20, coolend, temp, slope, constant);
   temp2 := slope * mid + constant; X1 := X0 + Xrange;
   Y1 := Y0 + (slope * 120 + constant - VSS) * Yscale;
   X2 := X0 + mid * Xscale; Y2 := Y0 + (temp2 - VSS) * Yscale;
   graphslope := (Y2 - Y1) / (X2 - X1);
   FOR n := (ROUND(X2) DIV tick) TO (ROUND(X1) DIV tick) - 1 DO
   BEGIN
      X := n * tick;
      drawline(X, Y2 + graphslope * (X - X2),
               X + tick/2, Y2 + graphslope * (X - X2 + tick/2));
   END;
   drawcross(X2, Y2);
   cleos(21 * chsize); cursorto(0, 4 * chsize);
   printtext('Initial temperature = ');
```

```
   formprint(temp1, 7, 3); printtext('K'); newline;
   printtext('  Final temperature = '); formprint(temp2, 7, 3);
   printtext('K'); newline; graphcont;
   cleos(21 * chsize); cursorto(0, 4 * chsize);
   printtext('Total heat input = ');
   formprint(heat, 7, 3); printtext('J'); newline;
   printtext('Temperature rise = ');
   formprint((temp2 - temp1), 7, 3);
   printtext('K'); newline; printtext('Mean temperature = ');
   formprint((temp1 + temp2) / 2, 7, 3); printtext('K');
   newline; printtext('Press any key to review these results');
   flush; dummy := getkey;
   clrtext; cursorpsn(1, 2);
   WRITELN('Heater voltage = ', vol:7:3, 'V');
   WRITELN; WRITELN('Heater current = ', amp:7:4, 'A');
   WRITELN; WRITELN('Heating period = ', tim:7:2, 'sec.');
   WRITELN; WRITELN;
   WRITELN('Total heat input = ', heat:7:3, 'J');
   WRITELN; WRITELN;
   WRITELN('Initial temperature = ', temp1:7:3, 'K');
   WRITELN; WRITELN('  Final temperature = ', temp2:7:3, 'K');
   WRITELN; WRITELN;
   WRITELN('   Temperature rise = ', (temp2 - temp1):7:3, 'K');
   WRITELN;
   WRITELN('   Mean temperature = ', (temp1 + temp2)/2:7:3, 'K');
   WRITELN; WRITELN;
   WRITELN('Press RETURN for the main MENU or any');
   WRITELN; WRITELN('other key to review the graph again');
   dummy := getkey;
   IF dummy <> CHR(13) THEN BEGIN
      repeating := TRUE; hgraph(repeating);
      FOR dash := 0 TO (ROUND(Yrange) DIV tick) - 1 DO
         drawline(X0 + stoppts * Xscale, Y0 + dash * tick,
                  X0 + stoppts * Xscale,
                  Y0 + dash * tick + tick / 2);
      FOR n := 0 TO coolend DO
         plotpoint(X0 + n*Xscale, Y0 + (temp[n] - VSS)*Yscale);
      calc; {NOTE recursive call of procedure}
   END;
END; {calc}

{*******  MAIN PROGRAM *******}

BEGIN {SPECIFIC HEAT PROGRAM}
initialize; clrtext; setpower(0);
```

```
cursorpsn(9, 1); WRITELN('SPECIFIC HEAT EXPERIMENT');
cursorpsn(5, 9); WRITELN('Please check that all electrical');
WRITELN; WRITELN('          connections are secure');
cursorpsn(1, 15);
WRITELN('Ensure that all controls on the power');
WRITELN; WRITELN('     supply are turned down to zero');
WRITELN; WRITELN; WRITELN; cont; clrtext; cursorpsn(1, 4);
WRITELN('   The thermometer can be calibrated in');
WRITELN; WRITELN('   two  ways.'); WRITELN; WRITELN; WRITELN;
WRITELN('1: By entering the appropriate');
WRITELN; WRITELN('   constants in the equation:'); WRITELN;
WRITELN; WRITELN('   Temperature = M x Voltage + Constant');
WRITELN; WRITELN; WRITELN('2: By calibrating the thermometer');
WRITELN; WRITELN('   at two or three known temperatures');
WRITELN; WRITELN; WRITELN;
WRITE('Enter the number of your choice: ');
REPEAT
   choice := eval(getkey);
UNTIL (choice = 1) OR (choice = 2);
CASE choice OF
1:BEGIN
    {THERMOMETER CAL. BY EQUATION}
    clrtext; cursorpsn(1, 5); {clrtext CLEARS SCREEN}
    WRITELN('Please enter the constants in the'); WRITELN;
    WRITELN('equation:'); WRITELN; WRITELN;
    WRITELN('Temperature = M x Voltage + C');
    WRITELN; WRITELN; WRITELN; WRITELN;
    cursorpsn(1, 15); WRITE('M = '); READLN(M); WRITELN; WRITELN;
    WRITE('C = '); READLN(C); cursorpsn(1, 23); cont;
  END;
2:BEGIN
    {CALIBRATION OF THERMOMETER AT 2 OR 3 POINTS}
    clrtext; cursorpsn(4, 10);
    WRITELN('CALIBRATION OF THE THERMOMETER'); WRITELN; WRITELN;
    WRITELN('   Do you wish to use two or three'); WRITELN;
    WRITE('   calibration points? (2/3): ');
    REPEAT
       noofpoints := eval(getkey);
    UNTIL (noofpoints = 2) OR (noofpoints = 3);
    FOR pointno := 0 TO 3 DO BEGIN
       cal[pointno, 0] := 0; cal[pointno, 1] := 0;
    END;
    clrtext;
    FOR pointno := 1 TO noofpoints DO
       calreading(pointno, cal); {calibration reading}
```

```
REPEAT
   REPEAT
      cursorpsn(1, 7); WRITE('Do you wish to change ');
      IF noofpoints=2 THEN WRITE('either') ELSE WRITE('any');
      WRITELN(' of        '); WRITELN;
      WRITE('these points ? (Y/N)'); spc(20);
      yn := getkey; IF yn = 'y' THEN yn := 'Y';
      IF yn = 'Y' THEN BEGIN
         REPEAT
            cursorpsn(1, 7);
            spc(38); WRITELN; WRITELN;
            WRITE('Which point do you wish to change? ');
            pointno := eval(getkey);
         UNTIL (pointno > 0) AND (pointno <= noofpoints);
         calreading(pointno, cal);
      END;
   UNTIL yn <> 'Y';
   Sx := 0; Sy := 0;
   FOR pointno := 1 TO noofpoints DO BEGIN
      Sx := Sx + cal[pointno, 0]; Sy := Sy + cal[pointno, 1];
   END;
   Sx := Sx / noofpoints; Sy := Sy / noofpoints;
IF noofpoints = 2 THEN BEGIN
   Sxx := SQR(cal[2, 0] - cal[1, 0]);
   Syy := SQR(cal[2, 1] - cal[1, 1]);
   Sxy := Sxx * Syy; END
ELSE BEGIN
   Sxy := 0; Sxx := 0; Syy := 0;
   FOR pointno := 1 TO 3 DO BEGIN
      Sxy := Sxy + (cal[pointno MOD 3 + 1, 0]
         - cal[pointno, 0]) * (cal[pointno MOD 3 + 1, 1]
         - cal[pointno, 1]);
      Sxx := Sxx + SQR(cal[pointno MOD 3 + 1, 0]
         - cal[pointno, 0]);
      Syy := Syy + SQR(cal[pointno MOD 3 + 1, 1]
         - cal[pointno, 1]);
   END;
END;
IF (Sxx = 0) OR (Syy = 0) THEN BEGIN
   M := 0; chk := 0 END
ELSE BEGIN
   M := SQRT(ABS(Syy / Sxx)) * Sxy/ABS(Sxy);
        {Sxy/ABS(Sxy) GIVES SIGN OF Sxy}
   chk := Sxy * Sxy / (Sxx * Syy);
END;
```

```
   C := Sy - M * Sx;
   IF (M <= 0) OR (chk < 0.9) THEN BEGIN
      cursorpsn(1, 5);
      WRITELN('Please check these points for errors!');
   END;
UNTIL (M > 0) AND (chk >= 0.9);
{NOW PLOT CALIBRATION CURVE}
setscales1(X0,Y0,Xscale,Yscale,Xrange,Yrange);
Graphics;
drawline(X0, Y0, X0, Y0 + Yrange);
drawline(X0, Y0 + Yrange, X0 + Xrange, Y0 + Yrange);
drawline(X0 + Xrange, Y0 + Yrange, X0 + Xrange, Y0);
drawline(X0 + Xrange, Y0, X0, Y0);
FOR n := 0 TO 7 DO BEGIN
   itemp := n * 50;
   drawline(X0, Y0 + itemp * Yscale, X0 - tick,
            Y0 + itemp * Yscale);
   cursorto(X0 - 2 * tick - 4 * chsize,
            Y0 + itemp * Yscale + chsize / 2);
   formprint(itemp, 4, 0);
END;
FOR n := 0 TO 5 DO BEGIN
      volts := 0.4 * n; v10 := n * 4;
      drawline(X0 + volts * Xscale, Y0, X0 + volts * Xscale,
               Y0 - tick);
      cursorto(X0 + volts*Xscale - 1.5*chsize, Y0 - 2*tick);
      formprint((v10 DIV 10), 1, 0); printtext('.');
      formprint((v10 MOD 10), 1, 0);
   END;
   verticalprint(X0 - 13 * chsize / 2, Y0 + Yrange / 2
                 + 13 * chsize / 2, 'Temperature K ');
   cursorto(X0 + Xrange / 2 - 19 * chsize / 2,
            Y0 - 2 * tick - 2 * chsize);
   printtext('Thermometer Voltage');
   FOR pointno := 1 TO noofpoints DO
      drawcross(X0 + cal[pointno, 0] * Xscale,
                Y0 + cal[pointno, 1] * Yscale);
   IF C < 0 THEN BEGIN
      X1 := X0 + (-C / M) * Xscale; Y1 := Y0 END
   ELSE BEGIN
      X1 := X0; Y1 := Y0 + C * Yscale;
   END;
   IF (M * 2 + C) > 350 THEN BEGIN
      X2 := X0 + ((350 - C) / M) * Xscale; Y2 := Y0 + Yrange END
   ELSE BEGIN
```

```
      X2 := X0 + Xrange; Y2 := Y0 + (M * 2 + C) * Yscale;
    END;
    drawline(X1, Y1, X2, Y2);
    cursorto(0, chsize); graphcont;
  END; {second case}
END; {case}

clrtext;
cursorpsn(1, 7); WRITELN('   The temperature will be evaluated');
WRITELN; WRITELN('from the thermometer output voltage');
WRITELN; WRITELN('using the following equation:');
WRITELN; WRITELN; WRITELN; WRITELN;
WRITE('  T = ', M:6:2, 'V ');
IF C > 0 THEN WRITE('+ ')  ELSE IF C < 0 THEN WRITE('- ');
IF C <> 0 THEN WRITELN(ABS(C):6:2);
WRITELN; WRITELN; WRITELN; cont;
clrtext; cursorpsn(1, 12);
WRITELN('Do you want to save your results');
WRITE('to disc (Y/N)? ');
REPEAT
   yn := getkey; WRITE(yn);
UNTIL (yn = 'y') OR (yn = 'Y') OR (yn = 'n') OR (yn = 'N');
IF (yn = 'y') OR (yn = 'Y') THEN save := TRUE ELSE save := FALSE;
IF save THEN BEGIN
   REPEAT
      cursorpsn(1, 16);
      WRITE('Enter name of the output file: ');
      READLN(filename);
      IF LENGTH(filename) > 7 THEN
         WRITE(CHR(7), 'FILE NAME TOO LONG!');
   UNTIL LENGTH(filename) < 8;
END;

REPEAT
   oldT := -1; oldV := -1; oldI := -1;
   clrtext; headings;
   cursorpsn(11, 8); WRITELN('MAIN MENU'); WRITELN; WRITELN;
   cursorpsn(1, 11); WRITELN('1: Manual temperature set');
   WRITELN; WRITELN('2: Automatic temperature set '); WRITELN;
   WRITELN('3: Manually set heating curve'); WRITELN;
   WRITELN('4: Automatically set heating curve'); WRITELN;
   WRITELN('5: Exit the program'); cursorpsn(1, 22);
   WRITE('Please choose the option you require: ');
   REPEAT
      readall(M, C, T, V, I); prvals(T, V, I, oldT, oldV, oldI);
      mmchoice := eval(checkkey);
```

```
   UNTIL (mmchoice > 0) AND (mmchoice < 6);
   CASE mmchoice OF
1: BEGIN
      clrtext; headings;
      cursorpsn(8, 8); WRITELN('MANUAL TEMPERATURE SETTING');
      cursorpsn(1, 11); WRITELN('0: Power supply off');
      WRITELN('1: 2.5V on power supply');
      WRITELN('2: 5.0V on power supply');
      WRITELN('3: 7.5V on power supply');
      WRITELN('4:  10V on power supply');
      WRITELN('5:  To return to main menu');
      REPEAT
         readall(M, C, T, V, I);
         prvals(T, V, I, oldT, oldV, oldI);
         cursorpsn(39, 25); choice := eval(checkkey);
         IF (choice >= 0) AND (choice < 5) THEN setpower(choice);
      UNTIL (choice = 5) OR (T >= 350);
      setpower(0);
   END;
2: BEGIN
      {AUTO TEMP SET}
      clrtext; headings;
      REPEAT
         cursorpsn(6, 8);
         WRITELN('AUTOMATIC TEMPERATURE SETTING ');
         WRITELN; WRITELN;
         WRITELN('Please  enter  the temperature you wish');
         WRITELN; WRITE('to reach: '); READLN(reqtemp);
         OK := NOT ((reqtemp > 350) OR (reqtemp < 4));
         IF NOT OK THEN WRITE(CHR(7));
      UNTIL OK;
      IF T <= reqtemp THEN BEGIN
         cursorpsn(5, 17);
         WRITELN('Now heating sample up to ', reqtemp:6:1, 'K');
         WRITELN; WRITELN('          *** Please wait ***');
         WRITELN; WRITELN;
         WRITELN(' To stop heating program press space bar');
         diff := 40; OK := FALSE;
         FOR pwr := 4 DOWNTO 1 DO BEGIN
            diff := diff / 2; IF pwr = 1 THEN diff := 0;
            IF ((reqtemp - T) >= diff) AND NOT OK THEN BEGIN
               setpower(pwr);
               REPEAT
                  readall(M, C, T, V, I);
                  prvals(T, V, I, oldT, oldV, oldI);
```

```
                IF checkkey = ' ' THEN OK := TRUE;
             UNTIL ((reqtemp - T) < diff) OR OK;
          END;
       END;
    END;
    setpower(0);
 END;
3: BEGIN
    {MANUAL HT CURVE}
    REPEAT
       clrtext; cursorpsn(1, 8);
       WRITELN('1: 2.5V'); WRITELN;
       WRITELN('2: 5.0V'); WRITELN;
       WRITELN('3: 7.5V'); WRITELN;
       WRITELN('4: 10V'); WRITELN; WRITELN; WRITELN;
       WRITELN('Please choose the power supply voltage');
       WRITELN; WRITE('you wish to use: ');
       choice := eval(getkey);
    UNTIL (choice >= 1) AND (choice <= 4);
    repeating := FALSE; hgraph(repeating);
    FOR n := 1 TO 20 DO temp[n] := 0;
    vol := 0; amp := 0; pow := 0; heatpts := 0;
    cursorto(0, 3 * chsize);
    printtext('* Now reading temperature: please wait *');
    waitnext(3);
    FOR n := 0 TO 19 DO BEGIN
       readall(M, C, T, V, I); printtemp(T); temp[n] := T;
       plotpoint(X0 + n * Xscale, Y0 + (temp[n] - VSS)*Yscale);
       waitnext(3);
    END;
    cursorto(0, 3 * chsize);
    printtext('Press "S" to start the heating curve    ');
    newline;
    REPEAT
       readall(M, C, T, V, I); printtemp(T);
       FOR q := 1 TO 19 DO temp[q - 1] := temp[q];
       {SHUFFLE POINTS BACK}
       temp[19] := T;
       FOR q := 1 TO 19 DO BEGIN
          ruboutpoint(X0 + q * Xscale,
                      Y0 + (temp[q - 1] - VSS) * Yscale);
          plotpoint(X0 + q * Xscale,
                    Y0 + (temp[q] - VSS) * Yscale);
       END;
```

```
      waitnext(3); key := checkkey;
   UNTIL (key = 'S') OR (key = 's');
   {*** START OF HEATING CURVE ***}
   setpower(choice); starttime := time; flush; n := 20;
   cursorto(0, 3 * chsize);
   printtext('Press "E" to end the heating curve     ');
   newline;
   REPEAT
      readall(M, C, T, V, I); temp[n] := T;
      vol := vol + V; amp := amp + I; pow := pow + I * V;
      heatpts := heatpts + 1; printtemp(T);
      plotpoint(X0 + n*Xscale, Y0 + (temp[n] - VSS)*Yscale);
      n := n + 1; waitnext(3); key := checkkey;
   UNTIL (key = 'E') OR (key = 'e') OR (T > (inittemp + 15))
         OR (time >= (starttime + 120));
   {*** START OF SECOND COOLING CURVE ***}
   setpower(0); stoptime := time; stoppts := n;
   cursorto(0, 3 * chsize);
   printtext('Now plotting cooling curve: please wait ');
   newline;
   FOR dash := 0 TO (ROUND(Yrange) DIV tick) - 1 DO
      drawline(X0 + stoppts * Xscale, Y0 + dash * tick,
               X0 + stoppts * Xscale,
               Y0 + dash * tick + tick / 2);
   REPEAT
      readall(M, C, T, V, I); printtemp(T); temp[n] := T;
      IF n = (stoppts + 25) THEN BEGIN
        flush; cursorto(0, 3 * chsize);
        printtext('Press space bar to end cooling curve    ');
        newline;
      END;
      plotpoint(X0 + n * Xscale,
      Y0 + (temp[n] - VSS) * Yscale);
      n := n + 1; waitnext(3);
   UNTIL (n > 120) OR (checkkey = ' ');
   coolend := n - 1; tim := (stoptime - starttime);
   vol := vol / heatpts; amp := amp / heatpts;
   pow := pow / heatpts;
   calc;
   IF save THEN
      writedata(filename, coolend, temp1, temp2, heat, temp);
 END;
4: BEGIN
   {AUTO HT CURVE}
   clrtext; cursorpsn(1, 10);
```

```
WRITELN('What temperature increment would you');
WRITE('like? '); READLN(tempinc);
repeating := FALSE; hgraph(repeating);
vol := 0; amp := 0; pow := 0; hpower := 1; rdgs := 0;
heatpts := 0;
cursorto(0, 3 * chsize);
printtext('* Now plotting the first cooling curve *');
newline; waitnext(3);
FOR n := 0 TO 19 DO BEGIN
   readall(M, C, T, V, I); printtemp(T); temp[n] := T;
   plotpoint(X0 + n * Xscale,
   Y0 + (temp[n] - VSS) * Yscale);
   waitnext(3);
END;
{*** START OF HEATING CURVE ***}
starttime := time; starttemp := T; setpower(1);
n := 20; cursorto(19 * chsize, 3 * chsize);
printtext('heating curve *      '); newline;
REPEAT
   readall(M, C, T, V, I); printtemp(T); temp[n] := T;
   vol := vol + V; amp := amp + I;
   pow := pow + I * V; heatpts := heatpts + 1;
   plotpoint(X0 + n*Xscale, Y0 + (temp[n] - VSS)*Yscale);
   n := n + 1;
   IF (rdgs >= 5)
       AND ((temp[n] - temp[n - 5]) < (tempinc / 10))
       AND (hpower < 4) THEN
   BEGIN
      hpower := hpower + 1; setpower(hpower); rdgs := 0;
   END ELSE
      rdgs := rdgs + 1;
   waitnext(3);
UNTIL (T > (starttemp + tempinc))
      OR (time >= (starttime + 120));
{*** START OF SECOND COOLING CURVE ***}
setpower(0); stoptime := time; stoppts := n;
cursorto(15 * chsize, 3 * chsize);
printtext('second cooling curve *'); newline;
FOR dash := 0 TO (ROUND(Yrange) DIV tick) - 1 DO
   drawline(X0 + stoppts * Xscale, Y0 + dash * tick,
            X0 + stoppts * Xscale,
            Y0 + dash * tick + tick / 2);
REPEAT
   readall(M, C, T, V, I); printtemp(T); temp[n] := T;
   IF n = (stoppts + 25) THEN BEGIN
```

```
            flush; cursorto(0, 3 * chsize);
            printtext('Press space bar to end cooling curve');
            newline;
         END;
         plotpoint(X0 + n*Xscale, Y0 + (temp[n] - VSS)*Yscale);
         n := n + 1; waitnext(3);
      UNTIL (n > 120) OR (checkkey = ' ');
      coolend := n - 1; tim := (stoptime - starttime);
      vol := vol / heatpts; amp := amp / heatpts;
      pow := pow / heatpts;
      calc;
      IF save THEN
         writedata(filename, coolend, temp1, temp2, heat, temp);
   END;

5: {end of program}
END; {case}

UNTIL mmchoice = 5;
setpower(0); clrtext;
END.
```

7. Computer-Controlled Observations of Surface Plasmon-Polaritons

A.D. Boardman, A.M. Moghadam, and J.L. Bingham

7.1 Introduction

The interaction between polarization excitations of condensed media and electromagnetic waves has attracted a lot of attention. This is because an electromagnetic wave travelling through a polarizable medium is modified by the polarization it induces and becomes coupled to it. This coupled mode of excitation is called a polariton. If the polarizable medium is identified, then the polariton is qualified. Thus, for example, in the case of an electron plasma the coupled modes are called plasmon-polaritons.

There has been a steady growth of interest in surface polaritons, which are modes associated with some form of guiding surface and are completely bound to it in the sense that their associated fields decay exponentially in the normal directions. Although the theoretical interest in surface polaritons, and the means of experimentally examining them, arose in the last two decades, a great deal of related work on electromagnetic surface wave propagation along a reactive surface originated much earlier than this. It was done from an electrical engineering point of view and the period, beginning at the turn of the century, leading up to the modern work is very interesting in its own right. In this chapter, however, attention is restricted to the plane surfaces of a metal whose high frequency behaviour can be realistically modelled as an electron plasma moving against a smoothed out compensating positively charged background.

For an electron plasma, such a surface mode has both a photon and a plasmon content simply because the total energy must be shared out over the whole system. It is possible to investigate plasmon-polaritons in their extreme states when they have either a very strong photon or a very strong plasmon content. A pure plasmon mode corresponds to a jelly-like plasma oscillation at the angular frequency ω_p. A surface plasmon corresponds to surface plasma oscillations with angular frequency $\omega_p/\sqrt{2}$. This classic result was first theoretically produced by *Ritchie* [7.1] and subsequent verification of it was soon obtained with the aid of fast electron spectroscopy.

An understanding of how surface plasma oscillations couple to incoming electromagnetic radiation stems from a major development by *Otto* [7.2] who pointed out that surface modes can be generated using attenuated (or frustrated) total reflection (ATR). It is interesting to note, however, that the technical application of frustrated total reflection goes back at least to 1947 with the work of *Leurgans* and *Turner* [7.3] and has since been widely used to couple light beams to thin-film guides. It is equally interesting that the type of ATR experiment proposed by Otto was actually performed in 1959 by *Turbadar* [7.4], but went unnoticed and was not properly interpreted. The major contribution by Otto was to link ATR to the launching of bound (surface) modes and to demonstrate that it is a direct and extremely sensitive way of measuring the wavenumber range in which the surface polariton has a fairly high photon content. This method was soon developed by *Kretschmann* [7.5] and *Raether* [7.6] into an alternative configuration. Interest in surface polaritons, following the spur given by Otto, has gathered momentum and the base of activity has broadened throughout the world.

The surface plasmon-polariton is a TM wave (magnetic field in the plane of the surface and normal to the propagation direction). If a polariton propagates along a plane surface between two semi-infinite media occupying $z > 0$ (medium 1) and $z < 0$ (medium 2), then the field components have space-time dependence $\exp[i(kx - \omega t)]\exp(-\beta_1 z)$ for $z > 0$ and $\exp[i(kx - \omega t)]\exp(\beta_2 z)$ for $z < 0$, where k is the wavenumber, ω is the angular frequency and β_1 and β_2 are decay coefficients. The relationship between angular frequency ω and wave number k is called the dispersion equation. This name arises because such an equation determines how a wave packet (pulse) will spread out.

β_1 and β_2 are determined by Maxwell's equations, under the TM mode assumption $\boldsymbol{E} = (E_x, 0, E_z)$, $\boldsymbol{B} = (0, B_y, 0)$. For a medium with a dielectric function ϵ the relevant components of Maxwell's equations are

$$\begin{aligned} &\frac{\partial E_x}{\partial z} - ikE_z = i\omega B_y\,, \quad \frac{\partial B_y}{\partial z} = \frac{i\omega}{c^2}\epsilon E_x\,, \\ &ikE_x + \frac{\partial E_z}{\partial z} = 0\,, \quad kB_y = -\frac{\omega}{c^2}\epsilon E_z\,, \end{aligned} \tag{7.1}$$

from which it follows, after suppressing the common factor $\exp[(ikx - \omega t)]$, that if $E_x = A_1 e^{\pm\beta z}$,

$$E_z = \pm i\frac{k}{\beta}A_1 e^{\pm\beta z}\,, \quad B_y = \mp i\frac{\omega\epsilon}{\beta c^2}A_1 e^{\pm\beta z}\,, \tag{7.2}$$

where

$$\beta^2 = k^2 - \epsilon\omega^2/c^2\,.$$

Hence, associating a dielectric constant ϵ_1 with an upper passive medium and a frequency-dependent dielectric function ϵ_2 with the lower medium leads to the definitions

$$\beta_1 = \left(k^2 - \epsilon_1\frac{\omega^2}{c^2}\right)^{\frac{1}{2}}\,, \quad \beta_2 = \left(k^2 - \epsilon_2\frac{\omega^2}{c^2}\right)^{\frac{1}{2}}\,. \tag{7.3}$$

The boundary conditions at the interface between the two media are that E_x and B_y are continuous. The imposition of these gives

$$\frac{\beta_1}{\beta_2} = -\frac{\epsilon_1}{\epsilon_2} > 0 \,, \tag{7.4}$$

and

$$k^2 = \frac{\omega^2}{c_2} \frac{\epsilon_1 \epsilon_2}{\epsilon_1 + \epsilon_2} > 0 \,. \tag{7.5}$$

where (7.5) is the dispersion equation of the surface waves. Equations (7.4) and (7.5) also show that the existence of such waves requires ϵ_1 and ϵ_2 to be of opposite sign and their sum to be negative. This can be understood as follows: (7.4) can be satisfied only if $\epsilon_1 > 0$, $\epsilon_2 < 0$, or vice versa; if this is so then $\epsilon_1 \epsilon_2 < 0$, hence $\epsilon_1 + \epsilon_2 < 0$, for (7.5) to be satisfied.

A medium with a positive dielectric function is termed a 'passive medium', and commonly used examples of such media include air, vacuum, glass, or similar media whose dielectric functions are fairly constant over the frequency range of interest. The dielectric function for the passive medium will be labelled ϵ_1, for this work, and has the value unity for air or vacuum and 2.25 for typical glasses. A medium with a negative dielectric function is called an 'active' medium and include materials and semiconductors at frequencies below the plasma frequency ω_p. This fact permits the study of materials with the simple model dielectric function

$$\epsilon_2(\omega) = \epsilon_L \left(1 - \frac{\omega_p^2}{\omega(\omega + i\nu)} \right) \,, \quad \omega_p^2 = \frac{Ne^2}{\epsilon_L \epsilon_0 m} \,, \tag{7.6}$$

where ϵ_L is the high-frequency dielectric constant, which is unity for metals but between 10 and 20 for semiconductors; N is the free electron density, e is the electronic charge, m is the effective mass, ϵ_0 is the permittivity of free space. ν is a damping term describing electron-photon collisions, and it is quite small, compared to ω, in the optical range of frequencies. If ν is included, the dielectric function of an active medium is a complex quantity with a large negative real part, provided $\omega < \omega_p$. $\epsilon(\omega)$ is, approximately, real and negative for $\omega < \omega_p$. The plasma frequency for metals is typically $\sim 10^{15}$-10^{16} and in semiconductors 10^{13}-10^{14}, therefore surface plasmon-polaritons might be expected to be generated at infrared frequencies for semiconductors and at optical frequencies for metals.

For real materials ν must be included in the dielectric function and, as a consequence, (7.5) may then be interpreted in two different ways that correspond to a certain choice of experimental procedure. Firstly, the wave number may be assumed to be real so that (7.5) may be solved for a complex frequency, corresponding to a wave that is temporally damped. Alternatively, the frequency may be taken as real so that (7.5) yields a complex wavenumber, corresponding to spatial damping. In either case the surface wave is not a true, long-lived, normal mode and leaks energy into the active medium. This can be seen from the fact that β_1 and β_2 are complex and hence the fields may have a small propagating component normal to the surface.

The dispersion curves obtained by these two choices are quite different but the curve obtained for real wave number and complex frequency is almost indistinguishable from the curve calculated with $\nu = 0$ and has a large wave number limit given by

$$\omega = \omega_p \left(\frac{\epsilon_L}{\epsilon_L + \epsilon_1} \right)^{\frac{1}{2}} \tag{7.7}$$

where, for a metal bounded by air, $\epsilon_L = 1$, $\epsilon_1 = 1$.

For small wave numbers the surface polariton is very 'photon-like' so the most suitable probe is light. The technique is known as **at**tenuated **t**otal **r**eflection (ATR).

ATR uses the evanescent wave that is set up at a medium-air interface when light in a high refractive index medium, such as glass, suffers total internal reflection. It is the reduction of the reflected intensity wave due to surface wave generation that is called ATR. If the weakening occurs by some other means, it is usually called frustrated total internal reflection (FTIR). The names, however, have sometimes been interchanged and, rather remarkably, FTIR can be traced back to *Newton* [7.7]. The knowledge and use of ATR has been widespread since the early 1960s due to pioneering work by *Fahrenfort* [7.8] and *Harrick* [7.9]. As we pointed out earlier, the fundamental idea of using ATR to generate surface polaritons was first published by Otto as late as 1968.

An inspection of the dispersion curve for a (air) vacuum/metal interface, shown in Fig. 7.1, reveals that in the photon-like (small k) region, the curve lies *below* the light line. For a plane electromagnetic wave, incident through the air on the interface between the ϵ_1 and ϵ_2 at some angle θ, the component of wave number parallel to the surface is

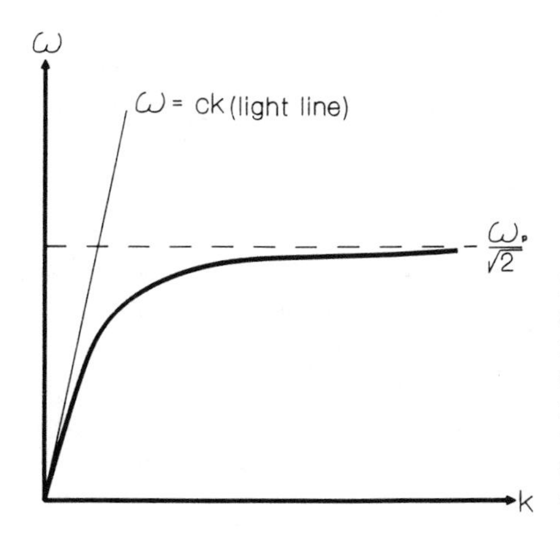

Fig. 7.1. Surface plasmon-polaritons dispersion curve, for surface waves on the boundary between semi-infinite air (vacuum) and a semi-infinite metal. The curve is asymptotic to $\omega_p/\sqrt{2}$ where ω_p is called the plasma frequency. $\omega = ck$ is the dispersion curve of light in air and is called the *light lines*

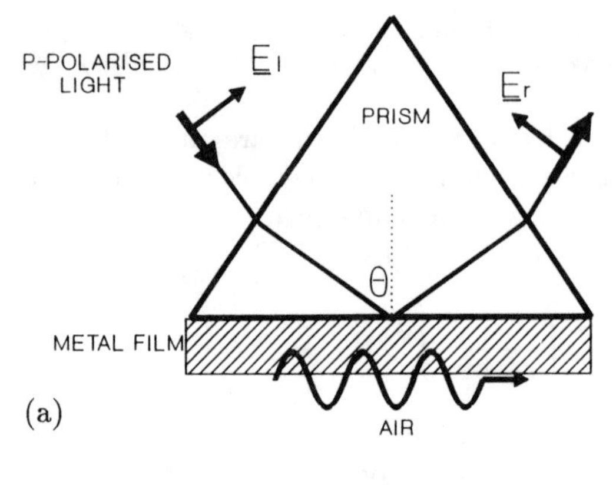

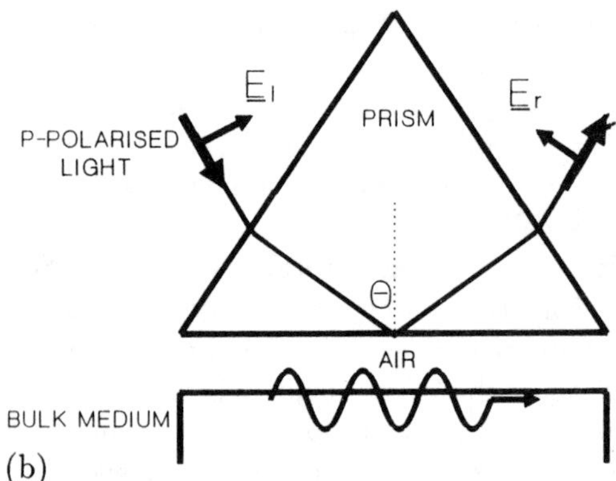

Fig. 7.2a,b.
(a) Excitation of surface plasmon-polaritons in the Kretschmann-Raether geometry. The surface wave is excited at the air-metal interface. **(b)** Excitation of surface plasmon-polaritons in the Otto geometry. The surface wave is excited at the air-metal interface

$$k = \frac{\omega}{c}\epsilon_1^{\frac{1}{2}} \sin\theta \,. \tag{7.8}$$

This corresponds to a line *steeper* than the light line of Fig. 7.1 and, hence, to the region *above* the light line. Indeed, only under the impractical condition of grazing incidence ($\theta = 90°$) does k approach a surface-wave value. The awkward conclusion is that surface polariton waves cannot be generated by propagating plane electromagnetic waves, in a semi-infinite passive medium, on to a *single* interface separating it from a semi-infinite active medium. If surface modes are to be generated, then an ATR system with more than one surface, first proposed by *Otto* [7.2] (Fig. 7.2(a)) and subsequently by *Kretschmann* [7.5] (Fig. 7.2(b)), must be used.

A prism-air-medium system is the Otto configuration. It consists of a prism (sometimes hemicylindrical) with a dielectric constant ϵ_1 separated from a thick sample of the active medium by a small air (vacuum) gap of dielectric constant $\epsilon_2 < \epsilon_1$. Light incident, through the prism on to the prism-air interface, at angles greater than the critical angle θ_c is normally totally reflected back out through the prism. However, under total internal reflection conditions, there will always be an exponentially decaying evanescent field extending into the air gap. This totally reflected wave has a component of wave number parallel to the surface, given by (7.8). The dispersion curve (Fig. 7.1) for surface polaritons on a *metal-air interface* lies *below* the air (vacuum) light line ($\omega = ck$) but an evanescent wave, created by the total reflection in the prism, exists in the range $\theta_c \leq \theta_1 \leq 90°$ for which

$$\frac{1}{\sqrt{\epsilon_1}} < \sin\theta < 1 . \qquad (7.9)$$

The range of surface wave number associated with this evanescent wave is

$$\frac{\omega}{c} < k < \sqrt{\epsilon_1}\frac{\omega}{c} . \qquad (7.10)$$

This is a region *above* the prism light line but *below* the air/vacuum light line. As shown in Fig. 7.3, the evanescent wave has a larger k than the corresponding wave vector, in air. Hence some wave numbers of the air/metal surface polariton become directly accessible with an ATR system.

From a practical point of view, the air gap must be small enough for the evansecent field from the prism to be significant at the air-medium interface. Energy from the incident wave is used to stimulate a surface wave, resulting in a *weakening* to a sharp *minimum*, of the reflected intensity returned by the prism. Hence the name '**a**ttenuated **t**otal **r**eflection'. The second system, developed by *Raether* and *Kretschmann* [7.5] is more attractive because it consists of a thin film of an active medium deposited onto the prism. This is much easier to do experimentally. The surface wave is excited in just the same way, except that the fields from the prism must now penetrate the film to reach the medium-air interface. In both cases the excitation of a surface plasmon-polariton occurs at the second interface and is observed as a sharp minimum in the reflected intensity of (TM) p-polarised light. In fact the purpose of an ATR experiment is to measure, for a given frequency, the real angle at which the reflected intensity is a *minimum*, or to measure for a given angle the real frequency at which a *minimum* occurs.

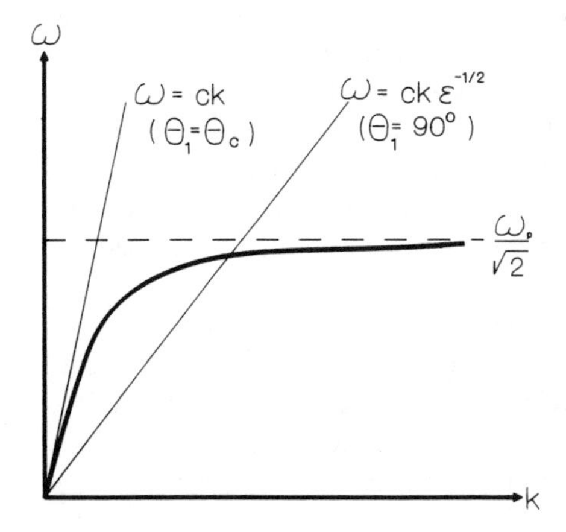

Fig. 7.3 The ATR configuration permits some of the surface plasmon-polariton dispersion curve, for a metal-air interface, to be observed. The observed wave numbers, for a given frequency, of a surface wave lies between the light lines $\omega = ck$ and $\omega = ck/\sqrt{\epsilon}$

The study of the characteristics of surface plasmon-polaritons by the method of attenuated total reflection (ATR) is also stimulated by the fact that surface plasmons are a very sensitive probe for the study of surface properties. For example, the ATR minimum shifts in a measurable way if the metal/air surface receives a layer of foreign atoms. The ATR method is a non-destructive technique and it is by far the most suitable and versatile technique for planar geometries. It has been used to measure the dielectric permittivity of metals and magneto-optic materials, and has also been employed as a tool for monitoring molecular absorption at a surface of an interface. Surface plasmon-based devices are finding roles as chemical sensors because of their fast response and very high sensitivity. It is important, therefore, to investigate the design and construction of an advanced, and fully computerised, ATR system that can yield direct observation of surface plasmon-polaritons. The method yields, for a fixed frequency, the *whole* reflectivity curve as a function of angle of incidence and it is the quality of the minimum created by surface plasmon generation that is so useful as a tool, i.e. the width and depth of the minimum (resonance) are sensitively dependent on the dielectric function and the film thickness of the metal film. For decreasing thickness of the film (increasing radiation damping), the resonance position is shifted to larger angles and the minimum becomes broadened. The minimum also becomes broader as the radiation wavelength decreases. The primary reason for this is that the internal damping becomes greater.

7.2 A Computer-Controlled ATR Experiment

An advanced system will now be described that can automatically perform an ATR experiment and allow the direct observation of surface plasmon-polaritons. Some physical features and the technical virtues of the specific prism geometry that is adopted in the system will be explained, and the full ray geometry of the special prism used here as it is scanned during the measurement process will be discussed. The Kretschmann-Raether geometry and an angle scan method is used since, as has already been pointed out, this is the most robust and versatile one. Brief mechanical design aspects of the system are presented and the computer interface and the computer-control strategy for the ATR measurement is fully described. The experimental results are then presented and compared with theory in order to establish the validity of the system.

The computer procedure that controls the system is so versatile that it enables the experimenter to employ any right-angle prism with different angles or refractive indices. It also provides probes to check for optical alignment within the system. Moreover, it performs the angular scan as many times as the experimenter wishes and averages the results in order to reduce environmental and other sources of noise within the system, thus enhancing the signal-to-noise ratio.

7.2.1 Prism Geometry

In order to perform optical measurements with glass prisms, the prism angles and the refractive index must be accurately known. The prism employed in this ATR experiment is 60° right-angled and has its large non-hypotenuse face coated with a *thick* silver mirror to make a totally reflecting surface. The thin metal film that supports surface plasmon is coated onto a glass slide that has the same refractive index as that of the prism. The glass side of the slide is then brought into optical contact with the prism base using index-matching liquid oil. Effectively, the prism, the index-matching oil and the glass slide are a single medium because they all have the same refractive index. This feature is advantageous in the sense that a number of films of different thickness or different materials may be readily prepared. In addition, it allows a single prism to be used in all the experiments. In all the figures throughout this chapter that show the prism geometry, the index-matching liquid oil and the glass slide have not been shown, for clarity. It is a geometry that makes the detection of the reflected light from a low-powered laser beam a straightforward matter. For the range of incidence angles of interest, the beam refracted out from the prism hypotenuse face is *always parallel* to the corresponding incident beam on the hypotenuse. This feature reduces the number of optical components within the system and greatly facilitates the detection of the reflected light. Indeed, only one beam splitter and one focussing lens is needed to capture the reflected light. Employing such an interesting prism geometry provides a wide range of incidence angle without motion of the reflected light over the aperture of the detector, thereby obviating any further detector function. A wide range of incidence angle is essential to obtain the ATR angular spectrum of highly absorbing metals, such as iron, nickel and cobalt, whose ATR resonance curves are rather broad. Figure 7.4 shows the details of the prism. It is, initially, positioned on a turntable such that a laser beam strikes the 60° corner while it is tangential to the hypotenuse face. The two possible arrangements for the incident and reflected beams are shown in Figs. 7.5(a) and 7.5(b), while Figs. 7.6(a) and 7.6(b) illustrate the ray geometry of the prism as the prism rotates.

It is the angle of incidence θ on the prism-metal interface that is required, but the angle that is actually measured *by the computer* during the measurements is the one that the incident beam makes with the hypotenuse face. This angle is defined as β and the angle defined as γ is simply

$$\gamma = \sin^{-1}\left(\frac{\cos\beta}{n_p}\right) = 30 - \alpha\ , \quad \text{(Fig. 7.5(a))}\ , \tag{7.11}$$

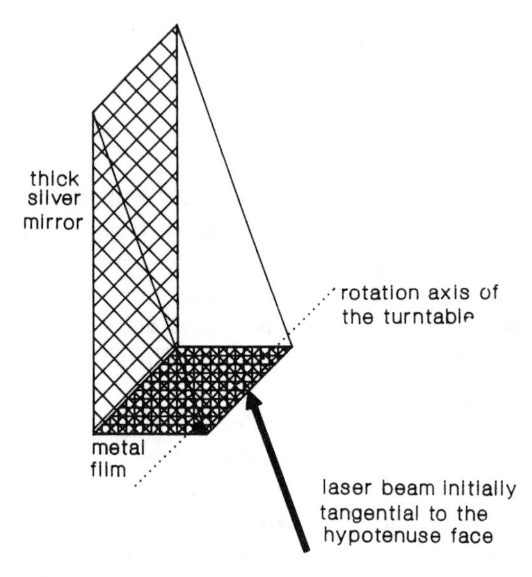

Fig. 7.4. Geometry of the prism showing the initial position of the laser beam. The rotation axis of the turntable is perpendicular to the triangular faces of the prism. Surface waves are generated on the metal film at the air/metal interface

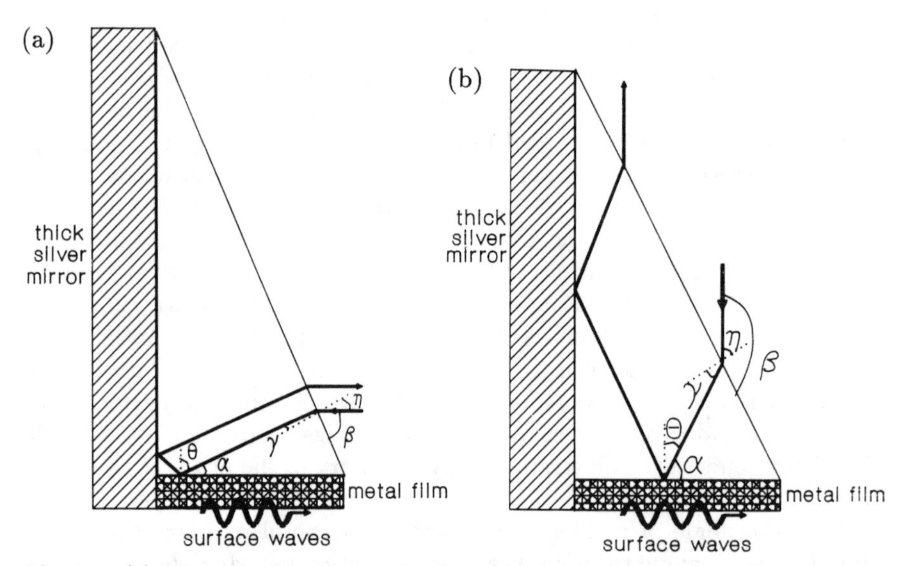

Fig. 7.5. (a) Direction of incident and reflected beams for large angles of incidence. In this case, $\alpha + \gamma = 30°$. In practice, the point of incidence is very much closer to the 60° corner to guarantee a wide range of incidence angles. The rotation axis passes through the 60° corner, perpendicular to plane of the figure. **(b)** Direction of incident and reflected beams for small angles of incidence. In this case, $\alpha - \gamma = 30°$. In practice, the point of incidence is very much closer to the 60° corner (rotation point) to guarantee a wide range of incidence angles

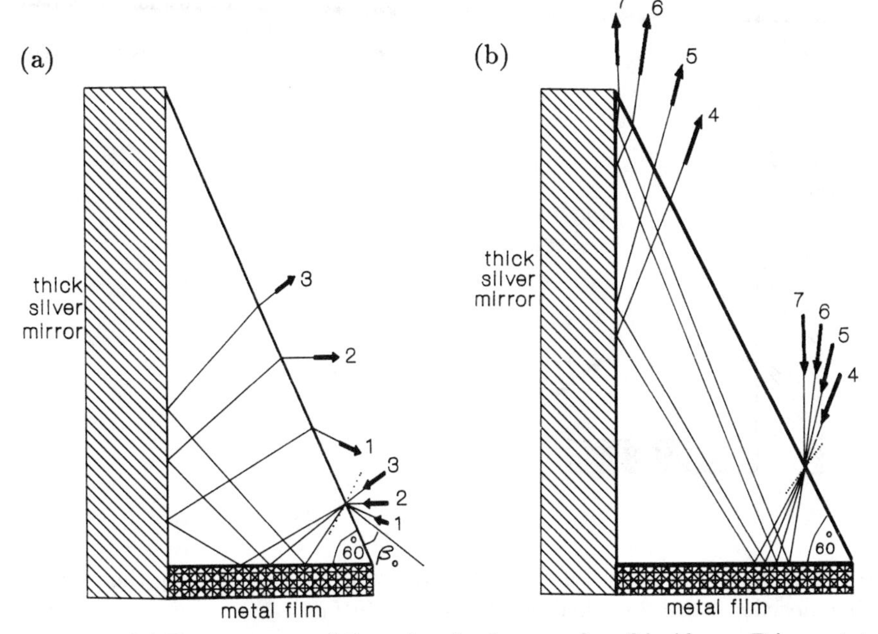

Fig. 7.6. (a) Ray geometry of the prism for large angles of incidence. Prism rotates about the 60° corner. **(b)** Ray geometry of the prism for small angles of incidence. Prism rotates about the 60° corner

$$\gamma = \sin^{-1}\left(\frac{-\cos\beta}{n_p}\right) = \alpha - 30 \ , \quad (\text{Fig. 7.5(b)}) \ , \tag{7.12}$$

so that

$$\theta = 60 + \sin^{-1}\left(\frac{\cos\beta}{n_p}\right) \quad (\text{Fig. 7.5(a)}) \ , \tag{7.13}$$

or

$$\theta = 60 - \sin^{-1}\left(\frac{-\cos\beta}{n_p}\right) \quad (\text{Fig. 7.5(b)}) \ . \tag{7.14}$$

Obviously, in Fig. 7.5(a), it is possible to have $\gamma = 30°$, which corresponds to the refracted beam being parallel to the base of the prism. This refracted beam hits the totally reflecting face of the prism, reflects back along its own path and refracts out along its original path. This is shown in Fig. 7.7.

As shown in Fig. 7.7, this is the only ray path for which the light travels back exactly parallel to the base of the prism and then back along its *original* path. This is a unique position for the prism, and for angles of $\beta > \beta_0$ the rays are reflected back along paths that are parallel to the incoming ray paths. The detector is only placed to receive such rays. Rays at an angle $\beta < \beta_0$ are reflected and refracted in other directions and are allowed to

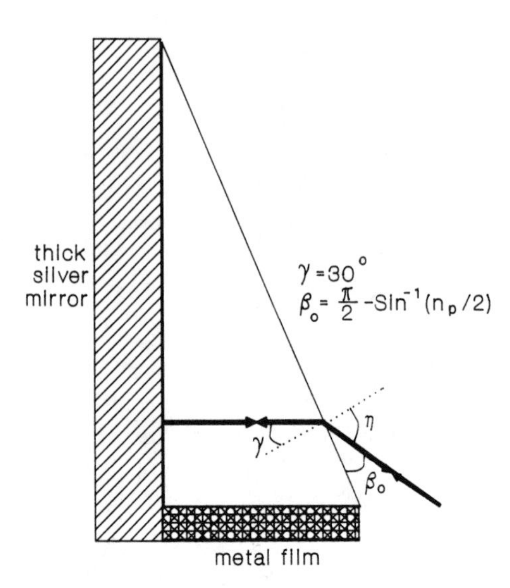

Fig. 7.7. Reference position. This position is searched for by the computer and the reference angle β_0 is calculated

escape undetected. Hence, if the computer can find β_0 precisely then every other position can be worked out from this. In fact

$$\sin\left(\frac{\pi}{2} - \beta_0\right) = (\sin\gamma)n_p = \sin(30)n_p = \frac{n_p}{2}, \tag{7.15}$$

and β_0 is called the 'zero position' value. This value is displayed by the computer. It then calculates the range of angle β for which the refracted beam inside the prism hits the metal film on the base of the prism. From these data the accessible range of incidence angles, θ, is calculated using (7.13) and (7.14).

The range of β that can be achieved depends on the relative position of the 60° corner of the prism and the point at which the laser beam hits the hypotenuse face. In practice, the prism must be positioned on the turntable where the laser beam just enters at the 60° corner during the whole of the prism rotation and ensures the greatest range of θ. As the turntable rotates, the angle β is swept through π but, of course, only the angles producing a θ are the ones required. Note that this result is independent of the position of the axis of rotation of the turntable relative to prism, but the axis must be perpendicular to the triangular face of the prism shown in Figs. 7.6(a) and 7.6(b). A uniform translation of the prism in the plane of Figs. 7.6(a) and 7.6(b) does not affect the calculation of β.

Instead of a triangular prism it is possible to use a half-cylindrical prism in ATR measurements. It must be precisely positioned, however, because its axis of rotation is at the centre of the half-cylinder base. An advantage would be that a laser beam always crosses the half-cylinder circumference at normal incidence, as shown in Fig. 7.8. A 90° rotation of the turntable corresponds to a 90° change in θ for the half-cylinder prism. A rotation of 180° of the turntable corresponds to a change in θ of about 80° in a

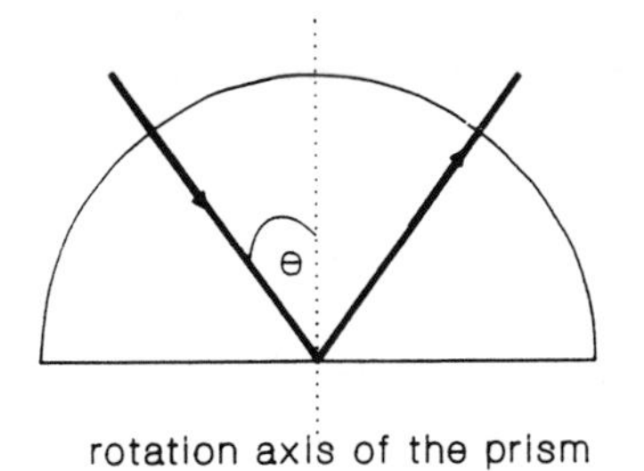

Fig. 7.8. Half-cylinder prism occasionally used in ATR experiments. The axis of rotation of the turntable passes perpendicularly through the centre of the prism base. The angle of rotation of the turntable is equal to the angle of incidence θ

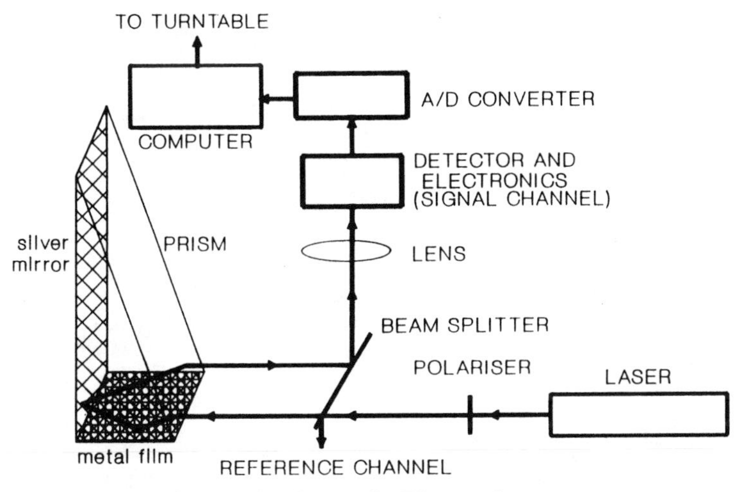

Fig. 7.9. Optical set-up for the novel ATR experiment

triangular prims, however, hence triangular prisms are more sensitive. A difficulty with the half-cylinder prism is that if the axis of rotation of the half-cylinder prism is not located accurately, serious errors will occur. The optical set-up used here involving a triangular prism is, in contrast, very efficient. The general layout of the experiment is shown in Fig. 7.9.

7.2.2 Computer Control of ATR Measurements

The block diagram shown in Fig. 7.10 illustrates the computer procedure and options that are available during a reflectivity measurement in the ATR experiments discussed here. Several programs are involved in the automated ATR experiment, both to obtain reflectivity measurements and to analyse the data. A brief description of each program will now be given.

a) PLASMON

This is the "start-up" master program that is called whenever the experiment is to be carried out. It provides access to the other programs, as

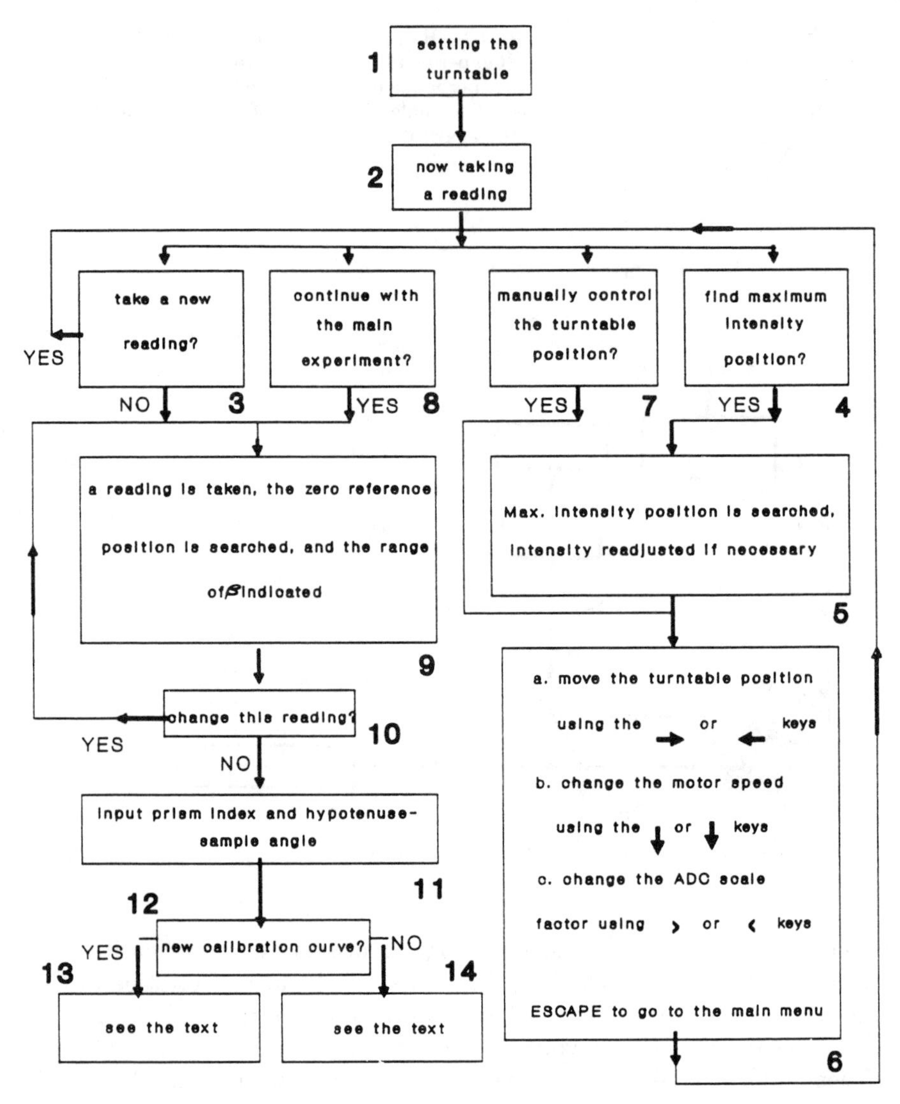

Fig. 7.10. Block diagram of the computer-controlled ATR procedure

required, by following on-screen prompts. It also sets up, if required, any redefined (new) characters used in the screen displays.

This program is called by a generic command (in our case "PLASMON"). The first screen display is the title page identifying the experiment as the "SURFACE PLASMON EXPERIMENT" with a prompt "press any key to continue".

The next page requests a choice of either using the apparatus to take a reading or of analysing results already held on file. If the choice is to analyse previously obtained results, then the program automatically calls a program called PLAS-SD.

If the choice is to perform a new experiment, then the program continues by displaying a graphical area on which the results will be plotted along with movable pointers (arrows) drawn graphically. Also shown are digital results of the relative intensity of reflected light and the angular position of the prism, namely β.

The experiment begins with the computer driving a turntable via an electric motor. The turntable may be in an arbitrary position after the last experiment but on starting again it is always turned to the same position. The optics is placed around the turntable to use this position by placing the prism on the turntable so that the laser beam is parallel to the hypotenuse. The 60° corner of the prism is also placed at the centre of the turntable so that prism rotation occurs about this vertex. At this stage, some degree of misalignment can be tolerated since the computer will determine the exact angular position of the prism later.

The computer then makes a fast 0 to 180 degree scan of the turntable, at low resolution, to show the profile of the reflected light. After the scan is completed a menu is displayed, with a choice of options for optimising the scan profile. These are options 3, 4, 7 and 8 shown in Fig. 7.10. After all parameters are set up to the satisfaction of the operator, the program continues by automatically calling another program called "PLASCAL". Before that, however, the calibration data are placed in "PCALDAT". This file is updated after each calibration run.

b) PLASCAL

This program is used to determine accurately the position of the prism on the turntable and to store a new calibration standard or recall an existing one. After entering this program the computer again makes an initial fast 0 to 180 degree scan of the turntable. It then analyses the reflectivity profile obtained to find the approximate position of the zero reference. By this we mean $\beta = \beta_0$. The turntable scans values of β ranging from the position β_0, for which the light is exactly parallel to the base and is totally reflected from the thick silver mirror, over all the $\beta > \beta_0$ range for which the returning rays are parallel to the incoming rays. For this stage of the experiment a very thick CALIBRATION metal film is used on the base of the prism to form yet another totally reflecting mirror. As shown in Fig. 7.11, a large spike occurs at $\simeq 41°$ and this is the $\beta = \beta_0$ value. After that, for $\beta > \beta_0$ the light refracted into the prism falls rapidly into the right-angled corner and then reflects off the base of the prism over a good β rangle. An ideal profile of this scan is actually shown in Fig. 7.11. The computer moves the turntable to a position two degrees to one side of the approximate β_0 position and then takes a slow scan to two degrees on the other side. This permits an accurate value of β_0 to be found. All subsequent values of β are measured from β_0. The β_0 value is made more accurate by taking derivatives of the

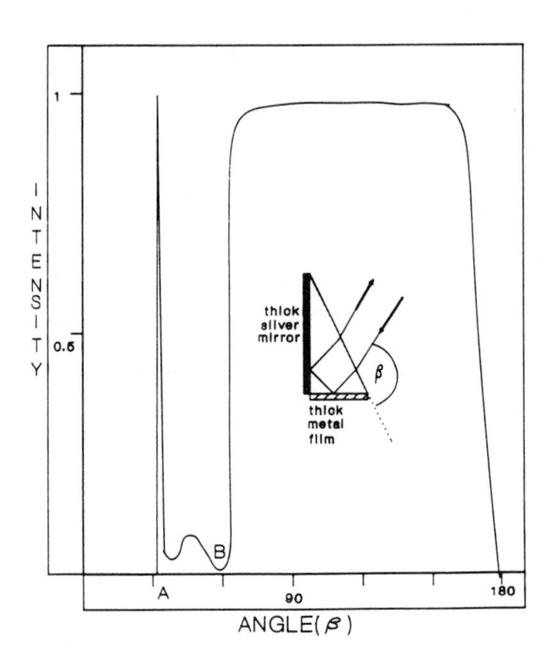

Fig. 7.11. This is a typical computer output to the monitor during calibration phase. The reflected intensity from a thick metal film is used. It is coated on a glass slide and the glass is placed in optical contact with the prism base using index-matching oil. Point A represents the zero (β_0) position, while point B shows that the laser beam is at the 90° corner

profile and then using a technique used in edge enhancement. Clearly, at this stage the computer can calculate the accessible range of θ.

Upon completion of this stage, the computer requests a choice of whether to retake the reading or set up again for a full calibration run. Any retake proceeds as above with a new evaluation of β_0 and an update of PCALDAT takes place, i.e. the parameters held in this file may now be changed and the option is available either to proceed to take a reading with a metal film on the base of the prism or a new reference (i.e. a thick material as fully internally reflecting reference). If the choice is to proceed with the existing calibration, the program re-saves the calibration data to PCALDAT, locks the file to prevent accidental erasure, and then proceeds to call a program called PLASRUN.

If the choice is to take a new calibration reading, then the program re-sets its scan parameters to take high accuracy reading only within the "useable" β range of the thick material (metal) on the base of the prism. In the previous readings the angular position was referred to the turntable position, but the readings in this high-accuracy mode are now all referred to the zero reference position (i.e. referred to the glass/sample interface).

The program then proceeds to take the first scan at low motor speed and will continue to take additional scans until instructed to stop. The purpose of this is to reduce any system noise by signal averaging of the successive scans, since signal theory suggests that system noise reduces as the squareroot of the number of results averaged. In this way a very accurate calibration profile may be obtained. After the calibration scan has been completed to the satisfaction of the experimenter, the program saves this new calibration standard to the file PCALDAT, locks the file and proceeds to call the program "PLASRUN".

c) PLASRUN

This program begins by, once again, asking the operator to ensure that the prism is correctly position on the turntable and then proceeds to take a fast scan to ascertain the position of the zero reference (β_0) as in the PLASCAL program. Once the zero reference has been satisfactorily determined the program then takes a slow, high accuracy scan of the sample over the range of angles that were used when the thick calibration metal film was attached to the prism base. This reading may be repeated as often as required, as in the calibration reading, to ensure the highest accuracy and lowest noise.

Once the reading has been completed, the full scan is compared with the previously taken calibration standard and the result is shown on the screen. These data may now be saved or repeated, to minimise error. The program, if instructed to save the data, asks the operator for information about the metal film being measured, the wavelength of the light used and the required file name. This information is saved and the file is again locked to prevent accidental erasure. Further readings may now be taken if required.

A program exists to display the names of all the appropriate files on the disc and ask for the required file name to be entered. The data are then displayed graphically on the screen and a set of options may be selected: measurements may be taken from the graph; the graph may be extended to show detail around an area of interest; the graph and its primary data may be sent to a printer. If the graphical data are shown in an expanded form, a simple polynomial fit is used to give a smooth curve through the data.

Figure 7.12 shows typical experimental results for a thin metal film on the base of the prism. The minimum created by the generation of a surface plasmon-polaritons is clearly shown. It also shows that these experimental results are in agreement with theory. The theoretical results can be obtained from a relatively straightforward electromagnetic wave calculation.

7.3 Comments on the Mechanical Design and the Computer Interface

The experiment consists of a turntable supporting the prism assembly, and a drive motor connected to a servo-potentiometer, whose output operates the turntable. It also sends a signal to the computer to indicate accurately the turntable position. The motor is a 12v dc motor, plus a gear box, which has been specifically made for the system, with a nominal speed of 1 rev per second. It is energised by an output port of the computer using two output lines only. Figure 7.13 shows a representation of the whole system. The computer automatically controls an angular scan ATR and plots the graph of the reflectance of the sample. Figure 7.14 shows the electronic detail of

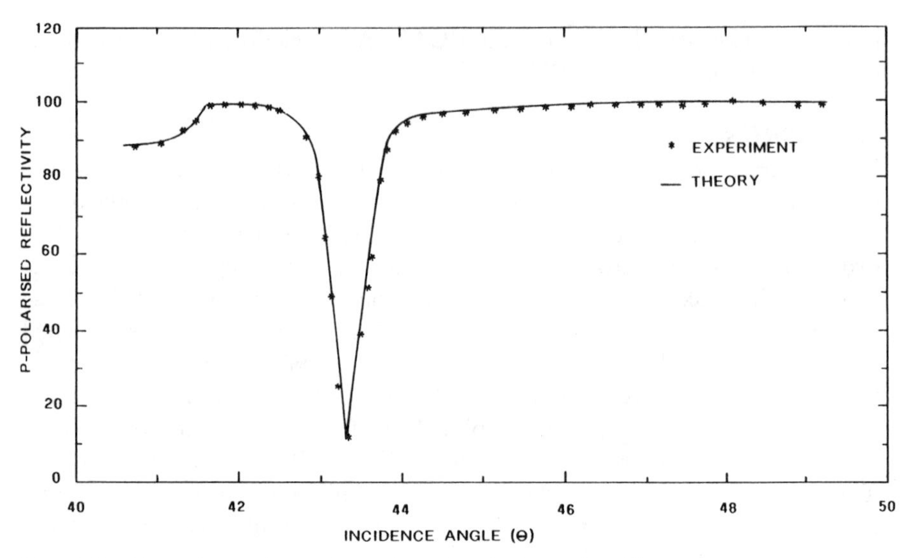

Fig. 7.12. Typical experimental results and their theoretical fit

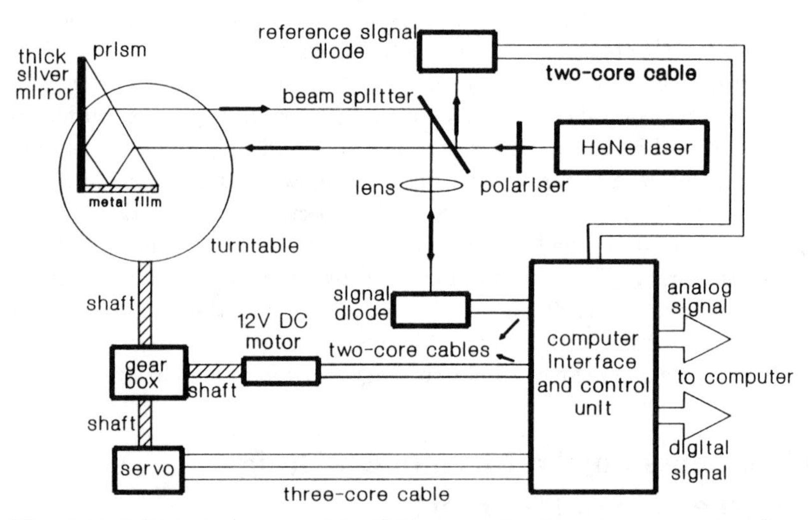

Fig. 7.13. Schematic diagram of the ATR system showing the computer interface, motor, gear box and servo-potentiometer

the control unit. The interface unit uses an H-bridge switch formed by four bipolar power transistors and two small signal MOSFETS to convert the digital signal from the computer into a voltage drive for the motor. A logic "1" on one output line from the computer sets the motor "forward" and the other output engages "reverse". In the event of a short circuit or high current drain of the motor (e.g. when starting up), or of a contention of *commands from the computer* (both forward and reverse lines being high together), the circuit protects itself automatically by limiting the output current to approximately 300mA. This current can be sustained indefinitely under fault conditions and is shown on the interface front panel by the illumination of a "motor fault" L.E.D. Since the motor is highly inductive it is also essential to protect the circuit against over-voltage spikes when the drive is switched off. This protection is provided by a 1A bridge rectifier across the motor output which clamps any voltage excursions to the ground and +12v supply lines.

The computer used provided a reference voltage of approximately 1.8v at its analogue input port. This is the reference for the multi-bit analogue-to-digital converter (ADC) of the computer and corresponds to a full-scale reading. The interface unit buffers this reference (see Fig. 7.14(a) with an operational amplifier to prevent any overloading of the computer and provides this voltage for a servo-potentiometer. This is buffered further by another operational amplifier and is then returned to the computer analogue input port. Since the servo-potentiometer output is always proportional to the computer reference voltage, no compensation needs to be made for variations in this reference (due to temperature, for example). The operational amplifiers that were chosen for buffering have extremely low offset voltages and currents and do not, therefore, contribute to system errors. The primary source of any error is therefore the ADC of the computer. In this case it is specified to 12 bits (equivalent to 0.024). Both the signal and reference channel photodiodes are amplified by operational amplifiers (see Fig. 7.14(b)) to convert their current outputs into a voltage signal. Again these operational amplifiers are chosen because any errors contributed by them to the signals will be well below the resolution limit of the ADC. Both channels also use another sort of operational amplifier to limit the output signal so that it cannot exceed the computer ADC reference voltage and thus cause an overload. These operational amplifiers are used as comparators that normally play no part in the signal conditioning unless an overload is detected. In this case, the output is limited to a reference voltage to prevent damage to the ADC and an LED on the front panel indicates the overload condition. In normal use the photodiode reference channel is taken directly from the laser via a beam splitter and the signal channel is then referred to this. This compensates fully for any fluctuations in the intensity of the laser by simply measuring the ratio of the two signals. Any variation in the computer reference voltage is also eliminated since this becomes a common factor for both signals. The interface unit also provides a fixed voltage reference of 12.4v for use when a reference photodiode is not employed. The interface unit also employs a simple visual indicator of four LEDs to enable the correct setting up of reference channel gain. Figure 7.14(c) shows the H-bridge switch to drive the motor.

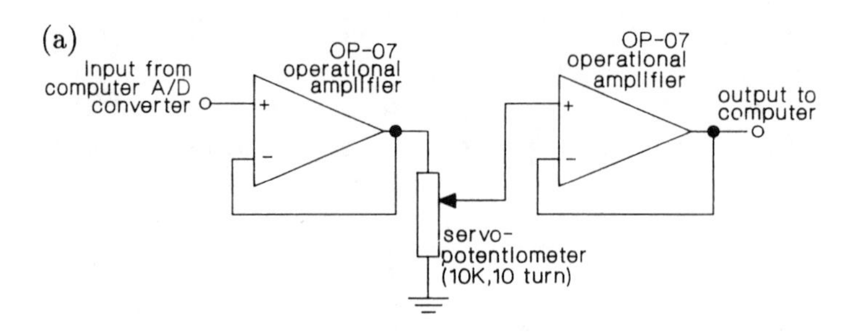

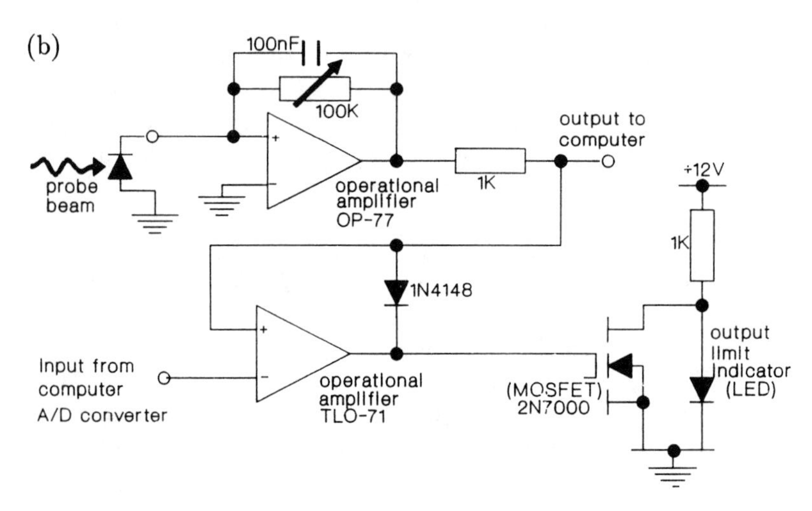

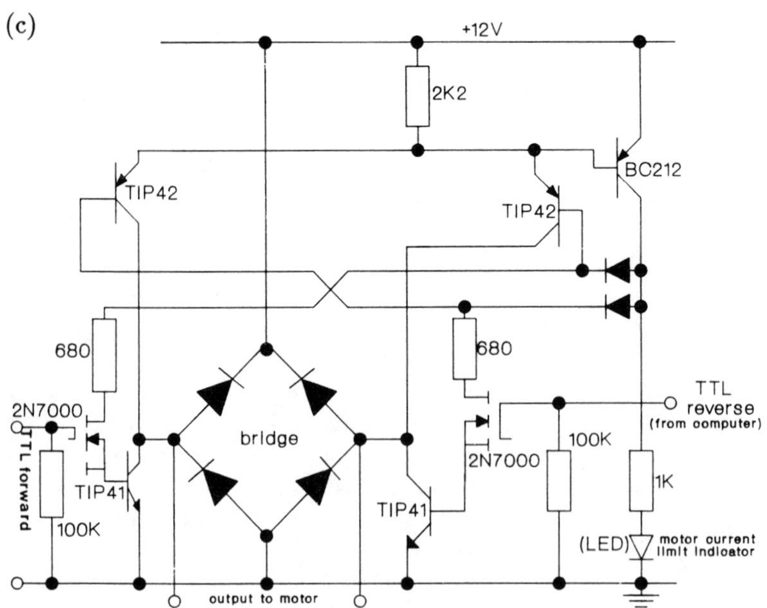

Fig. 7.14. (**a**) Servo-potentiometer interface circuit. (**b**) Electronics showing the amplifier circuit. (**c**) Electronics showing the H-bridge switch to drive the motor

7.4 Conclusion

This chapter deals with a computer-controlled experiment to observe surface plasmons. These surface excitations are generated on the surface of a metal film and this film is on the base of a prism. In the past, many experiments have been performed by manually shining a laser through the prism at various angles of incidence to the base. The reflectivity was monitored until an angle was found at which the reflectivity dropped dramatically due to the generation of surface waves on the metal/air interface of the film. In this case we devise a *computer-controlled* scanning technique that spins the prism around on the turntable. As the prism, directed by the computer, turns, the reflected laser beam is collected and returned through an analogue-to-digital converter to the computer. At this point the information is processed, firstly to set up the initial parameters and, secondly, to collect a complete scan of the reflectivity as a function of angle of incidence. All this is done completely accurately and automatically, and enables the presence of the surface plasmon-polariton to be detected directly with minimum operator intervention. The software is comprehensive and is designed to give good user-friendly graphics. The experiment is a beautiful example of the use of a computer in physics instruction.

References

[7.1] R.H. Ritchie: Surface Sci., **34**, 1 (1973)
[7.2] A. Otto: Z. Phys., **216**, 398, (1968)
[7.3] P. Leurgens and A.F. Turner: J. Opt. Soc. Am. **37**, 983 (1974)
[7.4] T. Turbadar: Proc. Phys. Soc. (London) **73**, 40 (1959)
[7.5] E. Kretschmann: Z. Phys. **241**, 313 (1971)
[7.6] H. Raether: Phys. of Thin Films **9**, 145 (1971)
[7.7] I. Newton: Optiks II, Book 8, 97 (1817)
[7.8] J. Fahrenfort: Spectrochim. Acta., **17**, 698 (1961)
[7.9] N.J. Harrick: Phys. Rev. **125**, 1165 (1962)

Part IV

Optics and Atomic Physics

8. Molecular Spectroscopy of I_2

U. Diemer, H.J. Jodl

8.1 Introduction

The experiment described in this chapter is used to teach the students some specific problems of molecular spectroscopy. This is important, because students do not touch this topic within standard lectures, and the special lectures on this subject are attended only by a few of them.

To start learning molecular physics, it would be a good idea to choose a system which is not too complicated in theory and practice. The hydrogen-like diatomic molecules are such systems. Looking for a suitable molecule which is easily handled, not too reactive (as alkali metals are), possessing transitions in the visible region of the spectrum, and a reasonable vapour pressure at not too high temperatures, iodine is a very good candidate.

The transition $X^1\Sigma_g^+ \longrightarrow B^3\Pi_{0u}^+$ of the I_2 molecule can be treated as a prototype example with which the students learn that molecular spectra, even in the case of the bandspectra of a diatomic molecule, are much more complicated than atomic spectra (line spectra), because of the enormous amount of lines appearing due to different excitations within the molecule.

On the experimental side, the students will become familiar with the classical method of absorption spectroscopy and recognize that this technique can be used to get information about the groundstate and the excited states of the molecule. At the same time it will be clear that this advantage will be balanced by the disadvantage of the large amount of information included in such a spectrum, which makes the analysis pretty complicated.

In addition, the students will learn the modern technique of laser-induced fluorescence (LIF), and its advantages and disadvantages. The analysis of such a spectrum is much easier than in the case of an absorption spectrum, because only one upper level is excited (in most cases well known), and there exist selection rules for allowed transitions. So there are only a few lines in the spectrum, and the advantage of easy interpretation of the spectrum is purchased for the price of less information (only about the groundstate).

In connection with those types of measurements where the computer is used as a tool to record data, the students will use the computer as a powerful aid in analysing data, although they use classical methods such as *Deslandres tables* and *Birge–Sponer plots*[1] to deduce molecular constants

---

[1] See the next section for a short explanation of these methods.

from their spectroscopic data. The students will also model a *Morse potential* containing the studied molecular states by using their calculated molecular constants.

If there is some time left for pleasure at the end of all that work, the students may do some optional exercises applying the computer. Here they can use a lot of useful tools available in a software pool at our physics department to do their investigations. For example, they can use a nonlinear fit routine (*NILFIT*) to fit a theoretical curve to the measured line shape of a fluorescence line, or to fit the shape of an absorption band system.

If they wish, the students can also use a special program to calculate a potential curve based on their measured data using the RKR (Rydberg, Klein, Rees) method and compare this *RKR potential* with the *Morse potential* they calculated before; a typical procedure, well established in spectroscopic research laboratories.

8.2 Some Basic Physics of the Diatomic Molecule

Because the recorded spectra are doppler limited, no rotational structure is resolved. Therefore it is sufficient to describe the diatomic molecule I_2 by the model of an anharmonic oscillator, explaining the obtained spectra. The energy levels of the anharmonic oscillator are given by

$$G(v) = \omega_e(v + 1/2) + \omega_e x_e(v + 1/2)^2 + \cdots \tag{8.1}$$

(see [8.1]). ω_e and $\omega_e x_e$ are called vibrational constants of the molecule. Notice that the sign of $\omega_e x_e$ is negative, therefore this value decreases the energy gap between two vibrational levels, referred to by the vibrational quantum numbers v and $v + 1$. To distinguish the constants and quantum numbers of the groundstate and the excited state, it is usual to refer the groundstate by a double prime ($''$) and the excited state by a single prime ($'$).

The energy of an observed line in the spectrum is due to the difference of the energy of the ground and the excited state:

$$\tilde{\nu}(v'', v') = G(v') - G(v'') =$$

$$[T_0 + \omega_e'(v'+1/2) + \omega_e x_e'(v'+1/2)^2 + \cdots] - [\omega_e''(v''+1/2) + \omega_e x_e''(v''+1/2)^2 + \cdots] \,. \tag{8.2}$$

T_0 is the energy difference between the minima of the two potentials, representing the pure electronic transition. Usually these energies are expressed in cm^{-1}, i.e. the reciprocal of the wavelength in centimeters. As one can see from dimensional considerations, this value is proportional to the energy by the factor of hc (c expressed in cm/s).

Because the anharmonicity constant $\omega_e x_e$ is in general much smaller than ω_e, one can see from (8.2) the following:
If the measured energies are arranged in a matrix form, such that the number of the columns[2] corresponds to v'' and that of the rows to v', the difference of two adjacent data in two rows will be ω_e', and the difference in two columns will be equal to ω_e''. Such a matrix scheme is called a *Deslandres table*. It is a very useful tool to find the correct assignment of the quantum numbers to the measured transition energies. Only in the case of a correct assignment, will the differences be a constant value, or decrease with increasing quantum numbers with respect to the anharmonicity constant $\omega_e x_e$.

Using (8.2) one can calculate the energy gap between two vibrational levels v and $v+1$:

$$\Delta G_{v+1/2} = G(v+1) - G(v)$$

$$= \omega_e(v+3/2) + \omega_e x_e(v+3/2)^2 - \omega_e(v+1/2) - \omega_e x_e(v+1/2)^2$$

$$= \omega_e + 2(v+1)\omega_e x_e \; . \tag{8.3}$$

Thus the plot of $\Delta G_{v+1/2}$ versus $v+1/2$ leads to a linear dependence of $\Delta G_{v+1/2}$ with v. The intersection with the ordinate yields ω_e, the slope of the curve is due to $2\omega_e x_e$. The intersection with the abscissa gives the highest still-bound vibrational level v_{max}. Therefore one can calculate the dissociation energy (D_e) by calculating the $G(v)$ of v_{max}. The energy of this level is the sum of all $\Delta G_{v+1/2}$. Because $\Delta G_{v+1/2}$ decreases by $2\omega_e x_e$ every step, there will be $\frac{\omega_e}{2\omega_e x_e}$ vibrational levels. This sum of all $\Delta G_{v+1/2}$ is the average value of G_v (which is $\omega_e/2$) times the number of vibrational levels:

$$D_e = \frac{\omega_e}{2\omega_e x_e} \cdot \frac{\omega_e}{2} = \frac{\omega_e^2}{4\omega_e x_e} \; . \tag{8.4}$$

This method of determining the molecular constants basing on the linear dependence of $\Delta G_{v+1/2}$ on v is called the *Birge–Sponer extrapolation.*

There are several methods modelling an analytical form of the molecular potentials. The most used form is the *Morse potential*:

$$U(\xi) = D_e(1 - \exp^{-\beta\xi})^2 \; , \tag{8.5}$$

where $\xi = R - R_e$ (R_e : equlibrium distance of the two nuclei) and $\beta = \sqrt{\frac{c\mu}{D_e\hbar}}\,\omega_e$ (μ : reduced mass of the molecule)[3].

This form of a potential curve has some disadvantages, i.e. the slope for large distances of the two nuclei is too weak. But within the region where the values of our measurements are lying (all nearby the equilibrium distance R_e), the *Morse potential* is a sufficiently good approximation.

---

[2] Start counting at 0!

[3] Pay attention to the dimensions of the constants! In this form of the equation one has to express the molecular constants in cm^{-1}, the molecular dimensions (R and R_e) in cm and the speed of light in cm/s!

For readers who want a more precise description of the spectra of diatomic molecules, it will be given here:
To include rotational structure in the model, one has to use the concept of a vibrating rotator by adding a further term $F(v, J)$ to the vibrational energy given by (8.1):

$$F(v, J) = B_v(J(J+1)) + D_v(J(J+1))^2 + \cdots ,$$

where J is the rotational quantum number and B_v and D_v are rotational constants. D_v, in common with $\omega_e x_e$, has a negative sign. The subscript v indicates that these constants are slightly different for each vibrational state, due to a slightly different moment of inertia. The whole energy of a level will now be given by:

$$T(v, J) = T_0 + G(v) + F(v, J) .$$

Figures 8.1 and 8.2 show two potentials schematically and also some energy levels. The arrows indicate possible transitions and dashed lines connect these transitions to the resulting structures in the spectrum. Notice in

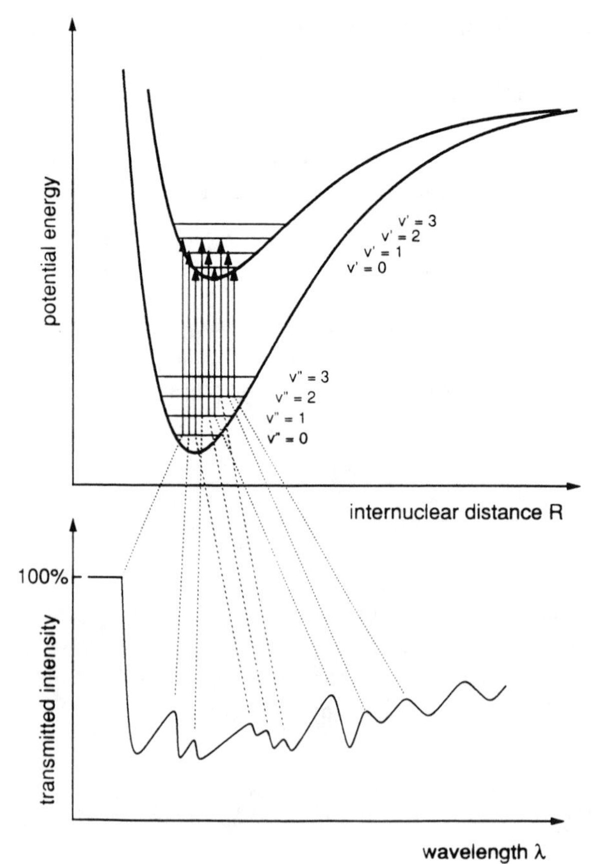

Fig. 8.1. Energy levels and resulting absorption spectrum. The shown transitions in the upper part of the figure yield the absorption spectrum shown below the termscheme

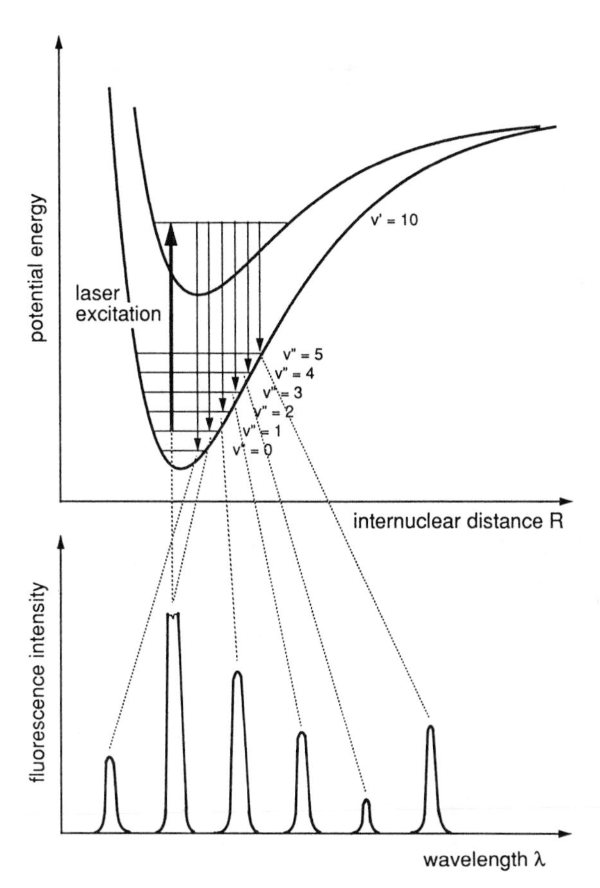

Fig. 8.2. Energy levels for LIF. On the left side of the scheme a laser excitation from one lower level to one upper level is shown. Some possible transitions resulting from the upper level are also shown. Below the energy levels the resulting LIF spectrum is plotted

Fig. 8.1 that only a few of all possible transitions are plotted, so the absorption spectrum will be very dense. In Fig. 8.2, a (nearly) realistic amount of transitions is shown, which proves that LIF spectra include less information but are easy to analyse. It is also indicated that stray light of the excitation laser enhances the fluorescence intensity of the excited transition.

At the end of this short introduction to molecular spectroscopy we will give some books in which the students may find further information about this subject. The book of *Herzberg* [8.1] was already mentioned. It is a complete reference also for advanced spectroscopists, but the amount of information may be too much for a beginner. The books of *Steinfeld* [8.2] and *King* [8.3] are suitable as textbooks and contain basic introductions to molecular spectroscopy. In these books the students may find any information necessary to understand our experimental description. For German readers, the book of *Hellwege* [8.4] is recommended for an overview.

8.3 Experimental Setup

The students may use different setups:

- the classical version (without computer)
- an off–line computerised version (only data analysis by computer)
- an on–line computerised version (the data are taken with the assistance of a computer)

8.3.1 The Classical Arrangement

There are two (slightly) different setups for the experiment, depending on whether one uses absorption spectroscopy or LIF. Both setups are shown in Fig. 8.3.

a) Absorption Spectroscopy The I_2 is contained in a stainless steel tube of about 10 cm diameter and 1 m length sealed by two glass windows. To increase the vapour pressure of the I_2, the tube can be warmed up by a heating coil.

An intense halogen lamp is placed at one end of the tube, and lenses are arranged at the other end, in front of the entrance slit of the monochromator. Here the students have to understand the meaning of the focal length of the monochromator and how to focus the light onto the entrance slit to achieve maximum resolution of the instrument. They learn that they will achieve this if they focus the light with the aperture ratio of the monochro-

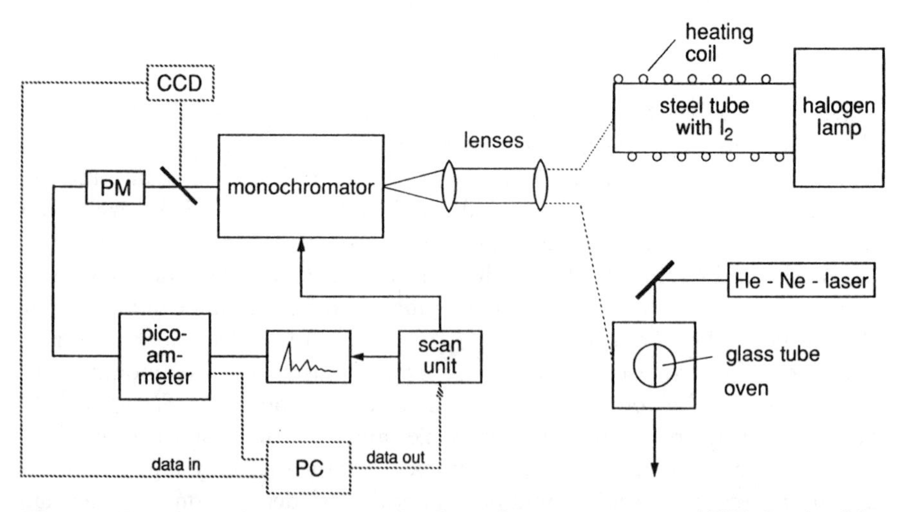

Fig. 8.3. Experimental arrangements for absorption spectroscopy and LIF. On the right side of the figure there are two different setups shown: The upper one is used to do absorption spectrosocpy, the lower one to do LIF. The left side is identical for both setups. The dashed lines indicate that these components are used by the computerised versions of the setup (compare Sect. 8.3.2)

mator to the entrance slit. This is the ratio of the beam diameter at the lens (in the case of imaging a fluorescence spot, the diameter of the lens (why?)) and its focal length. In this case the grating will be totally illuminated.
The monochromator unit is a 350 mm single–pass Czerny–Turner monochromator. The grating, with 1180 lines per mm and a blaze wavelength of 250 nm, allows a resolution of about 0.1 nm. A control unit scans the monochromator and also drives the strip chart recorder. To detect the light, a photomultiplier is used. The signals of the multiplier are amplified by a pico–ammeter and recorded by a strip chart recorder.

To calibrate the whole system, the students replace the halogen lamp by a mercury lamp and record part of the spectrum. (Of course, the I_2 cell is not heated during calibration.) Then they compare the wavelength readings of the monochromator scale to the known wavelengths[4] of the Hg spectrum.

b) Laser–Induced Fluorescence To record a LIF spectrum, the setup for absorption spectroscopy is slightly modified (compare Fig. 8.3). The steel tube is replaced by a glass tube of about 5 cm in diameter and 10 cm in length. This tube is placed inside an oven to heat it up to a desired temperature. The halogen lamp is exchanged by an HeNe laser (1 mW). The laser is placed above the cell in such a way that the resulting fluorescence line is parallel to the entrance slit of the monochromator. To enhance the intensity of the emitted fluorescence light, the laser beam is focussed inside the cell by a lens of 10 cm focal length.

The difficulty of this setup is that one cannot optimize the image of the fluorescence line by just watching the signal amplitude when the monochromator is tuned to the wavelength of the HeNe laser line. In such a case, one will only optimize the image of some stray light or some reflexes, therefore it is necessary to consider where a fluorescence transition beyond the laser wavelength could be and one firstly needs the results of absorption spectroscopy. Even with the knowledge of the wavelength of a real fluorescence line, it is not an easy task to optimize the signal by watching the output of the pico–ammeter and shifting the lenses and the laser. This disadvantage should be avoided by replacing the detector system by a CCD array to achieve instant recording of the spectrum (see also Sect. 8.3.2b)!

8.3.2 Extensions: Online Use of a Computer

In this section we will describe some extensions of the experiment which are indicated in Fig. 8.3 by dashed lines. These extensions are different ways of including a computer to record the data. If necessary, the disadvantages of the setups will be mentioned.

[4] The students have to go to the library and look up these wavelengths in some specific handbooks.

a) Hardware Requirements All our programs run on IBM AT/XT or compatible computers. We use MS–DOS as an operating system and TURBO PASCAL 5.5 as the programming language. For doing the calculations described in the next section, it would be advantageous to use a fast system, e.g. a mathematical coprocessor, but it is not necessary. To display the graphics EGA, CGA, VGA or HERCULES hardware may be used. The necessary drivers for the hardware are included in our software package.

Any AD/DA device may be used to acquire the analogue data, and any parallel I/O card to read external digital signals, or to output digital signals. To adapt the software to one's special hardware (e.g. the addresses of the interfaces), there are only minor changes in some subroutines[5]. To read data from the CCD device, an additional self–made control unit was developed[6].

b) Recording Data If one replaces the exit slit by a **CCD** array (**C**harge **C**oupled **D**evice), it is possible to have a real–time display of the spectrum. The resulting system may then be called an **OMA** (**O**ptical **M**ultichannel **A**nalyser). Such OMA systems are commercially available, but are quite expensive ($\approx$ 60000 DM), whereas our system (as described in [8.5]) costs only a few hundred DM. It is a very good alternative for students laboratories; in the beginners laboratories a typical setup costs about 5000 – 8000 DM, in the advanced laboratories 10000 – 30000 DM. Therefore, our technical solution is well suited for such an application because this setup, including the computer, costs less than 10000 DM.

The only disadvantage of this CCD device is its sensitive width of only 1 inch. The reciprocal dispersion of the monochromator is 2 nm per mm, so the array may register only part of the spectrum (50 nm in first order). The students have to think about this problem, and discuss their solutions. They may wish to use another array (not available), and (or) other orders of the grating (not possible in our case). They also have to understand the limits of the resolution of their optical setup, e.g. to choose a suitable combination of spectral resolution and linear dimension of a single diode of the array.

Due to the given equipment, the students should realise that they can inspect only limited spectral ranges of interest using a CCD device.

The easiest way to record the spectrum, therefore, is to tune the monochromator to a wavelength which is about in the middle of the spectral range one intends to detect. Using a mirror in front of the exit slit (see Fig. 8.3), one can choose if the CCD array or the photomultiplier is active to record data. If one uses a spectrograph, the photographic plate has to be replaced only by the array[1]. The rest of the work will be done by our program *LAmDA* (**L**icht **A**nalyse **m**it **D**ioden **A**rray, [8.6]). This program

[5] Please see the *READ ME* files on our discs for additional information.

[6] If you use the same array as we do (TC100 by Texas Instruments), please contact us, and we may send you the drawings to build up this unit.

[7] Of course, if one chooses the 'CCD setup', the calibration has to be done after placing the array, and the grating and mirror must not be scanned or moved!

controls the array, produces all necessary pulses, and reads the data from the array. The read data may be displayed in real time on the screen. This aspect of the program makes it a very powerful tool for the spectroscopist. If he is adjusting the experimental setup, e.g. optimizing the image of the fluorescence to the entrance slit, he can always control his success by looking at the screen, and doesn't have to scan the monochromator to do this. This advantage will be estimated by everyone who did the experiment the previous way, shifting the lenses some mm around, scanning the monochromator, ...
Of course, one saves not only time in adjusting the setup. Because all channels of the CCD device are read out several times per second, one needs only a fraction of the time that one would need in the classical way to record the same spectral range.

In addition, the programm *LAmDA* allows one to process the data, as will be described in the next section.

Considering all aspects previously mentioned, one will realise that the CCD device is a powerful tool for recording spectra. The disadvantage of the limited spectral range is balanced by the rapidity of the recording.

c) Computer Control of the Monochromator System For some applications (e.g. for recording a spectrum over a wide range) it is better to use a different setup. In this version, the computer will control the scanning of the monochromator and also read the signals of the photomultiplier.

The output of the pico–ammeter is stored by the computer and also recorded by a strip chart recorder. The advantages are three–fold:

1. To achieve an overview of the spectrum and the relative intensities. Instead of looking at the screen for parts of the spectrum, the strip chart can be used.
2. To find special structures in detailed spectra, one should use the stored data by the computer.
3. To reduce noise, it is possible to measure several times at the same wavelength and store only the average value of the signal.

To scan the monochromator, the computer is connected to the scan control unit. The computer is now in control, producing pulses to step to the next wavelength. The step rates have to be selected, as usual, on the control unit by hand. At the beginning of each run, one has to indicate to the program the initial wavelength of the monochromator, the step rate of the control unit, and the spectral range to be scanned. The computer will start the scan, and one can do more important things than watching the machines.

Of course, it is necessary that the students are aware of parameters such as scan rate, spectral range, number of datapoints or number of measurements at one wavelength, to find a compromise between a very good spectrum and the time needed to record it.

8.4 Measurements

8.4.1 Calibration of the System

To control the accuracy of the wavelength reading of the monochromator, the students have to calibrate this scale before it can be used to analyse the recorded spectra. The calibration is done by recording a mercury spectrum. Therefore, the lamp at the end of the stainless steel tube (compare Fig. 8.3) is replaced by a mercury lamp. Of course, the tube is held at room temperature to avoid absorption, which would contribute background to the recorded spectra and complicate the interpretation. The wavelength reading is then calibrated by comparing the so-determined Hg lines to the tabulated wavelength published in handbooks. At this point, the students are confronted with the fact that some books publish the wavelength in vacuum and others in air. They also have to understand that wave numbers are given always in vacuum (why?).

During this calibration procedure, the students will get a feeling for the accuracy of the monochromator, i.e. how good is the scale of the instrument in displaying absolute wavelength, how good can a single wavelength be reproduced, and which of these introduces the major error. (How can the latter be tested?)

The students also have to recognize that the same accuracy in the wavelength scale of about 0.1 nm will give very different accuracies in the energy scale, from 1.3 cm^{-1} to 3.3 cm^{-1} depending on the absolute wavelength within the I_2 spectrum.

In the computerised versions, a fit is done to achieve a correct assignment between reading, true wavelength and the corresponding parameter stored by the computer (number of steps, channel number of the CCD device ...). If the CCD array and the program *LAmDA* are used, the calibration is done selecting the desired line with the cursor and typing in the accurate wavelength taken from the literature. Of course, this calibration will be lost if the grating or the CCD array are moved in any way after this step.

While in the computerised versions the fit of the internal values to the wavelength is absolutely necessary, in the classical version the students can decide if it has to be done or not. The scale of the monochromator is accurate enough for the necessary precision, but a fit to the line positions may give a feeling for the measuring errors. In addition, the students should consider this effect on the potential (see Sect. 8.5.1e).

8.4.2 Recording the Absorption Spectra

After this calibration procedure the students start recording the I_2 absorption spectra. They have to study the influence of slit width on the spectrum, and find a compromise between high intensity (wide slits) and high resolution (narrow slits).

The students then try to record spectra at different temperatures of the I_2 gas, so they recognize that the form of the spectrum (the shape of the band systems) depends on the temperature of the gas. Rising temperature increases the absorption strength and the number of details in the spectrum. The first contribution is due to the fact that as more molecules are present in the gas phase, more molecules absorb light. The second contribution yields from the population of energetically higher levels in the molecular ground-state at higher temperatures, and therefore more states can absorb and will rise up in the spectrum. This temperature dependence of the population is given by a Boltzmann distribution:

$$N(T) \propto \exp^{-\frac{E}{kT}} ,$$

where N is the number of molecules at the level, E is the energy of the level, k is the Boltzmann constant and T is the temperature of the gas.

Here the students also have to find a compromise, because a lot of structure means a lot of information, but it also makes the interpretation of the spectrum more complicated.

All these studies are executed within a small spectral region containing only a few intense band systems. In the CCD version the students can watch in real time how the structures (band heads of higher v'' bands) grow with respect to other bands, when the temperature is increased. Two spectra at different temperatures are shown in Fig. 8.4.

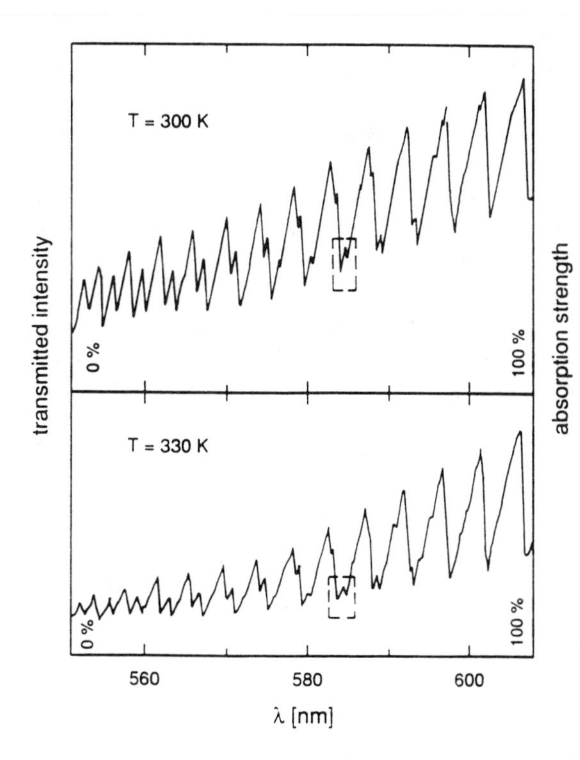

Fig. 8.4. Two absorption spectra at different temperatures. Both spectra are recorded at the same sensitivity and no offset of the intensity or absorption scale. Note that the small change in temperature (only 30 K) yields an increase in structure and an increase of the total absorption strength (compare dashed boxes)

After these preliminary studies, the students have to choose a specific temperature and a suitable slit width and then have to take a spectrum over the whole range where significant structure occurs. This will be the range from about 500 nm to 650 nm. In this part of the spectrum, band heads with quantum numbers $v'' = 0$ to $v'' = 6$ (max.) and $v' \sim 2$ to $v' \sim 50$ can be observed.

8.4.3 Recording the Fluorescence Spectra

In the groundstate it is possible to detect vibrational levels only up to $v'' \approx 6$ using the technique of absorption spectroscopy. All higher levels are barely populated at the reachable temperatures, and so cannot absorb. Therefore, another technique has to be used to determine some higher levels of the groundstate, which may be used to model the potential curve with a reasonable precision. This technique is **L**aser **I**nduced **F**luorescence (LIF).

The experimental setup for doing LIF is shown in Fig. 8.3. As mentioned already in Sect. 8.3.1b, it is rather difficult to image the fluorescence line to the entrance slit of the monochromator. But if this has been done correctly, it will be very easy to record the spectra. The use of the CCD device will, in this case, be even a bigger problem than in the case of absorption spectroscopy, because the spectrum is extended over a range of about 250 nm, from $\approx$ 600 nm to $\approx$ 850 nm. This range is much larger than the spectral range that the CCD can monitor at one time.

The CCD will be a nice system for recording some special features of the spectrum. The temperature dependence of the spectrum can be analysed directly, but increasing temperature will only increase the signal up to a certain limit depending on the special setup. In addition, the influence of the slit width to the signal form is easily studied: i.e. narrow slits yield low intensity, but one can detect the very narrow splitting of the lines into two rotational components. Remember that the great advantage of the CCD setup is that one can watch the influence of its actions in real time on the screen! The line shapes may also be stored for a fit procedure later on (see Sect. 8.5.2). Figure 8.5 shows such a section of a LIF spectrum.

Doing fluorescence spectroscopy, one can determine the levels of the groundstate up to $v'' \approx 25$. Although there are some lines missing (because of very small *Frank–Condon factors*[8]), there are enough levels to describe the groundstate with reasonable precision.

Remember that there is only one upper level excited, so by doing LIF one finds only information about the groundstate!

Note that if the HeNe laser (fixed wavelength at 632.8 nm) would be replaced by a tuneable laser (e.g. a dye laser), then this setup is identical to basic setups for doing LIF in every spectroscopic research laboratory.

---

[8] See Sect. 8.5.2d for a short explanation.

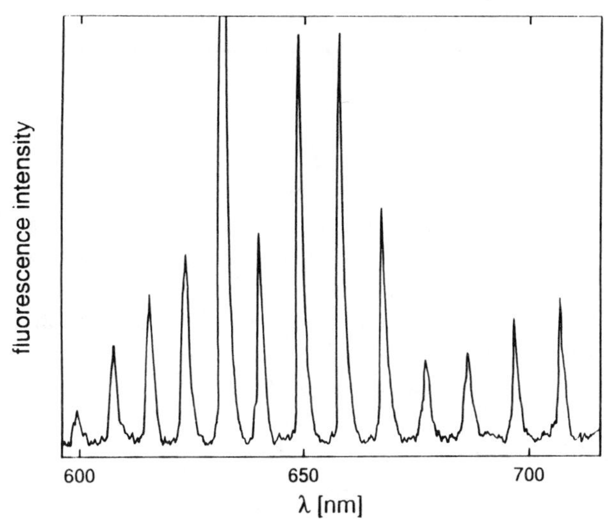

Fig. 8.5. Section of a LIF spectrum of I_2. Note that the huge line (632.8 nm, out of scale) is the line excited by the HeNe laser directly

8.4.4 Some Additional Features of the Program *LAmDA*

In this section, some features of the program *LAmDA* shall be described. They are not necessarily used to do the measurements in this experiment, but they are rather useful to record and analyse spectra in general.

LAmDA was not especially built to record spectra. It is a universal tool to study all kinds of intensity distributions. To do this at rest, the recorded distributions can be stored. To work later with them, *LAmDA* can also read in data again. These data may not only be analysed, but can also be manipulated in various ways. The most important features of the program are:

- **Scaling:** The two axes may not only be scaled in relative intensities (between 0 and 1) and channel numbers (depends on the number of diodes of the used CCD array), but can be scaled in every desired way. It is, for example, no problem to connect the x axis with a wavelength scale.
- **Inspecting:** The recorded data can be scanned by a cursor. For each data point, the coordinates of this point are displayed depending on the actual scaling of the axis.
- **Marking:** By pressing the enter key, the data of the selected point can be stored for further use. One can easily analyse a spectrum by scanning the cursor to the middle of the lines, and store their wavelength position by one keystroke.
- **Zooming:** To inspect special regions of the whole spectrum in detail, it is possible to zoom into a selected range.

- **Printing, Plotting:** Of course it is possible to print or plot the spectrum, (or a zoomed range of it), after doing some manipulations.
- **Data Manipulations:** The following options may be used to manipulate the recorded data:
- **Filtering:** If the data are too noisy, there are two ways to smooth the spectra. The first is to use a digital low pass filter to subtract the higher frequencies of the spectrum. The original data and the filtered data are displayed on the screen, and one can see that this kind of filter broadens and shifts(!) the lines. If one is not satisfied with the effect, one can reject the filtered data, and can use the original one's for further manipulations (e.g. filtering with a filter of another strength).
- **Smoothing:** Secondly, the data points are averaged by their neighbouring data points. To avoid shifting the data, the averaging is done by weighting the points by a Pascal triangle. Here also, the original and the smoothed data are shown, so one can study the effect of this procedure.
- **Combining:** This allows the user to combine different measurements by simple mathematical functions such as to add, subtract, multiply or divide two curves. It is also possible to do this not by a second measured dataset, but, for example, by a constant value. Therefore, it is possible to subtract some offset, or to normalize the spectrum by multiplying it by the sensitivity curve of the detector system, etc. .
- **Fourier Transformation:** Besides the possibility of storing data for use in an additional Fourier–transform program (e.g. [8.7]), it is possible to do this transformation within the program. Therefore one can, for instance, directly inspect the effect of the digital low pass filter.

8.5 Analysis of the Spectra Using the Program JOD

8.5.1 Analysis of Absorption Spectra

The first step of the analysis will be to obtain the wavelength of each band system. This work will be quite different if the students work with the computerised experiment or with the classical version. In the classical version, the students have to determine the wavelength by using ruler and calculator. In the on–line version, the spectrum (or part of it) is displayed on the screen, then the cursor is moved to the structure or line of interest. Simultaneously the computer displays the adjacent wavelength, which is calculated by using the stored calibration data. These wavelengths may also be stored for further use.

The next step will be the same for all versions. In this part of the work the students are assisted by the program *JOD* [8.8]. This is a very powerful tool to analyse the band spectrum of a diatomic molecule. Even though the program is named *JOD*, there are no assumptions made about the molecule,

Table 8.1. The entries of the *Jod* menu. Note that some of the entries lead to submenus, e.g. error handling, where the different errors are handled in a different way

Data	Assignment	Constants	Potential	Quit
help	help	help	help	quit program
edit data	compare progressions	show constants	show constants	
test table	Deslandres table	Birge–Sponer extrapolation	grafics	
print table		error analysis	error analysis	
load data				
store data				

so the program may be used for every diatomic molecule[9]. The options of the program are shown in Table 8.1. They allow the user, which may not only be a student in a laboratory course but also an advanced student in some research work, to analyse spectra of diatomic molecules. The only restriction is that no rotational structure analysis is included here. The major features of the program are:

- data manipulation
- finding the correct assignment of quantum numbers
- calculating molecular constants
- calculating a *Morse potential* by using these constants
- error analysis

a) Input of the Data The students start their analysis by typing in the measured data at the menu point edit data. It is also possible to read in data from a file. Because it is not easy to figure out the correct assignment of the vibrational quantum numbers without any information, there is some help for the students: they are told that the band head at wavelength 548.33 nm corresponds to the transition $v'' = 0 \longrightarrow v' = 24$. This information and the knowledge that the band system with $v'' = 0$ is the strongest system (because most of the molecules are in this energy level), allows the students to start their analysis. They only have to count up and down the strongest band heads starting at this given wavelength. For the other vibrational levels, a knowledge of the correct quantum number v' is not necessary at this point. They only have to find out the correct v'', which is not difficult because the strength of the band system decreases with increasing v'' (because population decreases). Therefore one can start at any desired point and one has to arrange the bands by increasing energy only. The correct assignment will be found later on at the menu point *Deslandres table*.

If the data of one progression (a progression is a series of band heads with the same v'') are completely typed in, they may be controlled afterwards. Therefore, they can be displayed as a test table (or printed out).

[9] If the program is used to analyse molecular systems other than I_2, one has to change the reduced mass of the molecule, which is necessary to calculate a *Morse potential.* The reduced mass of I_2 is included in the program as a default value. All other activities can be done without any change of the program for every diatomic molecule.

This table contains v', the measured wavelength, the energy (in cm^{-1}) and the energy difference between two successive vibrational levels ($\Delta\tilde{\nu}$). An abnormal variation of this difference may indicate an error in the analysis (or maybe only an error in typing).

b) Assignment and Deslandres Table After typing in all data and correcting errors, the next menu entry may be chosen, building up the *Deslandres table* of the band system. The columns already exist, which are the different progressions (series of the same v'') just typed in. Now one has to find the correct adjustment of the rows. This is the correct assignment of v'. Remember that if this assignment will be correct, the energy differences within one row would nearly be constant. To test this, the menu point *compare progressions* is very useful. Here, one can compare two progressions in the following way (compare Table 8.2):
On the left side of the screen the first progression is displayed, as it would be displayed in the test table after typing in data. Only the display of the wavelength is omitted. The assignments of this progression cannot be changed. The point of interest is now the column $\Delta\tilde{\nu}$, describing the energy differences. On the right side of the screen are firstly the energy difference and secondly the wave numbers of the respective transition displayed. Therefore one can directly compare the differences of the two progressions. To find the correct assignment, the progression on the right side can be shifted in v' up or down. One uses the cursor keys (up or down) to scroll up or down the correct column, and one compares the differences $\Delta\tilde{\nu}$. In an ideal case, all neighbouring differences should be the same, but this case (of course) never occurs, so the total sum of the difference of the differences ($\sum(\Delta\tilde{\nu}_{(left\ side)} - \Delta\tilde{\nu}_{(right\ side)})$ is displayed. A minimum of this number indicates a correct assignment. In our example shown in Table 8.2, it decreases from 103.6 cm^{-1} to 0.7 cm^{-1}.

Because the $v'' = 0$ progression is fixed (one knows the correct assignment of the $0 \longrightarrow 24$ progression!), it is recommended to start with this progression on the left side and then adjust all the others. If one is sure that $v' = 1$ has the correct assignment, one also can use this as a fixed progression (and so on).

When all progressions are correctly assigned, the complete *Deslandres table* is built up. One now can display the table on the screen (or part of it, because it is rather big) or print it out. It may be stored also in a file to calculate the molecular constants in the next step.

c) Calculating the Molecular Constants To determine the molecular constants from the data stored in the *Deslandres table*, a *Birge–Sponer extrapolation* is set up analytically (refer to Sect. 8.2). There will be no graph on the paper or the screen; only the resulting constants are displayed. To do this calculation, the stored differences are again read in. The mean value for each $\Delta G_{v+1/2}$ is then calculated and displayed on the screen. These values may also be printed out.

Table 8.2. This table is displayed on the screen if the menu entry *compare progressions* is selected. The upper part shows the table before the right side is correctly assigned. Pay attention to the large differences in the column $\Delta\tilde{\nu}$. The lower part shows the table after scrolling down the right side. Now the assignment is correct, and the differences are much smaller in this case

	CRSR UP – scroll up CRSR DOWN – scroll down		< ESC > – store table < SPACE > – next data	
	v'' =	0	v'' =	1
v'	$\tilde{\nu}$	$\Delta\tilde{\nu}$	$\Delta\tilde{\nu}$	$\tilde{\nu}$
15	17416.4	100.4	93.2	17587.1
16	17514.7	98.2	92.3	17679.4
17	17611.5	96.6	87.0	17766.4
18	17706.9	95.4	90.1	17856.5
19	17799.6	92.7	83.0	17939.5
20	17892.0	92.4	83.4	18022.9
21	17980.4	88.5	82.6	18105.4
22	18069.5	89.0	77.7	18183.1
23	18153.1	83.6	78.7	18261.8
24	18237.2	84.1	75.7	18337.5
25	18317.0	79.8		
26	18399.3	82.2		
27	18477.1	77.9		
28	18550.8	73.7		
29	18623.4	72.6		
30	18697.2	73.8		
total sum of differences: 103.6 [cm^{-1}]				

The table when the assignment is correct:

	CRSR UP – scroll up CRSR DOWN – scroll down		< ESC > – store table < SPACE > – next data	
	v'' =	0	v'' =	1
v'	$\tilde{\nu}$	$\Delta\tilde{\nu}$	$\Delta\tilde{\nu}$	$\tilde{\nu}$
15	17416.4	100.4	100.6	17202.5
16	17514.7	98.2	100.3	17302.8
17	17611.5	96.6	94.8	17397.7
18	17706.9	95.4	96.2	17493.8
19	17799.6	92.7	93.2	17587.1
20	17892.0	92.4	92.3	17679.4
21	17980.4	88.5	87.0	17766.4
22	18069.5	89.0	90.1	17856.5
23	18153.1	83.6	83.0	17939.5
24	18237.2	84.1	83.4	18022.9
25	18317.0	79.8	82.6	18105.4
26	18399.3	82.2	77.7	18183.1
27	18477.1	77.9	78.7	18261.8
28	18550.8	73.7	75.7	18337.5
29	18623.4	72.6		
30	18697.2	73.8		
total sum of differences: 0.7 [cm^{-1}]				

Table 8.3. The results of the Birge–Spooner extrapolation. The values given in parenthesis are taken from [8.1] and are not displayed on the screen. The students have to look these up in the literature. Notice the big difference in the D_e values. See text for explanation

groundstate		excited state
−1.419	slope	−1.970
214.8	intersection	129.8
$\omega_e'' = 214.8$ (214.57)		$\omega_e' = 129.8$ (128.0)
$\omega_e x_e'' = 0.710$ (0.6123)		$\omega_e x_e' = 0.985$ (0.834)
$D_e'' = 16257.30$ (12435)		$D_e' = 4273.45$ (3206.6)
	Transition energy (T_0)	
15745.2		(15641.67)

Using the $\overline{\Delta G_{v+1/2}}$ values, the molecular constants ω_e and $\omega_e x_e$ for every electronic state are evaluated by a linear regression procedure (Table 8.3). To determine the value of T_0, which is the energy gap between the minima of the two involved potentials, one could use the energy of any measured band head, subtract (according to (8.1)) the vibrational energy of the upper state and add the energy of the lower state. However, there is no information about the quality of the measured lines used in the program, so it is better to type in the wave number of a measured line, for which one is sure that it results from a reliable measurement. The calculation of T_0 is then done in the described way.

If the students compare their results with the values given in the literature (e.g. in [8.1], given in parenthesis in Table 8.3), they will find quite good agreement. However, there is one exception: The value of the dissociation energy given in the literature is much smaller than their value. If they calculate the D_e values by using (8.4) and the molecular constants given in the literature, the result will be comparable to their value of D_e. The reason for this behaviour is that the *Birge–Spooner extrapolation* is a linear extrapolation, which is quite good for lower vibrational quantum numbers. At higher quantum numbers however, the influence of the higher vibrational constants becomes bigger and bigger and yields a stronger (rather than linear) decrease of the vibrational energy gap. Because it is not possible to measure these higher constants directly within the accuracy of the experiment, there is no chance to determine the correct value of D_e.

d) Error Analysis The next point to figure out is the quality of analysis. This involves an error consideration which is assisted by the program. There are two different types of errors which may be considered: errors occurring from uncertainties of the measurement, and those which appear from incorrect assignment of the band heads.

To test the magnitude of errors according to incorrect assignment, one can vary the quantum numbers v' and shift them by a desired integer i. The calculation for the molecular constants will then be done again for values $v'_{\text{new}} = v'_{\text{old}} \pm i$. The results are displayed again, together with the molecular constants resulting from unshifted values.

To determine the error arising from the uncertainty of the measurement, one has to type in the precision of the wavelength determination, which is the reading error of the monochromator. The students should be able to estimate a correct value for this error after doing the calibration procedure. The influence of this error on the molecular constants is now calculated. Again, the primary constants and the possible maximum deviations are displayed.
The students can therefore assume different magnitudes of the errors, and will get a feeling for how the different errors influence their results. Of course, this is not the usual way to execute an error analysis, where one has to use the law of error propagation to calculate the magnitude of the errors. This allows the students to estimate the quality of their measurements, because they can see immediately the influence of their changes. It is also very easy to compare the standard deviation of their constants with the calculated errors, and to realize which errors possess the largest influence on their measurments.

e) Calculation of the Morse Potential This is the last task for the students. In this section, they can calculate a morse potential, as given in (8.5). In general, they will use the molecular constants calculated in the previous section. The resulting potentials are then displayed on the screen. If desired, the vibrational levels may also be plotted (Fig. 8.6).

The advantage of this method is that the students may vary the constants and see immediately how the potential is changed. If they vary only one constant, they can watch the special contribution of this constant to the form of the potential. They also may change the step width of the calculation to determine some special regions of the potential in detail, e.g. around the equilibrium distance of the nuclei.

In a last step, the students choose an uncertainty of their constants (considering the errors estimated in Sect. 8.5.1d), and the upper and lower limit of the potential curves (corresponding to the error limits of the constants) may then be plotted within the unaltered potential curve on the screen. Thus they can easily see where there is a maximum influence of these errors. As one can see in Fig. 8.6, the devation will increase with increasing potential energy, and may be neglected near the potential minimum.

There is also a possibility to print or plot the results of the calculations.

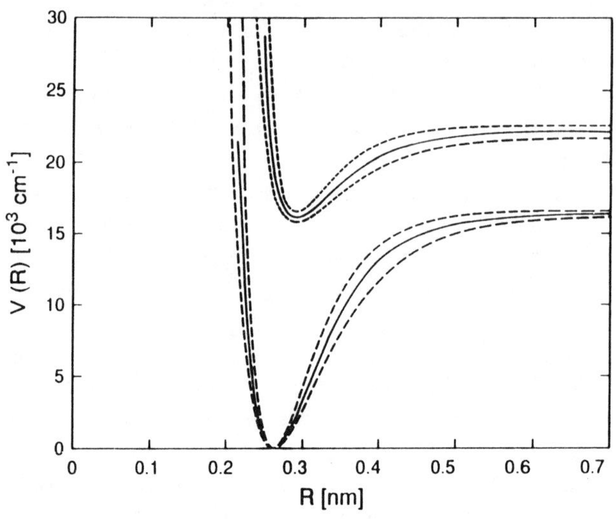

Fig. 8.6. The morse potentials calculated from the constants given in Table 8.3. The dashed lines indicate the upper and lower limits according to the errors of the constants

8.5.2 Some Optional Exercises

In this section, we will give some additional exercises, which can be done by the students if there is enough time. These exercises may be done voluntarily, so there is some kind of "playground" where the students are not forced to do something, but learn by following their own instincts.

a) Calculating an RKR Potential by the Program RKR In research work, one is interested in a potential that is already described by all measured values and not only by the three calculated constants, such as the *Morse potential.* Of course, a potential which is described by these measured data will be much closer to reality than the coarse model of a *Morse potential.* A famous method calculating a potential using measured energy values is that developed by *Rydberg, Klein* and *Rees* [8.9]; this method is therefore called the *RKR method.*

Unfortunately there is no simple mathematical explanation of this method available; the interested reader has to look to the original papers [8.9]. We will not explain the *RKR method* in this article, but we will give a short and obvious hint as to how one can calculate the potential using measured energy values.

If one looks at Fig. 8.7, one can see that the potential curve encloses an area A, which depends on the actual energy level $E_{\tilde{\nu}} = E(v, J)$ and the potential curve $V(R)$. This area A may be calculated in the following way:

$$A = \int_{R_1}^{R_2} (E_{\tilde{\nu}} - V(R))dR \,, \tag{8.6}$$

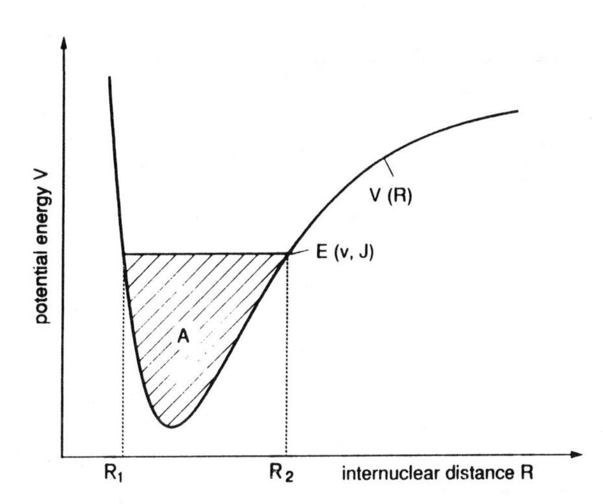

Fig. 8.7. A schematic diagram which visualizes the basic idea of the *RKR procedure*. The hatched area may be calculated by using (8.6)

where R_1 and R_2 are the inner and outer turning points of the potential at this energy level. One may trust the authors in that it is possible, in principle, to calculate the curve $V(R)$ from such an integral. However, it is necessary to have information about the rotational constants B_v of each vibrational level. This is a restriction to our experiment because it is not possible to resolve rotational structure in our case. Therefore the students may try to estimate B_v values from the small splitting observed in the fluorescence lines, or they can use rotational constants given in the literature, e.g. in [8.1].

Our program *RKR* [8.10] performs a calculation of a potential using either the typed–in discrete values of $E_{\tilde{\nu}}$ and B_v, or by calculated values using the constants given in the literature. The program displays the resulting potential on the screen. One may also plot the potential and store the results for further use in other programs.

b) Comparing the RKR Potential to a Morse Potential; the Program NILFIT The program *RKR* may be chained to the program *NILFIT* [8.11], which is a non–linear parameter fit routine. In this routine, a fit to the points of the RKR potential can be done. One has to select an analytical form of the potential which will be fitted to the data, such as the *Morse potential*, a *Lennard–Jones potential* or the potential of an *harmonic oscillator*. One may also type in one's own form of a potential to which the RKR data may be better suited.

After choosing some (more or less) meaningful initial values for the fit procedure, the program performs a desired amount of iterations and then displays the result of the fit. One can also plot the result as a potential curve in which the RKR data points are also included. Thus it is possible to compare the accuracy of different analytical potential forms to the RKR potential, which is a non–analytical potential (it consists of a set of discrete data points!).

Because *NILFIT* is a universal nonlinear fit routine, it may also be used to fit any other data. If the students used the CCD device and stored some line shapes or band shapes, they may try to fit these structures to some analytical expressions. They will learn that the shape of a fluorescence line yields a Gaussian profile and not a Lorentz profile, as expected from theory, (they are Doppler limited!). They may also model the shape of some absorption bands and find that these shapes are formed by the rotational distribution of the molecules. Therefore they may realize that there is some information about the rotational structure buried in the absorption spectra, even if this structure can not be resolved directly.

c) Calculating Wave Functions (Solving the Schrödinger Equation) The exercises described in this and the next section are only for those students who are familar with programming. A lot of mathematical routines are available in a software pool to which the students may have access. Some of these routines can be used to solve differential equations, such as the *Schrödinger equation.* Using these subroutines, the students have only to write a small program to organize the in– and output of their data, and the correct transfer of their data to subroutines.

If the students have previously calculated a reasonably good potential, they may calculate wave functions as solutions of the *Schrödinger equation* for some given vibrational–rotational levels. They will learn that such solutions only exist for discrete energy levels. They will also learn that the actual form of the wave function gives information about the vibrational quantum number (this is the number of nodes). If they plot the wave functions, they will recognize that the nuclei will be located near the classical turning points of the potential, most of the time at higher vibrational quantum numbers, because there the wave functions possess the largest amplitudes.

d) Determining Frank–Condon Factors (FCF) This proclivity to locate the nuclei near the turning points is the reason why some lines in the LIF spectra are not present, even if the neighbouring lines are quite strong. To understand and visualize this, the students may write a small program to calculate *Frank-Condon Factors.*

Within one electronic transition of a molecule, the strength of a line can be described by the *Frank–Condon factor* of this transition, which is directly related to measurable line intensities. This factor is the square of the absolute value of the overlap integral of the wave functions ($\Psi_{v,J}$) of the two involved energy levels:

$$FCF = | \int \Psi_{v'',J''} \Psi_{v'J'} \, dR|^2 \, .$$

Figure 8.8 shows two different cases, one with a strong line and one where the line almost disappears.

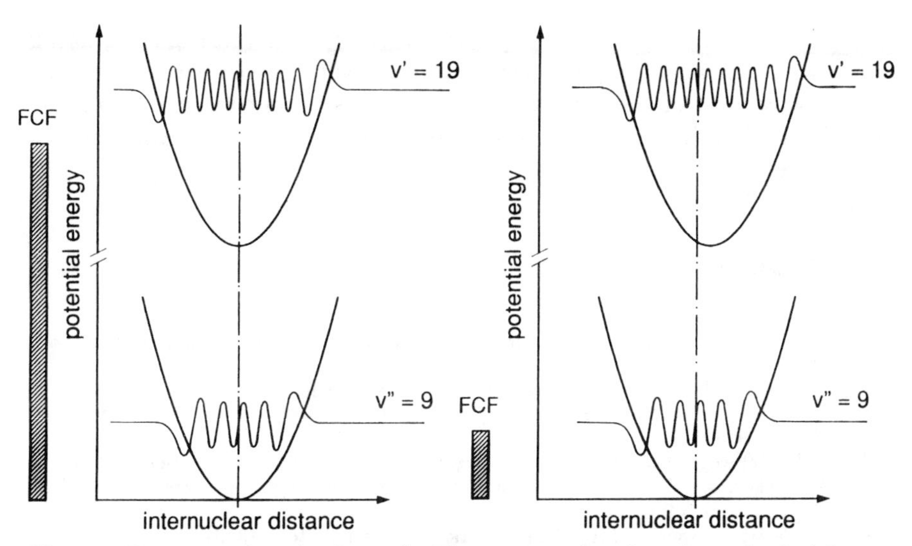

Fig. 8.8. Two cases of overlap integrals. The two wave functions shown in the left part yield a *Frank–Condon factor* which is shown as bars beside them. Note that the two wave functions on the right side are the same, but the upper potential is shifted a little bit to the right. The small bar indicates the dramatic decrease of the line strength

8.6 Pedagogical Aspects

Besides the basic knowledge that students learn within an experiment, such as experimental techniques or molecular physics, there are some additional aspects in our version. Of course, the students learn (by using it) that a computer is a powerful tool in physics. However, the main additional aspects are the following.
The students use the computer in the manner that is later used in research laboratories. They recognize that there exist, in general, different programs for different problems, and that one program alone cannot handle the whole problem. It would be no problem, in principle, to write a program that can do everything needed for this special case, but this will not be feasable. Therefore the students should realise that it is advantageous to work with different programs, and that they have to transfer their data between these.

The students are confronted with problems to be solved. They are given some hints on how they could solve them, but they are not told the whole solution. Thus the students learn how to analyze problems in, e.g., molecular physics. The individual solutions may be completely different. Some students like to go to the library and look up books or papers, if anybody else had already solved their problem. Others, perhaps, like to find their own solutions and learn to use the available hard- and software.

Last but not least, by introducing optional exercises, the students find a playground where they can, free of any compulsion, follow their own preferences. In these exercises they may find (without any stress) that exploring unknown facts can be very interesting and motivating. (Of course, this is the reason why most of them started studying physics, but most of them also lost this feeling during the years because there are too many restrictions in the studies and the curricula.)

References

8.1 G. Herzberg: *Spectra of diatomic molecules* (Van Nostrand Reinhold, New York 1950)
8.2 J.I. Steinfeld: *Molecules and radiation* (Harper and Row, New York 1974)
8.3 G.W. King: *Spectroscopy and molecular structure* (Holt, Rinehart and Winston, New York 1964)
8.4 K.H. Hellwege: *Einführung in die Physik der Molekeln* (Springer, Berlin 1974)
8.5 J. Becker, A. Färbert, H.J. Jodl: *CCD-Sensoren im physikalischen Praktikum* PhuD **2**, 163-169 (1990)
8.6 A. Färbert: *program LAmDA* (Universität Kaiserslautern 1989)
8.7 A. Schmidt: *program FFT (fast fourier transformation)* (Universität Kaiserslautern 1989)
8.8 K.D. Bier, A. Färbert, et al: *program JOD* (Universität Kaiserslautern 1986–1991)
8.9 R. Rydberg: *Graphische Darstellung einiger bandenspektroskopischer Ergebnisse* Ann. Phys. **73**, 367 (1931)
O. Klein: *Zur Berechnung von Potentialkurven 2–atomiger Moleküle mit Hilfe von Spektraltermen* Z. Phys. **76**, 226 (1932)
A.L.G. Rees: *The calculation of potential–energy curves from band–spectroscopic data* Proc. Phys. Soc. **A59**, 998 (1947)
8.10 U. Tönning: *program RKR* (Universität Kaiserslautern 1990/91)
8.11 M. Thoma: *program NILFIT* (Universität Kaiserslautern 1988)

9. Optical Transfer Functions

H. Pulvermacher

9.1 Introduction

If one asks a student to characterize the quality of a hi-fi device, he will take a sine wave generator, measure the ratio of the amplitude at the output and input, represent it as a function of frequency, and deduce the bandwidth of the device from that plot.

This possibility is a consequence of the fact that, according to Fourier, a general oscillating phenomenon can be decomposed in a series of sine-like oscillations. In an analogous way, a wave field may be decomposed in a series of plane waves. In a single plane each of these waves is represented by a sine-like oscillation. Thus the decomposition into plane waves may be found from a Fourier transform of the light disturbance (amplitude) in a single plane. As it is a two-dimensional Fourier transform, things are more complicated in optics than in electronics. If these plane waves are focussed in the posterior focal plane of a lens, one gets a Fraunhofer diffraction image and furthermore finds immediately that Fraunhofer diffraction may be described by a Fourier transform. This fact is the basis of Fourier optics, both in its classical form as Abbe's theory of the microscope and its resolution, and in its more refined modern variants.

For the standard formula for the resolution of the microscope one assumes spatially coherent illumination parallel to the optical axis. If one uses oblique illumination, one gets an improvement of resolution up to a factor of two. As spatially incoherent illumination may be interpreted as an illumination from different directions at the same time, one will expect the same improvement of resolution under these illumination conditions. The formula for the resolution of the telescope confirms this assumption: There one has to insert the diameter of the lens where the radius appears in Abbe's theory. A more refined theory for the case of spatially incoherent (the field will be temporally coherent in any case) illumination can be found in Duffieux's theory of optical image formation and the optical transfer function (see Sect. 9.2.5).

Parameters influencing the transfer functions are, firstly, diffraction and, secondly, the geometric aberrations of the imaging systems. Both will be studied in the following experiments; aberrations, however, only in the simple case of defect of focus. As can be seen from the preceding considerations, diffraction and aberrations are not the only parameters. The illuminating conditions play an important role too, and will be studied later on.

9.2 Mathematical Tools

9.2.1 Fourier Transforms

a) Definitions

1. In this paper, the Fourier transform is defined in the following way (bold-face letters designate vectors):

$$\tilde{f}(R) = \int_{-\infty}^{+\infty} f(x) e^{-2\pi \imath x R} dx \tag{9.1}$$

and the inverse

$$f(x) = \int_{-\infty}^{+\infty} \tilde{f}(R) e^{2\pi \imath x R} dR. \tag{9.2}$$

Fourier pairs are usually designated by the same small and capital letters or by a letter and the same letter with a ˜ (tilde) over it.

2. An integral transform of the form

$$h(x') = \int_{-\infty}^{+\infty} f(x'-x) g(x) dx \quad \text{or shortly} \quad h(x') = f(x') \star g(x') \tag{9.3}$$

is called a convolution.

3. Special functions:

$$\text{rect}(x) = \begin{cases} 1 & \text{for } -\frac{1}{2} < x < \frac{1}{2} \\ 0 & \text{elsewhere} \end{cases}$$

$$\text{sinc}(x) = \frac{\sin(\pi x)}{\pi x}$$

$$\text{circl}(r) = \begin{cases} \frac{4}{\pi} & \text{for } r < \frac{1}{2} \\ 0 & \text{elsewhere} \end{cases} \qquad \text{with } r = \sqrt{x^2 + y^2}$$

$$\text{beinc}(x) = \frac{2J_1(\pi x)}{\pi x}$$

$$\begin{aligned} \text{circon}(x) &= \frac{4}{\pi} \text{circl}(x) \star \text{circl}(x) \\ &= \frac{2}{\pi} \left\{ \arccos(x) - x\sqrt{1-x^2} \right\} \end{aligned}$$

b) Basic Theorems

1. Shift invariance of the modulus of the transform:

$$g(x) = f(x - x_0) \quad \rightarrow \quad \tilde{g}(R) = \tilde{f}(R)e^{-2\pi \imath x_0 R} \ . \tag{9.4}$$

2. Symmetry for real valued functions:

$$f(x) \quad \text{real} \quad \rightarrow \tilde{f}(-R) = \tilde{f}^*(R) \ . \tag{9.5}$$

3. Convolution theorem:

$$h(x) = f(x) \star g(x) \quad \rightarrow \quad \tilde{h}(R) = \tilde{f}(R)\tilde{g}(R) \ . \tag{9.6}$$

In two dimensions the formulae given until now are analogous.

c) Common Fourier Pairs In the following Table 9.1 x and R are corresponding coordinates in direct and Fourier space. In two dimensions r is the radial part of polar coordinates in direct space. As the radial part in Fourier space is identical with the spatial frequency corresponding to the x coordinate, R may also be used for it.

Table 9.1. Common Fourier pairs

function type	direct space	⇔	Fourier space
δ-"function"	$\delta(x - x_0)$		$e^{-2\pi \imath x_0 R}$
Dirac comb	$\sum_{j=-\infty}^{+\infty} \delta(x - ja)$		$\sum_{j=-\infty}^{+\infty} \delta\left(R - \frac{j}{a}\right)$
Rectangle	$\text{rect}\left(\frac{x}{a}\right)$		$\text{sinc}(aR)$
Triangle	$\begin{cases} \frac{1}{a}\left(1 - \frac{\lvert x \rvert}{a}\right) & \text{for } \lvert x \rvert < a \\ 0 & \text{elsewhere} \end{cases}$		$\text{sinc}^2(aR)$
Cylinder	$\frac{1}{d^2}\,\text{circl}\left(\frac{r}{d}\right)$		$\text{beinc}(Rd)$
Airy disc	$\frac{1}{d^2}\,\text{beinc}^2\left(\frac{r}{d}\right)$		$\text{circon}(Rd)$
Gaussian	$\frac{1}{\sqrt{2\pi}\sigma}\exp\left(-\frac{x^2}{2\sigma^2}\right)$		$\exp(-2\pi^2\sigma^2 R^2)$

9.2.2 Theory of Transfer Functions

a) Definitions

1. The *light disturbance* or *amplitude* of a light field is a complex valued quantity $A(\mathbf{x})$ with the following properties:
   a) $I(\mathbf{x}) = A(\mathbf{x})A^*(\mathbf{x})$ is the irradiance at point $\mathbf{x}$.
   b) $arg(A(\mathbf{x}_1)) - arg(A(\mathbf{x}_2))$ is the phase shift of the light disturbance between the points $\mathbf{x}_1$ and $\mathbf{x}_2$.

2. The *amplitude transmittance* t of a transparent object is a quantity with the properties:
   a) $tt^* = \tau$, where τ is the (irradiance) transmittance of the object.
   b) $arg(t)$ is the phaseshift introduced by the object.
3. The *point spread function* $d(\mathbf{x})$ is the irradiance in the image of a point source divided by this irradiance integrated over the whole image of the point source. In general the point spread function depends on the position of the point source.
4. The *line spread function* $l(v)$ taken in direction $\mathbf{s}_1$ is the point spread function integrated in direction $\mathbf{s}_1 : \int d(u\mathbf{s}_1 + v\mathbf{s}_2)du$ with $\mathbf{s}_1$, $\mathbf{s}_2$ normalized orthogonal vectors. The point spread function is a function of two variables, the line spread a function of one.
5. A *spatially invariant* or *isoplanatic* or *spatially stationary* system is a system with a point spread function independent of the position of the point source. For a spatially invariant system, the line spread function is the distribution of the irradiance in the image of a line source.
6. The *optical transfer function* (OTF) D($\mathbf{R}$) is the Fourier transform of the point spread function. The variable $\mathbf{R}$ in Fourier space is called *spatial frequency*. It has the physical significance of the inverse value of the periodicity constant of a grating and is measured in lines per mm: 1/mm. Generally, the optical transfer function is a complex valued function:

$$D(\mathbf{R}) = T(\mathbf{R})e^{\Phi(\mathbf{R})}.$$

7. The modulus $T(\mathbf{R})$ of $D(\mathbf{R})$ is called the *modulation transfer function* (MTF), the phase angle $\Phi(\mathbf{R})$ the *phase transfer function*. $D(\mathbf{R})$ resp. $T(\mathbf{R})$ for a given value of $\mathbf{R}$ is called the *optical* resp. *modulation transfer factor*.

b) Linear Transfer Theory The definition of point spread and transfer functions only is reasonable if a linear superposition law is valid for the quantity used to define the point spread function. In the case of a fully incoherently radiating or illuminated object, this quantity is the irradiance, as used above, for coherent illumination the light disturbance. We will first discuss items common to both situations and then those valid especially for coherent or incoherent light.
As for every Fourier transform of a real-valued function, the optical transfer function follows the well-known relation

$$D(\mathbf{R}) = D(-\mathbf{R})^*.$$

Let $k(\mathbf{x})$ and $K(\mathbf{R})$ be the point spread and the coherent transfer function of an imaging system. Then the incoherent point spread function is $k(\mathbf{x})k^*(\mathbf{x})$. Using the convolution theorem for the incoherent (optical) transfer function results in:

$$D(\mathbf{R}) = \int_{-\infty}^{+\infty} K\left(\bar{\mathbf{R}} + \frac{\mathbf{R}}{2}\right) K^*\left(\bar{\mathbf{R}} - \frac{\mathbf{R}}{2}\right) d\bar{R}^2. \tag{9.7}$$

9.2.3 Imaging with Space Invariant Systems

a) The Filter Equation Let $d(\mathbf{x},\mathbf{x}')$ be the point spread function for a linearly transferred quantity and an imaging system dependent on the position $\mathbf{x}'$ of the point source $\mathbf{x}$, the coordinate in object space. For the image $o'(\mathbf{x}')$ of an object $o(\mathbf{x})$ with Fourier transform $\tilde{o}(R)$, one finds:

$$o'(\mathbf{x}') = \int_{-\infty}^{+\infty} o(\mathbf{x})d(\mathbf{x},\mathbf{x}'))dx^2.$$

In the case of a space invariant system, the kernel has the form

$$d(\mathbf{x},\mathbf{x}') = d(\mathbf{x}' - \mathbf{x})$$

and so

$$o'(\mathbf{x}') = \int_{-\infty}^{+\infty} o(\mathbf{x})d(\mathbf{x}' - \mathbf{x}))dx^2 \;, \tag{9.8}$$

or the image is the convolution of the object with the point spread function. Using the convolution theorem yields immediately

$$\tilde{o}'(\mathbf{R}) = \tilde{o}(\mathbf{R})D(\mathbf{R}). \tag{9.9}$$

This is called the *filter equation* (see Hecht [9.1]). Its importance is rather far reaching: It does not only describe an imaging process, but also opens a large field of possibilities for image manipulations and is the basis for a variety of image processing procedures: If the image and the transfer function are given, (9.9) allows the direct reconstruction of an object from its image.

b) The Image of a Sine Grating Imagine an object with a sine-like irradiance distribution $1 + M\sin(Rx)$, where M is the *contrast* or *modulation* of the object. The Fourier transform of it is composed of two δ peaks symmetric to zero with weight $\frac{M}{2}$ (as can simply be deduced from Euler's relation and the given Fourier pairs) and a δ peak at zero. From this fact and the filter equation (9.9) it follows that the image is $1 + T(R)M\sin(xR - \Phi(R))$. The modulation in the image is reduced by the factor T(R), the modulation transfer factor of the imaging system. These considerations lead further to the result $M = \frac{a_1 + a_{-1}}{2a_0}$ for the modulation depth M, where a_j are Fourier components.

9.2.4 Coherent Optics

a) Fraunhofer Diffraction Assume a plane with an object of amplitude transmittance A($\mathbf{x}$) illuminated with a plane wave propagating parallel to the optical axis with wavelength λ. We may further assume coarse structures of the diffracting object, and so small diffraction angles. Then the light disturbance at a point $\mathbf{x}'$ in the posterior focal plane of a lens with focal length f is beside a constant factor that guarantees conservation of energy:

$$A'(\mathbf{x}') = \int_{-\infty}^{+\infty} A(\mathbf{x}) \exp\left(-\frac{2\pi i \mathbf{x}\mathbf{x}'}{\lambda f}\right) dx^2. \tag{9.10}$$

If the object is situated in the anterior focal plane of the transforming lens, the relation is valid for modulus and phase of the light disturbance, in other cases only for the modulus.

b) The 4—f Setup Consider a setup (Fig. 9.1) with an object of amplitude transmittance $o(\mathbf{x})$ in the anterior plane of a lens of focal length f illuminated by a plane coherent wave. At the distance $2f$ behind the first lens there is a second lens of the same focal length; its posterior focal plane is the image plane. The illuminating plane wave is focussed in the posterior focal plane of the first lens and is rendered again into a parallel beam by the second lens. Thus the intermediate focal plane is the plane of the aperture stop and is therefore called the pupil plane. There, essentially the Fourier transform $\tilde{o}(\mathbf{x}_p/\lambda f)$ of the object transmittance is found. Now there may be a filter $P(\mathbf{x}_p)$ with influence on modulus and phase of the light disturbance in the pupil plane. $P(\mathbf{x}_p)$ is called the ***pupil function*** of the system. Its influence on the modulus may come from a real absorbing structure like a stop, and its influence on the phase from a corrector plate, but usually from the aberrations of the optical system. The distance ΔW between the wave front in the posterior focal plane of the first lens and this plane is called the ***wave aberration***. The corresponding phase shift is $\Delta\Phi = \frac{2\pi\Delta W}{\lambda}$. Together with the absorption filter A, one finds for the complete pupil function

$$P(\mathbf{x}_p) = A(\mathbf{x}_p) \exp\left\{\frac{2\pi i}{\lambda} \Delta W(\mathbf{x}_p)\right\}.$$

Behind this filter the light disturbance is $\tilde{o}(\frac{\mathbf{x}_p}{\lambda f})P(\mathbf{x}_p)$. Thus one finds for the Fourier spectrum of the image, using (9.10), $\tilde{o}(R)P(R\lambda f)$. It is Fourier transformed by the following lens. From the convolution theorem it follows

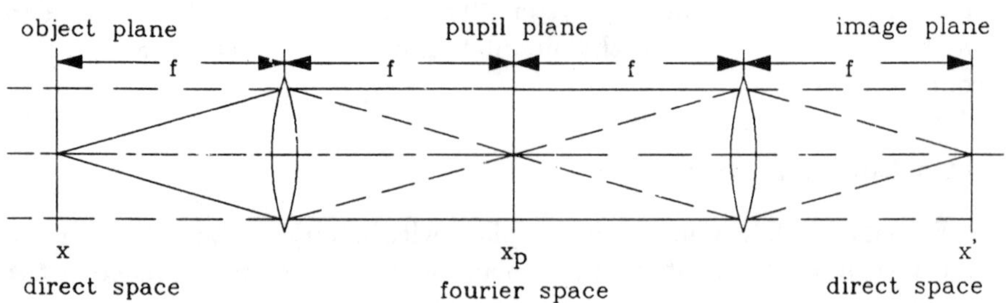

Fig. 9.1 4—f setup. f: focal length, $\mathbf{x}$, $\mathbf{x}_p$, $\mathbf{x}'$: coordinates in object, pupil and image plane

that the light disturbance in the posterior focal plane of the second lens is $o(\mathbf{x}') \star p(\frac{\mathbf{x}'}{\lambda f})$, where p is the Fourier transform of P. So $p(\frac{\mathbf{x}}{\lambda f})$ is the coherent point spread function. It is also obvious that in the case of coherent illumination the light disturbance is linearly transferred by the system and $P(R\lambda f)$ is the transfer function.

c) Objects of Low Contrast Now let us assume an object of low contrast $1 + o(\mathbf{x})$, where $o(\mathbf{x})$ is small. The irradiance in the image is

$$\begin{aligned} o'(\mathbf{x}') &= (1 + o(\mathbf{x}) \star p(\mathbf{x}))(1 + o(\mathbf{x}) \star p(\mathbf{x}))^* \\ &\approx 1 + \Re(o(\mathbf{x}) \star p(\mathbf{x}) + o(\mathbf{x}) \star p(\mathbf{x})) \\ &= 1 + 2(o_r(\mathbf{x}) \star p_r(\mathbf{x}) - o_i(\mathbf{x}) \star p_i(\mathbf{x})) \, , \end{aligned} \tag{9.11}$$

where the indices r and i designate the real and imaginary part of the respective quantities. That means that the real and imaginary part of the object are transferred independently. The appropriate point spread functions are the real and imaginary part of the point spread function of the system. For symmetric point spread functions, the appropriate transfer functions are again the real and imaginary part of the transfer function of the system. Objects with purely real or imaginary amplitude transmittances are called *absorption* resp. *phase objects*.

This leads to the conclusion that in the case of a symmetric point spread function, an absorption object is transmitted by the real part of the coherent transfer function and a phase object by the imaginary part.

9.2.5 Incoherent Optics

a) Duffieux's Integral Now let us assume spatially incoherent illumination and perfect temporal coherence. In this case the irradiance is transferred linearly. The irradiance is found from the light disturbance by multiplying it with its conjugate complex. Therefore using the convolution theorem shows that the incoherent transfer function $D(\mathbf{R})$ is found from the coherent transfer function $K(\mathbf{R})$ by a convolution of the coherent point spread function with its conjugate complex. If such a convolution integral is normalized, it is called an *autocorrelation integral*. Finally, the coherent transfer function must adequately replaced by the pupil function and the result is:

$$D(\mathbf{R}) = \frac{\int_{-\infty}^{+\infty} P(\mathbf{x} - \lambda f \mathbf{R}/2) P^*(\mathbf{x} + \lambda f \mathbf{R}/2)) dx^2}{\int_{-\infty}^{+\infty} P(\mathbf{x}) P^*(\mathbf{x}) dx^2} \, . \tag{9.12}$$

This formula is called *Duffieux's integral*.

b) Defect of Focus Now let us imagine that the object is not situated in the anterior focal plane of the first lens, but a distance Δx behind it. A diverging spherical wave is impinging on the pupil plane. Its radius is $r = f^2/\Delta z$, which can be seen using Newton's lens equation. Approximating the spherical by a parabolic wave, one finds for the pupil function (formulated in a one-dimensional model)

$$P(x) = \exp\left\{\frac{2\pi\imath}{2r\lambda}x^2\right\} = \exp\left\{\frac{\pi\imath\Delta z}{\lambda}\frac{x^2}{f^2}\right\}, \tag{9.13}$$

and so for the coherent transfer function, $\exp\left\{\pi\imath\Delta z\lambda R^2\right\}$. If we further assume a slit-like pupil of diameter a, the whole coherent transfer function is $\operatorname{rect}\left(\frac{R\lambda f}{a}\right)\exp\left\{\pi i\Delta z\lambda R^2\right\}$. The evaluation of the Duffieux integral gives, after the normalization to $D(0) = 1$ and an introduction of the numerical aperture $A = \frac{a}{2f}$,

$$D(R) = \left(1 - \frac{R\lambda}{2A}\right)\operatorname{sinc}\left\{2zRA\left(1 - \frac{R\lambda}{2A}\right)\right\}. \tag{9.14}$$

For the limiting cases of vanishing defect of focus and very small wavelength, we get the special cases of an OTF that is only affected by diffraction on the one hand or by geometric optics on the other. The result for pure diffraction is

$$D(R) = \left(1 - \frac{R\lambda}{2A}\right), \tag{9.15}$$

and for the geometric optics limit

$$D(R) = \operatorname{sinc}(2zRA). \tag{9.16}$$

This formula may be interpreted rather simply: The exit pupil of the system is illuminated uniformly and the point spread function is generated by a central projection of the exit pupil on the image plain. Comparing (9.14) and (9.16), one finds that the argument of the sinc function is, in the case of diffraction effect, diminished by the the factor $1 - \frac{R\lambda}{2A}$, and so diffraction effects are able to shift the zero of the transfer function to higher spatial frequencies.

Now it is interesting to study the situation for small spatial frequencies; that is, we take only terms linear in R:

$$D(R) = \left(1 - \frac{R\lambda}{2A}\right)\operatorname{sinc}(2zRA). \tag{9.17}$$

This is the product of the OTF in the geometric optical limit and the OTF for pure diffraction, which is in accordance with a theorem due to Hopkins. It must be considered that this approximation is only valid for small spatial frequencies.

9.2.6 Exercises and Questions

1. Prove the formula for the diffraction by a slit and by a circular hole using the table of common Fourier pairs.
2. Use the convolution theorem to find an approximation formula
   a) for the irradiance in the neighbourhood of the diffraction order of a grating,
   b) for the envelope of the diffraction orders of a grating.
3. Interpret the results of (2.a) as a double diffraction at grating and slit.
4. Problems concerning circular apertures:
   a) Find the OTF for a defocussed system with circular aperture in the geometric optical limit using the interpretation of the formula for the slit aperture under analogue conditions.
   b) Find the light disturbance in the diffraction image of a circular aperture using the table of common Fourier pairs, and from it the irradiance.
   c) Find from the result of (4.b) the MTF of a diffraction limited system using the table of Fourier pairs.

9.3 Experimental Set Up

9.3.1 Preliminary Considerations

a) The Method From a modern point of view, the only effective method for measuring an OTF would be the Fourier analysis of the image of a narrow slit or, better, a small hole. Such a measurement would be strictly within the modern definition of an OTF.

However such a measurement has some disadvantages from a didactical point of view: It is intuitively evident that the contrast in the image of a grating is relevant for the quality of an imaging process. For that reason, it is preferable to use the fact that the ratio of the contrast in the image of a grating to the contrast of the grating itself is identical with the modulation transfer factor for the spatial frequency of the grating used for the measurement.

However, there is a bigger disadvantage: A measurement sticking to the definition does not measure the OTF of an imaging but of a scanning process. The coherence properties of the process are not influenced by the aperture of the illumination, but of the detection system. To deal with those complications is far beyond the possibilities of an experiment that has to be performed within a limited time.

b) The Test Object The test object for the measurements must have a grating-like structure; that means it must contain only one spatial frequency and it must be possible to change the frequency easily. These are, approximately, the properties of a sector star, which Lindberg [9.3] used first for the measurement of transfer functions.

9.3.2 The Optics

From the preliminary considerations it is clear that one has to image a sector star by the system under test, to measure the irradiance in the image by an adequate detector system and to process the result to find the MTF.

a) The Imaging Optics As the MTF is a tool especially useful for characterizing photographic systems and in photography the object is often far away, the test object is first imaged to infinity by a collimator (Fig. 9.2). This image is further imaged by the system on the detector slit. As the collimator Cl_2 and the system under test are identical, they may form a 4-f system and so they are placed in the distance 2f apart. The whole system forms a telescope of angular magnification 1. Then the lateral magnification and the depth magnification are also 1. As the lateral magnification is independent of the position of the object, the depth magnification is also constant. That means that a certain defect of focus in the object space leads to the same defect of focus in the image space. This simplifies the evaluation of the experiments considerably.

In the adjustment procedure described later, it is necessary to know the positions of the focal planes of the systems relative to their mountings.

b) The Illumination The test object is illuminated with a collimator Cl_1 from the secondary image of the light source, the arc of a mercury high pressure lamp. The diameter of that image may be influenced according to the demands of the experiment by using a suitable condensor system. If a large diameter of the illuminating source is needed in order to illuminate a large entrance pupil uniformly, an illuminated ground glass must be used as the source.

For a measurement of the MTF for incoherent illumination, the aperture stop must be illuminated to a sufficient extent. That is the case if the radius of the image of the light source is an amount $R_{max}\lambda f$ (R_{max} maximum of used spatial frequency) larger than the radius of the pupil stop.

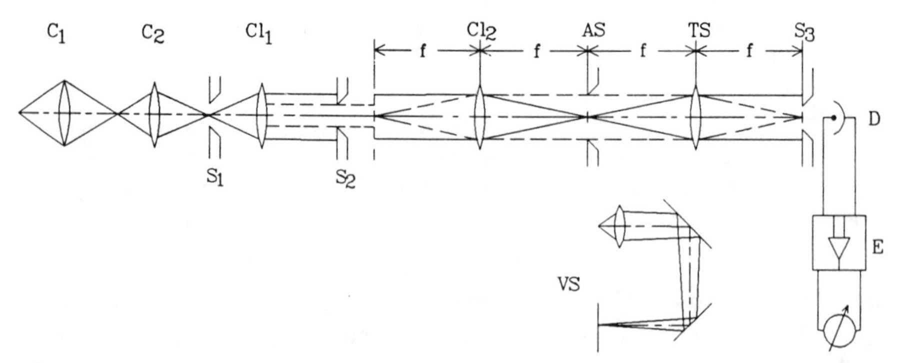

Fig. 9.2 Diagram of the setup. C_i: condensors, Cl_i: collimators, S_i: slits, AS: aperture stop, TS: system under test, D: detector, E: electronics, VS: viewing system

c) The Primary Image If the light source in the anterior focal plane of the first collimator is small, the collimators before and behind the test object form an optical system that produces a Fraunhofer diffraction image of the object in the back focal plane of the second collimator. It may be interpreted as the primary image of the object in the sense of Abbe's theory, and inspected by a viewing system.

d) The Detector The detector consists of an adjustable slit followed by a light ground glass, an interference filter and a photo multiplier. The ground glass is necessary to achieve a uniform illumination of the photo cathode. The slit may be turned to make it parallel to the grating under test.

9.3.3 The Test Object

a) Principal Problems As already mentioned, the test object is a sector star. At a definite distance from its centre, it represents a grating with a definite periodicity. If it is considered along a straight line directed tangentially, one finds immediately that the periodicity varies; one could speak of frequency modulation. Therefore small frequency bands and not single frequencies appear in the spectrum (see Appendix 9.A). This broadening is more pronounced for higher frequencies. As a consequence, the usage of the inner part of the sector star for studying the image of a grating with a high spatial frequency is limited.

As the spectrum of the sector star contains more than one spatial frequency, its image must not be scanned by a long slit but by a small hole, or at least a short one. This short slit is produced by projecting a slit with a width of some 0.1 mm perpendicular to the detector slit into the object plane. It acts as a field stop.

b) The Experimental Realization The sector star (Fig. 9.3) is mounted in the free aperture of a ball bearing of sufficient diameter and rotated by a motor. The rotating star is shifted horizontally perpendicular to the optical axis by a stepping motor. The actual position may be controlled by a dial gauge. The actual spatial frequency may be calculated from the distance between the centre of the star and the "projected centre" of the field stop.

The sector star is imaged on the detector slit. In a plane behind the slit, the diffraction images discussed in the appendix will appear. The vertical adjustment of the slit is achieved if the hyperbolae above and below the slit are symmetric. A preliminary focussing is achieved if the hyperbola on the right and left side are just tangent to one another (see Appendix 9.A).

If the field stop (Fig. 9.5) is projected into the object plane by a bundle of low divergence (small light source), the same diffraction image will appear. By symmetrizing the hyperbola-like structure, the position of the centre of the sector star relative to the field stop can be found, and so can be used correctly in calculating the spatial frequency.

The whole set up is mounted on a carriage and so its focal position may be manipulated with a screw and read on a dial gauge.

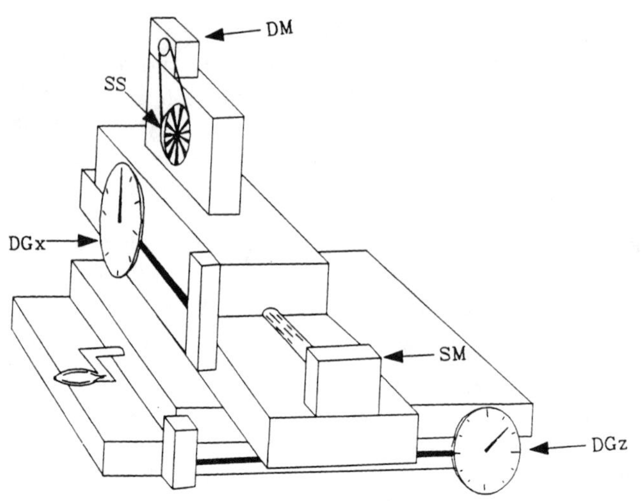

Fig. 9.3 Mounting of the sector star. DM: DC-motor, SM: stepping motor, AG_x, DG_z: dial gauges for x and z direction

9.3.4 The Electronics

The Fourier spectrum of a periodic object consists of equally spaced δ-functions of different weight. If the object is real valued and the origin of the coordinate system is placed adequately, the first positive and negative order are real and equal. If a_i designates the ith Fourier coefficient, the modulation depth is given by $\frac{2a_1}{a_0}$ (see Sect. 9.2.3) where a_i are the Fourier coefficients. Thus the measurement of the modulation of the grating is made possible by the measurement of the Fourier coefficients a_0 and a_1. These measurements are performed by two electronic filters: (Fig. 9.4) a low pass (integrator) and a band pass [9.4]. The band pass filter also has the effect of an improvement of the signal-to-noise ratio. The outputs of both filters have to be divided and adequately normalized to form the modulation transfer factor for the used spatial frequency.

This division is done by an analogue circuit. However, we used it only for the monitoring in the experiments. In order to achieve a higher accuracy, the signals of the low and the band pass output were measured separately and divided by the program. The overall offsets of the electronics for both signals were measured at the start of a measurement by shielding the detector. Between shielding and measurement there must be a delay longer than the tuning time of the band pass. The rotational speed of the sector star may be tuned by the voltage of its power supply. It may be controlled by the phase shift between input and output of the band pass.

If possible, the normalization constant was determined for each experiment, and at least twice for every series: at the beginning and at the end; the control at the end being usually identical with the first control for the next series.

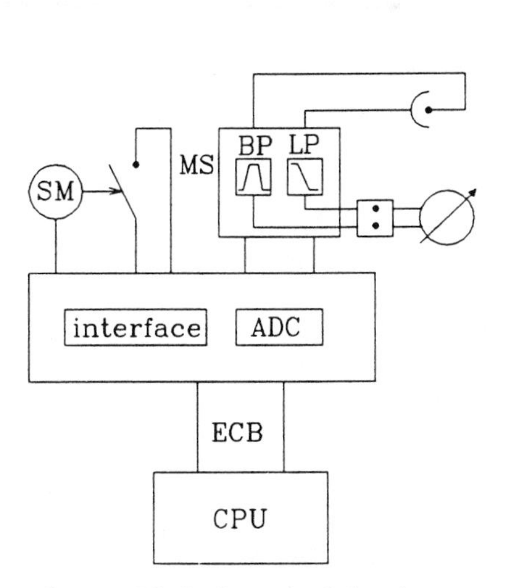

Fig. 9.4 Block diagram of the electronics. SM: stepping motor, MS: microswitch, BP: band pass, LP: low pass, ADC: analogue-digital converter, ECB: ECB bus

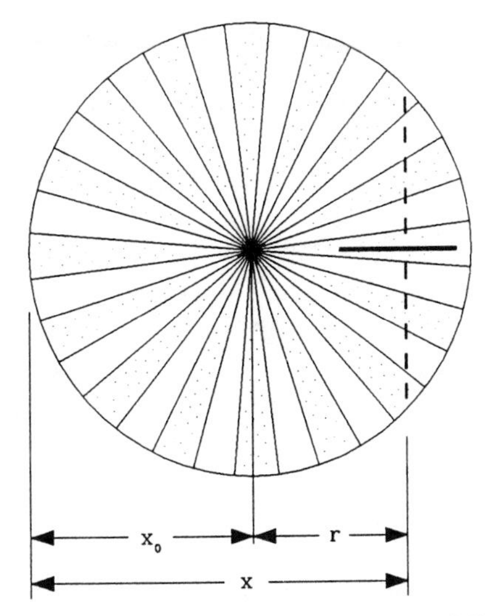

Fig. 9.5 The sector star. r: radius, full line: radial orientation, dashed line: tangential orientation

9.3.5 The Adjustment

a) Adjustment Procedures In the following discussions, positioning means finding the correct position of an element in the direction of the optical bench.

For the whole adjustment, the arc of the mercury lamp is imaged on a slit in the anterior focal plane of the first collimator. The second condensor is positioned toward the focal plane of the collimator (minimal magnification of the arc).

There are two adjustment procedures necessary for the set up: A general alignment procedure and the usual autocollimation procedure. For the alignment procedure, a screen with a number of concentric circles and a small cross in their centre is a valuable tool. At first a suitable diameter is chosen for the light bundle. Then the screen is placed directly behind the optical element or system to be adjusted and centered to the light bundle. Afterwards, the screen is moved away as far as possible and the bundle is centered on the screen by adjusting the system. Usually the alternating adjustment with the screen positioned immediately behind the system and far away must be repeated several times.

For the autocollimation a mirror is used. The diameter of the bundle is first chosen to be as narrow as possible. The mirror is first placed in front of the system (attached to the mounting) and the system is adjusted (its angular degrees of freedom) in such a way that the bundle is reflected in itself. Now the mirror is placed behind the system and the optical element is adjusted by shifting it perpendicular to the optical axes so that the bundle is again reflected in itself. If the bundle behind the system must be parallel, then the bundle must be opened as wide as possible and a focussing of the system may be achieved by reflecting the focus exactly in itself.

Usually some iterations will be necessary where the adjustment using alligning and autocollimation alternate.

b) Adjustment of the Setup

First Collimator, Object and Second Collimator The *first collimator* is adjusted by using autocollimation and alignment alternatingly.

Now the bundle (large diameter) is focussed into the plane of the aperture stop by positioning the *second collimator* adequately. Then the *object* is placed in its anterior focal plane (its position relative to the mounting should be known, so that a coarse adjustment can be achieved by choosing an adequate distance) and centered to the illuminating bundle. Next it is checked to see whether the centre of the star is projected exactly parallel to the optical bench. This control must be done very carefully as it influences the quality of the zeros of the transfer functions for coherent illumination in the case of defect of focus. Finally, the second collimator (its stop) is centered to the projected image of the star.

The Aperture Stop and the Control of the Primary Interference Image The viewing system for the inspection of the plane of the aperture stop (primary interference image) is now installed. It consists of a lens and two mirrors. These mirrors reflect the light in a way that it finally travels backwards parallel to its primary direction. There the plane of the stop is imaged on a screen and thus can be inspected. In our experiment the plane of the stop was fixed. This is important for the following details of the adjustment process:

Firstly the stop is focussed on the screen by positioning the lens of the viewing system. Then the slit of the first collimator is focussed on the screen by fine positioning of the second collimator, and subsequently the image of the slit is centered on the screen by adjusting the lens of the viewing system. Finally the stop is centered on the screen. Thus one has the entrance slit imaged in the plane of the stop and the stop centered to the image of the slit.

System Under Test, Detector, Field Stop and Focus The *system under test* is adjusted by a combination of the autocollimation and alignment procedure, as described for the first collimator. The light source is the image of the entrance slit in the plane of the system stop.

The *entrance slit* of the detector is first positioned in the right distance from the system under test. The slit is then adjusted vertically using the hyperbola-like diffraction image, as described before. An improvement of the position of the test object may be done also by using the diffraction image. The tilting angle of the slit can be controlled by shifting it horizontally (relative to the star).

The *field stop* (Fig. 9.5) is adjusted horizontally using the hyperbola-like diffraction image behind the sector star.

The *optimal focal position* of the star must be fixed with a large aperture stop. One uses a spatial frequency of about 4.5 l/mm and looks for the position of optimal contrast. Furthermore, one defocusses the star until the contrast is reduced to about half the optimal value in both directions. The mean of these two positions is the value for the optimal focussing.

c) Preparations for the Single Series of Experiments For the first two series of experiments, diffraction must be negligible. Therefore the aperture stop must be large as must the light source to illuminate it. In order to achieve that, a strongly defocussed image of the arc of the mercury lamp is projected on a ground glass. The field stop is positioned close to the sector star on the side facing the light. For adjusting the field stop with the diffraction image, the arc must be focussed on the ground glass. After the adjustment the previously determined defect of focus must be restored.

In the next series, a much smaller aperture is needed. A small auxiliary stop is placed in the anterior focus of the first collimator and the image is centered to the aperture stop. The field stop is placed between the second condensor and the first collimator and imaged on the sector star by the collimator. It is adjusted horizontally so that the auxiliary stop is illuminated symmetrically. Finally the centre of the star is adjusted horizontally using the diffraction image. The auxiliary stop must be withdrawn for the measurement.

For the last series one needs a slit-like illumination. To illuminate the slit as homogeneously as possible, the second condensor is placed away from the focal plane of the first collimator (maximal magnification of the arc). To avoid an enhanced resolution due to oblique illumination, the stop must be centered very carefully to the image of the slit vertically. If the spatial frequency is increased, both first diffraction orders must touch the margin of the stop simultaneously.

9.3.6 The Software for Experimentation and Evaluation

The software is written in PASCAL. Subroutines from "Numerical Recipes" [9.5] are incorporated. The menu structure is taken from "Tool box" [9.6].

The program for the data aquisition and evaluation is menu guided according to the SAA standard: (Fig. 9.6). The menu is activated with the "ALT" key. The single actions can be selected either by the cursor keys or by hot keys. If a menu action is selected, its function is briefly explained in the bottom line. During the evaluation of the data, a graphical representation is always available. A window is superimposed where the interaction with the program is performed. Using "F9" one may switch between a short and a detailed explanation of the possible actions. With the "ESC" key it is often possible to return to the main menu.

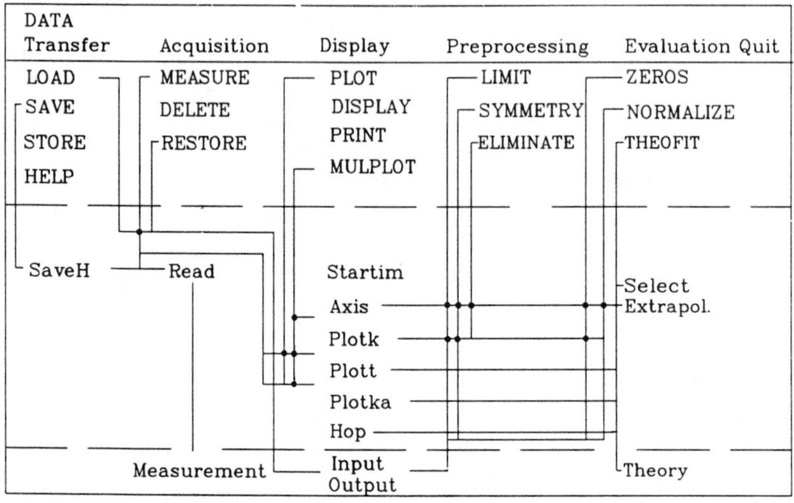

Fig. 9.6 Block diagram of the program

a) The Main Menu The main menu enables the activation of the following groups of actions: Data *transfer*, *acquisition*, *displaying*, *preprocessing* and *evaluation*. For an experiment, one first has to perform a data transfer and then the acquisition. For the first experiment of every series, an immediate preprocessing of the data is strongly recommended. Thus the parameters of the experiment are known more precisely and the handling of subsequent experiments of the series is simplified. The evaluation of the data should not be done immediately after their acquisition. It is preferable to do the evaluations with an overview of all the results.

b) Transfer of Data The data of the current experiment are represented in three ways: On the disc, in a large array together with all other data on the heap, and the data of the current experiment itself in a small array in the working area of the program. As far as the saving of the data is concerned, the following steps are necessary: transferring the current data to the heap and saving the whole set of data on disc. Preparing for a new experiment SAVEs first the data of the preceding one, LOADs the set of data from disc, loads the currently needed ones in the working array and displays the parameters.

c) Acquisition of Data DELETing data is done by setting the data count to zero. The data themselves are not destroyed. If the data count is reset to its original value, the old data are RESTOREd.

For the MEASUREment of the data itself, the parameters may be presented in two ways: Either a standard set may be used or the parameters of the preceeding experiment. These parameters are dispayed and a line editing of them is possible. The repetition rate for one measured value is a helpful mean to enhance the accuracy in the case of large noise. For the distance of the single spatial frequencies, a value of 0.1 is recommended.

The electronic offsets of the AC and the DC channels are now measured. The experiment itself starts at a position of the star destined by a microswitch. One has to control whether it coincides with the zero of the mechanical gauge; if not the gauge must be reset. The position of the star for the first measurement is selected automatically. If it is situated too near to the border of the star, an adequate offset for the first position may be chosen (There are two mechanical and two eletronical setups; don't mix them up!). During the measurement, the spatial frequency, the position of the star, the output of the AC and DC channels and the modulation are displayed. Measurements are taken on both sides of the centre of the star. Therefore the resulting curve consists of two branches, which are designated with different symbols in the graphical representation. Finally the star is brought back into the starting position, and the measured curve is displayed.

d) Displaying of Data For the displaying of data, two functions are available: A PLOT (Fig. 9.7) with a usual linear representation of a single curve, and a common representation of several curves (MULPLOT). If a theoretical curve is fitted, it is also drawn in both repesentations.

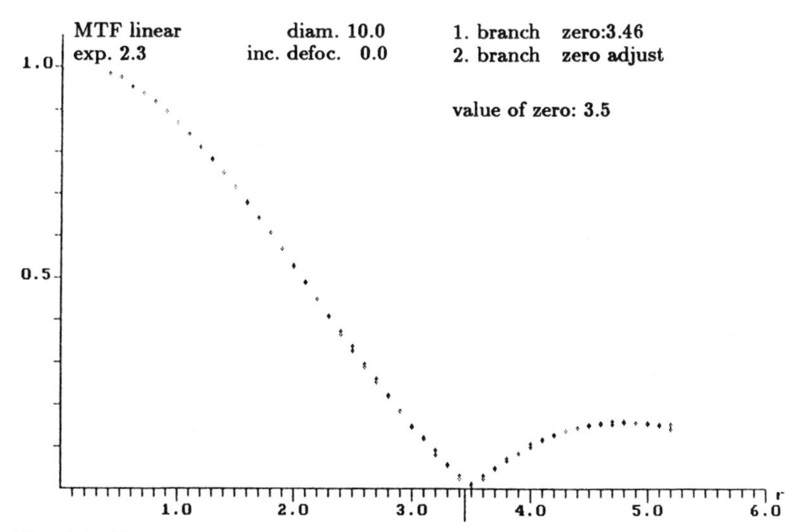

Fig. 9.7 View on the screen in a data processing situation: entering zero

e) Preprocessing of Data The preprocessing of data includes three procedures: There is the possibility for a refinement of the position x_0 (see Fig. 9.5) (SYMMETRY) of the centre of the star. If an inaccurate value for x_0 is used for the calculation of the spatial frequency, the two branches of the curve measured on different sides of the centre of the star will not coincide. Controlled by the horizontal cursor keys, the value of x_0 may be varied systematically and the new resulting curve is displayed. Thus an optimal value for x_0 may be found graphically interactively.

Furthermore, there is the possibility to do a preliminary normalization for spatial frequency zero. By visual extrapolation, an accuracy of a few % for the value of the normalization constant may be found, and so the curve is within the LIMITs of the screen. The same routine eliminates negative values of the MTF.

For a precise normalization it is necessary to ELIMINATE drastically deviating single data at low spatial frequencies.

f) Interpretation of Data For the final interpretation of data there are three functions available:

Firstly there is the possibility to mark the position of the first ZERO of the experimental curve.

Second is analytical NORMALIZation of the curves. The curve is approximated by a formula of the structure $a - bR^n$, where a is the normalization constant and the power n must be deduced from the expected analytical form for the curve. The region where this formula is fitted is defined by a lower bound for the values of the MTF; 0.8 is a good compromise. If diffraction and aberrations both play a role, the situation is more complicated. The curve may then be approximated by a product of a curve of the form $a - bR^2$ for the geometric aberrations and a circon function for the diffraction effects. The first zero of the purely diffraction-limited curve must be known. As this approximation is only valid for very low spatial frequencies, it is necessary to have enough measured values in this region.

The THEOretical curve is FITted graphically interactively. The formula to be fitted may be selected from a menu. The measured curve and the theoretical one are both drawn on a double logarithmic scale. Using the cursor keys, the experimental curve may be shifted relative to the theoretical one until an optimal fit between both is obtained. From the shifting distances, the correct scaling factors for the abscissa and ordinata on a linear scale may be calculated. With the "F1" key the correct scaling of the ordinata for the experimental and the abscissa of the theoretical curve is achieved. If a careful normalization of the experimental curve was performed before, a rescaling of the ordinata should not be necessary. A careful fixing of the first zero of the experimental curve is strongly recommended as it will facilate the fitting appreciably. The fitted experimental curve is plotted later with the usual linear representation of the experimental data.

9.4 Evaluation

9.4.1 The Tasks

In all experiments a MTF is measured under distinct experimental conditions. A theoretical curve is fitted to the experimental results and the first zero of it is determined. Its dependence on the experimental parameters is investigated.

The following general remarks concern all figures in this section: Examples of measured MTF are displayed on the left side, and the result of the whole series on the right. The single measured values of the two branches of the MTF are represented by rectangles and triangles. The solid curve is the theoretical curve. In the evaluation, the results of the single experiments are designated by circles, and the solid line represents the fitted straight line. The correlation coefficient of the fit is designated by r, the slope of the line by m, the diameter of the stop by D, the width of the detector slit by b and the defocus by z.

The software is designed for performing five series of experiments, namely: Investigating the influence of the detector slit and of defect of focus; investigating the combined influence of defect of focus and diffraction in two subseries, one for prevailing diffraction and one for prevailing defect of focus for incoherent illumination; and, for nearly coherent illumination, investigating the effect of pure diffraction and of defect of focus with negligible diffraction.

9.4.2 The General Procedure of Evaluation

The evaluation procedure is generally a combination of two steps: preprocessing and evaluation.

In the real evaluation there are again two tasks: The normalization for spatial frequency zero and the fitting of the theoretical curve.

One has to distinguish three situations for the normalization: The ideal case, if one may use an expansion of the form $D(R) = 1 - kR^n$; the more complicated situation, if one has to take diffraction into account: one must know the zero produced by diffraction effects and the power produced by the further influences, for instance 2 for geometric aberrations; and finally a situation where an analytical approximation in the surrounding of $R = 0$ is useless. In this situation the fitting of the theoretical curve on a double logarithmic scale may find the correct scaling factors for both axes.

If a good fit for low and high frequencies at the same time is impossible, a fit for low frequencies will usually lead to the correct result. At high frequencies a number of errors may occur.

9.4.3 Influence of the Detector Slit

In the first series of experiments, the influence of the detector slit is investigated. It may be described by a convolution of the point spread function with a rect-function describing the transmittance of the slit, and by a multiplication of the MTF itself by $sinc(bR)$, where b is the width of the slit. As the MTF itself is approximately 1, the measured curve may be fitted by $sinc(aR)$. The power for normalization is 2 and the first zero is characterized by

$$bR_0 = 1 \qquad \text{or} \qquad \frac{1}{R_0} = b.$$

As a consequence $1/R_0$ is plotted over b (Fig. 9.8). One expects a linear curve ascending with slope 1 and intersecting the abscissa at b=0. A slope significantly deviating from 1 may be explained by a magnification of the image different from 1. An intersection of the abscissa at a point a_0 may be interpreted as a misadjustment of the zero position of the screw used for tuning the width of the slit.

A magnification different from 1 must be taken into account in all further evaluations.

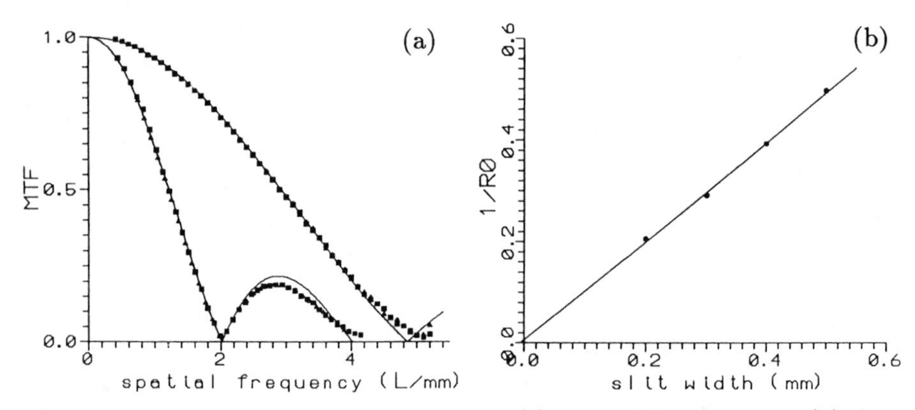

Fig. 9.8a,b. Variation of detector slit; D=10 mm. (**a**) slit width b=0.2, 0.5 mm; (**b**) plot of $1/R_0$ over slit width b; m=0.975, r=0.999

9.4.4 Pure Defect of Focus

In the second series, the influence of defect of focus is investigated. The theoretical formula is (see exercise 4.a in Sect. 9.2.6)

$$\frac{2J_1(2\pi AzR)}{2\pi AzR}.$$

The power for normalization is 2 and the first zero is characterized by

$$2AzR_0 = 1.22 \qquad \text{or} \qquad \frac{1}{R_0} = \frac{2A}{1.22}z.$$

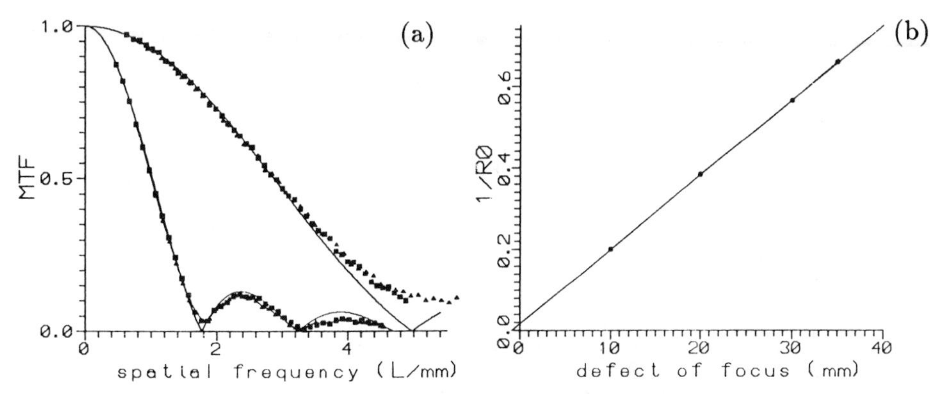

Fig. 9.9a,b.. Variation of defect of focus; D=10 mm. (a) z=100, 300 mm; (b) plot of $1/R_0$ over defect of focus z; m=0.0183, r=0.99996

$1/R_0$ is plotted over z and a linear relation results (Fig. 9.9). From its slope an effective value for the numerical aperture A may be found. If the illumination of the pupil is not uniform, the value of A derived from the parameters of the experiment may deviate significantly from that deduced from the plot – in our experience up to 5% or even more. This difference may be important before the form of the experimental curve deviates substantially from the theoretically expected one.

One needs a ground glass for uniform illumination of the pupil. Furthermore, the detector slit must be narrow. In this way the available light is low and the statistical errors are large. They may be reduced by multiple measurements.

9.4.5 Diffraction and Defect of Focus

a) Pure Diffraction The next series of experiments is dedicated to the investigation of diffraction effects. At first a single curve is measured for pure diffraction. The theoretical formula to be fitted is (see exercise 4.b of Sect. 9.2.6):

$$\operatorname{circon}\left(\frac{R\lambda}{2A}\right).$$

From the zero and the used wavelength the numerical aperture of the system may be found (Fig. 9.11a). The error is usually in the range of a few %.

b) Diffraction and Small Amounts of Defect of Focus If the aperture is very small even large amounts of defect of focus have only a limited effect on the MTF. The next series are aimed at demonstrating this. A fitting of a theoretical curve besides that for pure diffraction is senseless. It is recommended to use the normalization constant of Sect. 9.4.5a and to control it with that of the first curve of Sect. 9.4.5c (Fig. 9.11a).

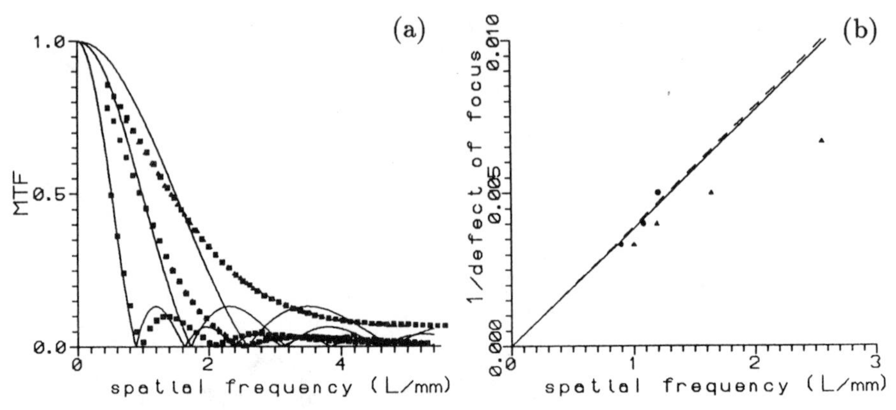

Fig. 9.10a,b. Combined influence of defect of focus and diffraction. (**a**) defocus z=100, 150, 300 mm; (**b**) plot of 1/z over first zeros R_0 of MTF; measured values: triangles, fitted values: circles, solid line: fitted by hand, dashed line: calculated from the parameters of the experiment

c) Diffraction and Large Amounts of Defect of Focus In the case of a larger aperture one has first to measure the curve for pure diffraction as one needs its zero for the normalization of part of the other curves or, preferable in our experience, its normalization constant for the rest of the series. In this case it is recommended to repeat the measurement of the focussed curve at the end of the series and to use it for a control of the normalization. Usually the zero will not appear on the display as it is situated at too high a spatial frequency. Its value may be guessed from the diameter of the pupil stop, and the zero of the curves of the preceeding series to shorten the fitting process.

The evaluation of the rest of the series is best started with the largest amounts of defect of focus. The inverse of the defect of focus is plotted over the zeros and a linear relation results, as in the second series of experiments (Sect. 9.4.4, Fig. 9.10). As the zero of the experimental curve is shifted more and more to the right of the zero of the theoretical curve for pure defect of focus, the fitted values of the zeros become more and more uncertain. For the last curves it is better to extrapolate from the zeros for large defect of focus and not to fit it to the experimental results.

As the experimental curve has, for small spatial frequencies, values less than the curve for pure defect of focus (a consequence of the theorem of Hopkins) but its zero is larger, there is an intersection of both curves. The result is a good demonstration of the fact that Hopkins theorem is only valid for small spatial frequencies.

9.4.6 Quasi-Coherent Illumination

The rest of the experiment is performed with a slit-like illumination. In the case of illumination with a small light source, the MTF can be separated into two factors. One is the MTF for perfectly coherent illumination, the second a damping factor that depends on the extension of the light source.

a) Pure Diffraction and the Effect of Partial Coherence In the case of pure diffraction, this damping factor can easily be calculated from Abbe's theory: The secondary image is built up from the primary one by interference. Interference is only possible between homologous points of the light source itself and both first diffraction orders. Therefore the portion of light that may interfere diminishes according to a function that is portrayed by a circle or an ellipse. If the energy that is diffracted into both first diffraction orders is small compared to the energy in the zero order, that circle-like function is identical with the MTF.

If the grating has a large contrast, the portion of light diffracted into the first two orders is no longer negligible. The denumerator of the expression for calculating the modulation decreases with increasing spatial frequency as the amount of light from the first diffracted orders transferred through the pupil diminishes with increasing spatial frequency. Therefore the calculated modulation is larger, as would be expected from the circle-like function for the MTF. For median spatial frequencies, this leads to values for the MTF larger than 1. Such values larger than 1 for the MTF are typical for the imaging of objects of large contrast with partial coherent illumination and characteristic for nonlinear effects.

In the outer part of the sector star, its Fourier spectrum is very similar to the spectrum of a grating: there are distinct diffraction orders. In the inner part the situation is different. Whereas for larger diffraction angles the first diffraction order has a sharp border, it is not the case for lower ones: The energy diminishes gradually (see Appendix 9.A). This fact may be observed in the primary interference image. As a consequence, the contrast will not vanish abruptly at the resolution limit, but there will appear a measurable modulation beyond it. The experimental curve has a small "tail" at its high-frequency end.

Due to coherent noise (speckles) the statistical errors for measurements with coherent illumination are larger.

On account of the large number of systematic deviations from the expected theoretical curve, the fitting is rather uncertain. As the experimental curve is curved upward for low spatial frequencies, an analytical normalization is possible but cumbersome due to special items of the program. The normalization is only practical in the double logarithmic plot. Due to the tail of the experimental curve, a precise fitting of the high-frequency part is also difficult. It is best to fit both curves tangent at the high-frequency turning point of the experimental curve. Usually the resulting zero of the fitted curve is too large (Fig. 9.11b).

b) Coherent Illumination and Defect of Focus For this situation the MTF is the real part of the pupil function or, according to Sects. 9.2.4 and 9.2.5, $\cos\left(\pi\lambda z R^2\right)$. The power for normalization is 4. Due to the damping that is

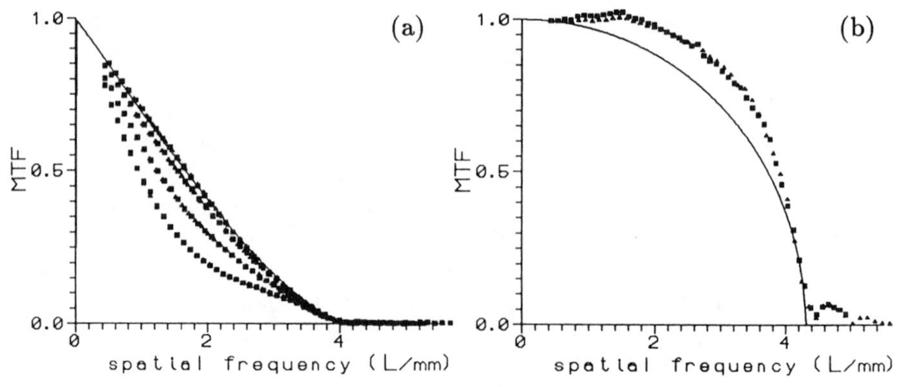

Fig. 9.11a,b. Influence of diffraction. (**a**) incoherent illumination; D=1 mm, R_0=4.076 l/mm, z=0, 100, 200, 300 mm; (**b**) nearly coherent illumination; D=2 mm, R_0=4.306 l/mm, z=0 mm

generally present for partially coherent illumination, the measured values are smaller than the theoretical ones. The best fit is achieved if the experimental curve includes the theoretical one symmetrically in the surrounding of the first zero.

The zeros are found from

$$\lambda z R^2 = \frac{1}{2} \qquad \text{or} \qquad \frac{1}{z} = 2\lambda R^2 .$$

Thus $1/R_0^2$ is plotted over z and a linear relation results (Fig. 9.12). From its slope λ may be derived. As all quantities involved in this calculation are known rather precisely, the zeros are well defined and the large amounts of defect of focus may be measured accurately, the result is precise and errors are usually less then 2%.

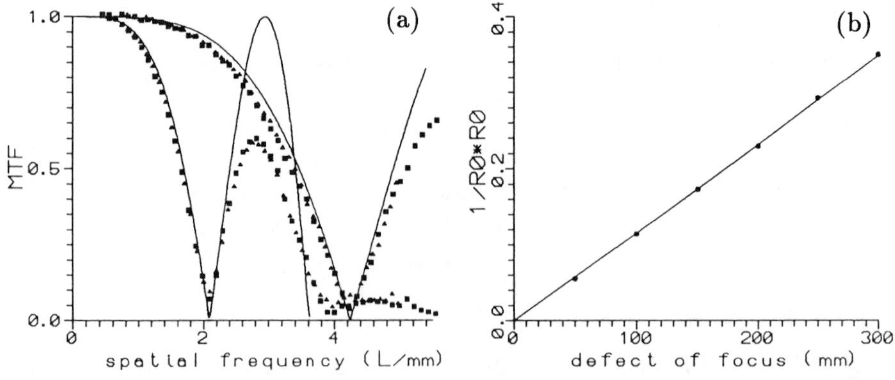

Fig. 9.12a,b. Influence of defect of focus, nearly coherent illumination; D=∞. (**a**) z=0, 200 mm; (**b**) plot of $1/R_0^2$ over z; m=0.00116, $\Rightarrow$ λ=578 nm, identical with theoretical value

9.5 Didactic and Pedagogical Aspects

9.5.1 Goals

The didactical goals of an experiment designed for teaching are usually two fold: repetition and confirming of subjects that should be known to the student, and the introduction of new fields.

Subjects of repetition are firstly diffraction optics with diffraction by a grating of infinite and finite length, by a slit, and by a circular opening, secondly the difference between Fraunhofer and Fresnel diffraction, and finally the resolution of an optical instrument, the Rayleigh criterium and Abbe's theory of the microscope.

New fields to be introduced are, generally, the problems of image quality and the possibility of image processing, Fourier optics as a tool for studying imaging processes, Duffieux's theory of the imaging of incoherently radiating objects, and the properties of transfer functions, details of Fresnel diffraction and the interaction of geometric aberrations and diffraction.

The experimental skills learned in this experiment are predominantly the adjustment of optical instrumentation and the usage of the autocollimation principle, also the handling of a photo multiplier and a band pass amplifier.

In none of these contexts is the computerisation of the experiments of importance. The advantage of the experiment is only that the acquisition of data is easier, and so a larger number of curves may be measured and their properties studied. In computer science the experiment in no way improves the student's programming skills. It trains him in handling a complex menu-guided program and it introduces the possibilities of graphically interactive data processing.

As the experiment demands a good overview over the field of classical diffraction optics and some knowlege of geometric optics, it is only suitable for graduate students.

9.5.2 Interpretation of Data

The perfect procedure for the interpretation of experimental data would be as follows: The student has a sufficient knowledge of the theoretical basis for the experiment, makes a thorough analysis of the situation and selects, or better deduces, the adequate formula for the evaluation of the data. Such a procedure would be called deductive. It can operate with a small number of experimental data.

If there is enough experimental material, another procedure could also be possible: The student studies the results for a series of experiments and tries to find their characteristic properties. Next he compares the properties of different series. Thus he may be able to select the correct theoretical interpretation even if he is not able to explain all relationships precisely. If he was successful, he may use the results to gain a better theoretical insight. Such a procedure will be called inductive.

A necessary prerequisite for the inductive procedure is a sufficiently large set of experimental data and a computerized experiment is very suitable for producing them.

There is no doubt that the deductive procedure is intellectually more delightful. However an experimental physicist must also be able to use the inductive procedure. In the special situation of the described experiment, the prerequisites for a successful application of the deductive procedure by an average student are not so good: His basis in diffraction and geometric optics must be very sound so that he is able to deduce the adequate point spread functions and from it the correct MTF. If this is not the case, there is great danger for him to mix things up, especially in connection with Fresnel diffraction: In the geometric optical limiting case, the Bessel function appears in the transfer function; in the limiting case of Fraunhofer diffraction it appears in the point spread function.

9.5.3 Presentation of Data

The logarithmic representation as a mean for linearising an exponential function will be well known to the average student. That it is generally useful for finding scaling constants in general relations is not well known. It is a powerful tool especially in the context of graphically interactive data interpretation.

9.5.4 Complications and Limitations of the Method

The experimental set up used in this experiment is rather complicated and a precise adjustment is a necessary prerequisite for good results. So there exists the danger that the student is controlled too strongly and feels led like a child. But this bothers only qualified experimentalists. A good compromise is to show the students how to adjust the first collimator and to use this opportunity to teach him the usual methods for adjustment. For the remaining part of the adjustment he will get more and more freedom, so that he finally should be able to adjust the system under test without any help. How autonomously the student works in the rest of the experiment is partly dependent on the time he is able to invest and on the number of curves for the single series the supervisor will demand. The ideal situation would be that the student will do the adjustment for the series and measure a characteristic curve without any help. Usually the supervisor is able to notice the consequences of a severe misalignment of the set up in the result, and so he may decide whether he has to control the adjustment or not.

Another danger common to the usage of menu-guided programs in didactical experimentation is that the student may know the necessary steps from his predecessors. No doubt this will reduce his profit from the experiment. However, there remains the benefit that he has to handle a large number of experimental results and from this training there will remain a profit for him in any case.

9.5.5 Applications of Fourier Optics

What is the general use of the knowledge that the student has acquired by doing the described experiments? Firstly he will be acquainted with the problems of image quality. He will know that resolution is only one figure of merit, but the problem as a whole is described by the transfer function. However, he will know that by using the filter equation, not only is a calculation of the image of an object possible but also by inversion of that relation, the object may be reconstructed from its image. This would be a first glance at the large field of image processing, where the principles of Fourier optics play a predominant role.

The author thanks his colleagues Mr. M. Seel for his help with TEX, Mr. K.H. Schmidt for developing the ECB-bus interfacing, Dr. J. Spahn for design work, and the graduate students Mr. Stölzle for realizing the preprocessing electronics, Mr. St. Enders for converting the software from BASIC to PASCAL, and Mr. H. Schulz for his help in dealing with the SAA standard.

Appendix 9.A: Diffraction by a Sector Star

9.A.1 Fraunhofer Diffraction If a slit is illuminated by a laser bundle and imaged on a sector star, a slit-restricted Fraunhofer diffraction image may be observed. If the slit is imaged on the outer part of the star, the Fraunhofer diffraction image will be very similar to the diffraction image of a grating. The spatial frequency of this grating is the local spatial frequency of the sector star at the middle of the slit. If it is imaged on an inner part, there will be no unique spatial frequency and the diffraction orders will be broadened. This effect may be seen in Fig. 9.13. One may notice that the broadening extends mainly to the low-frequency side of the order and that it is composed of several fringes parallel to the order. The fringes split up into single maxima. The existence of the fringes is experimentally confirmed. The splitting up into maxima may be due to aliasing.

9.A.2 Fresnel Diffraction As one can deal with a sector star as a composition of small gratings with a definite local spatial frequency, a narrow laser bundle is diffracted in the same way as by a grating of that spatial frequency. If the bundle scans the star radially, the diffracted bundles hit a plane somewhat behind on hyperbolae. If a radially orientated slit is imaged on the star with a very small aperture, the diffraction image behind the star is composed of these hyperbolae (see Fig. 9.14). If the slit image is

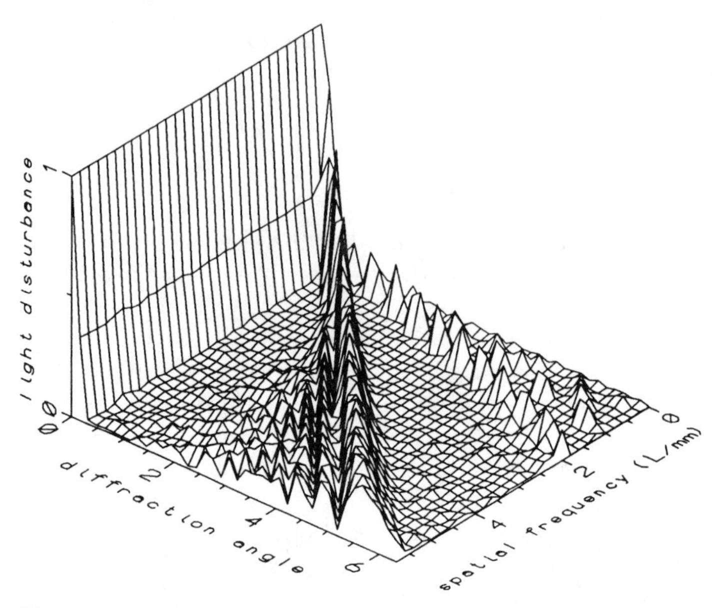

Fig. 9.13 Fraunhofer diffraction of a very narrow laser bundle by a sector star. The light disturbance is plotted over the diffraction angle and the spatial frequency at the centre of the bundle

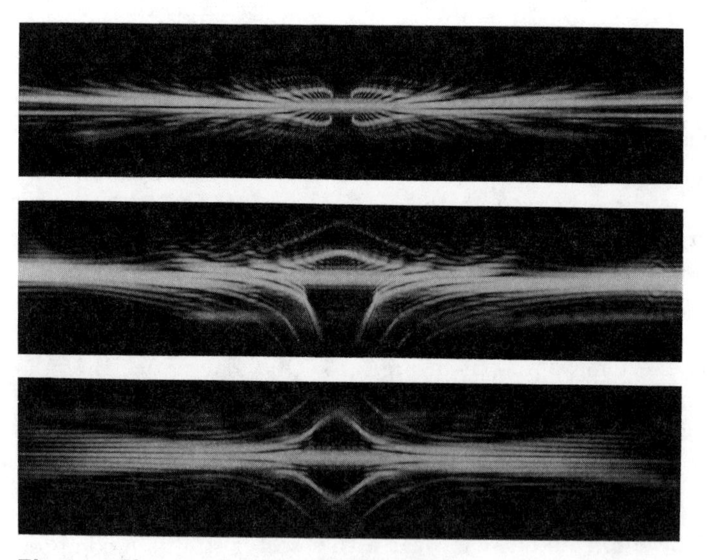

Fig. 9.14 Photographs of Fresnel diffraction at a slit imaged on a sector star; from top to bottom: defocussed, decentered, focussed and centered

decentered, the hyperbolae are deformed assymetrically (Fig. 9.14). If the slit image is defocussed, the hyperbolae are bent away from the centre of the star (Fig. 9.14).

References

1 E. Hecht: *Optics* (Addison-Wesley, Reading, Mass. 1974)
2 Duffieux, P.-M.: L'Intégral de Fourier et ses Applications a l'Optique, Masson et Cie, Paris 1970
3 P. Lindberg: Optica Acta **1**, 80 (1954)
4 H. Pulvermacher: Optik **23**, 88 (1965/66)
5 W.H. Press et al.: *Numerical Recipes in Pascal. The Art of Scientific Computing* (Cambridge University Press, Cambridge 1988)
6 R.G. Haslier and K. Fahnenstich: *Turbo Pascal Tool box.* Eine Einführung in den SAA-Standard mit einer umfangreichen Toolbox zur Realisierung von Benutzeroberflächen (Markt & Technik Verlag AG, Haar bei München 1989)

Part V

Nuclear Physics

10. Nuclear Spectrometry Using a PC Converted to a Multichannel Analyser

J.S. Braunsfurth

Conventional nuclear spectroscopy instrumentation is very costly. It requires trained operators to master the complex user interface to successfully perform even quite simple experiments. Both factors generally preclude introductory experiments beyond the most trivial ones for first year students, and limit considerably the possibilities to do experiments in advanced practicals with state-of-the-art equipment.

Viable concepts for converting a PC into a multichannel analyser for nuclear spectroscopy are discussed in detail, including hardware requirements, user software interface, and the minimal necessary software utility package required for the use of such a system with students not yet familiar with computers and sophisticated experimental equipment.

Sample experiments suitable for first year practical (physics and natural science students), and for an advanced undergraduate practical (second/third year physics students) are presented.

10.1 Introduction

10.1.1 Hardware Concept

Student instruction in introductory or advanced laboratory courses using state-of-the-art multichannel analyser (MCA) systems for nuclear spectroscopy experiments is plagued by prohibitive investment costs.

To circumvent this problem, a plug-in spectroscopy Analog/Digital Converter (ADC) board[1] for a suitable low-end personal computer (PC) has been developed, combined with an elaborate Pascal/Assembler software package.

At the time of the first implementation of the hard- and software described here[2], accelerated Apple IIe[3], and later, Apple IIgs PCs yielded the best price/performance ratio. They achieve event handling capabilities

---

[1] A 100 MHz 4096 channel- (12 bit) Wilkinson ADC.

[2] First version 1982/3, essential features of actual version approx. stabilized since 1987/8.

[3] Apple IIe, resp. IIgs are registered trademarks of Apple Computer Inc., Cupertino, CA., US.

and response times comparable to medium class state-of-the-art laboratory MCA systems. Therefore, they are used here for all kinds of conventional nuclear spectrometry experiments, in every day laboratory work and introductory or advanced student's practicals.

10.1.2 Target Group

This report will concentrate on two examples of experiments which may be configured flexibly either for first year physics, natural sciences and medical students, or for advanced physics students in the last phase of their practicals. The reduced versions for groups of two or three first year students are intended to be done in a time frame of half a day. The full versions meant for groups of two third year physics students will take two or three practical days to complete.

10.1.3 MCA Design Alternatives

Looking at the hardware structure, there are three possible general MCA design concepts, which for this purpose shall be named "classical", "modern", and "recent".

a) The "Classical" MCA Concept It is a hard-wired MCA system, consisting of a switch-controlled spectroscopy ADC, a hard-wired MCA memory array combined with a dedicated arithmetic/logic unit (ALU), and a hard-wired screen display of the data histogram. The ALU is used for data histogram generation and run time management. For data evaluation, the MCA is linked to a minicomputer or PC through an output-only interface. In many cases, the ADC is designed as a separate plug-in unit that may be exchanged.

b) The "Modern" MCA Concept Here, the ADC, the MCA memory array, and the interface to the outside world are managed by a firmware controlled single board computer. Usually, in order to minimize management overheads and system dead time, the ALU, including the run time management, and the screen display of the MCA system still are hard-wired.

The firmware may contain sophisticated calibration, display and pre-evaluation options. As these functions have to be adapted to the hard-wired ALU and MCA display, they are not compatible with – or portable to – the computer used for data evaluation. Although the data interface to host computers usually works in both directions, there are severe restrictions on the options applicable to data transferred back into the MCA system.

c) The "Recent" MCA Concept This is in fact a modification of the "modern" concept. The firmware-controlled single board computer is replaced by a small PC system. This PC also has to handle the data display and ALU functions, including the run time management. The latter topic proves to be the essential bottleneck of this concept.

Employing a PC instead of the ROM-based controller of the "modern" concept has several obvious advantages. Firstly, the machine language routines controlling the ADC and the experiment are freely accessible and may be altered to comply with changing experimental requirements. For this, quite sophisticated program development tools are at hand. Secondly, the operating system of the PC usually provides an almost complete program shell containing the necessary functions for data display, evaluation and communication to outside computers, and to peripherals.

To qualify for this task, the PC has to satisfy several hardware criteria.

1. The hardware structure has to be open, so that the ADC board can be interfaced directly to the processor bus. The overhead of a standard I/O interface between ADC and CPU would slow down the MCA system intolerably – the response time required is $\lesssim 20\mu s$, (preferably $\lesssim 10\mu s$).
2. The CPU preferably should support random access hardware handshaking with peripherals. This shortens the ADC device driver code by up to 40%, with a corresponding improvement in ADC event handling response time.
3. CPU-compatible LSI interface circuits should be available, which contain at least one fast hardware event counter and one timer of sufficient capacity. The minimum capacity needed is 16 bit – 24 bit or 32 bit are preferable. Otherwise the run time handling overhead becomes excessive.

Of no less importance is the structure of the PC's operating system.

1. It should support at least one complete high level programming language (Pascal, Fortran, C) including a Macro Assembler, service routine libraries ("Toolboxes"), etc., and data transfer to/from other computers. The last feature is important, as it allows free access to all MCA data structures, making available to the host computer all display and data handling capabilities of the MCA system.
2. For efficient use of the CPU as random access event handler, the operating system management overhead has to be either very small, or it must be possible to circumvent it altogether during time-critical operations.
3. The operating system should be user-fault tolerant, flexible, and easy to become acquainted with – at least to the standards of the rather ancient but comfortable UCSD Pascal[4] environment. [10.9–11]

Taking into account all six conditions mentioned, the range of choices becomes deplorably restricted. Although today the Apple II family of PCs must be considered obsolete, it appears difficult to find an adequate substitute yielding at least comparable MCA system specifications for a comparable price tag.

[4] Trademark of The Regents of the University of California.

10.2 Basic Physics

The experiments deal with phenomena accompanying β, γ and bremsstrahlung emissions from radioactive sources.

One of the basic goals of a practical should be to teach the students to retrieve – on their own – the necessary information from scientific literature. However, for a beginner's practical, there are limits to the extent a student is able to scan specialized literature in addition to the standard textbooks. Important topics relating to the experiment in question should be supplemented, if they are either too widely scattered through the literature, or there is only limited access to the sources.

For these experiments in the practical course, the students are provided with an exerpt based on the Nuclear Data Sheets [10.8] containing relevant decay schemes, some charts[5] relating to the interaction of γ radiation with matter, and a short review of concepts important in this context, to supplement the standard textbooks. In order to give an impression of its extent, this chapter presents a sketch of this supplement.

10.2.1 Interaction of Electromagnetic Radiation with Matter

In the energy range between 50 keV and 3 MeV, the interaction of electromagnetic radiation with matter is mediated by three mechanisms. In all of them, the electrons bound in the atomic shells play an essential role. They are:

a) Compton Effect The scattering of a γ photon by the weakly bound outer shell electrons of atoms is called *Compton effect*. The photon ionizes the atom and transfers part of its energy to the outgoing electron, which – like a low energy β particle – has a much higher absorption probability in the surrounding matter than the residual γ photon. After Compton scattering, the photon has lower energy E_γ^{c} (i.e. longer wavelength), depending on the scattering angle θ according to

$$E_\gamma^{\mathrm{c}} = \frac{E_\gamma}{1 + \dfrac{E_\gamma(1-\cos\theta)}{m_{\mathrm{e}}c^2}} , \tag{10.1}$$

while the outgoing Compton electron carries the energy $E_{\mathrm{e}^-}^{\mathrm{c}} = E_\gamma - E_\gamma^{\mathrm{c}}$.

$m_{\mathrm{e}}c^2 = 511\,\mathrm{keV}$ is the rest mass of an electron.

b) Photoelectric Effect Here, the incident photon transfers its total energy E_γ mostly to inner shell electrons of atoms in a tightly localized multiple ionization process. The range of these photoelectrons is so low that usually they will be absorbed near their point of origin. The probability of an ab-

[5] Based on the *Harshaw Radiation Detectors Catalogue*, Harshaw Chemicals Comp., Solon/Ohio USA, and De Meern, Holland.

Table 10.1 Dependence of the linear absorption coefficients on Z of the absorber element and on the photon energy E_γ, for the three interaction types. The total linear absorption coefficient is given by $\mu = \tau + \sigma + \pi$

Interaction	symbol	Z-dependence	energy-dependence
Photoelectric effect	τ	$\propto Z^4 \dots Z^5$	$\propto E_\gamma^{3.5} \dots E_\gamma^{-1}$
Compton effect	σ	$\propto Z$	$\propto E_\gamma^{-1}$
e^--e^+-pair production	π	$\propto Z^2$	$\propto \ln(E_\gamma - 2m_e c^2)$

sorption by photoelectric effect depends very strongly on the atomic number Z of the absorber material and the incident photon energy E_γ (see Table 10.1).

c) e^--e^+-Pair Production Electron-positron pair production may occur inside the electric field of an atomic nucleus, if the energy of the incident photon exceeds two electron rest masses, i.e. $E_\gamma > 2m_e c^2 = 1022\,\text{keV}$. Only for photon energies $E_\gamma > 1.5\,\text{MeV}$ does this effect play a notable role.

10.2.2 Absorption of Electromagnetic Radiation in Matter

In the most simple case of monoenergetic photons being attenuated by an homogenuous absorbing layer of thickness x, the outgoing intensity $I(x)$ as a function of the incident intensity I_0 is given by the absorption law

$$I(x) = I_0\, e^{-\mu x}, \quad \mu = \text{linear absorption coefficient in cm}^{-1}, \tag{10.2}$$

or, for radiation containing $k \geq 1$ different monoenergetic components,

$$I(x) = \sum_{k=1}^{n} I_{0k}\, e^{-\mu_k x}. \tag{10.3}$$

The linear absorption coefficient μ for electromagnetic radiation varies by orders of magnitude with the absorbing element's Z and the photon energy E_γ.

Instead of the linear absorption coefficient μ (in cm^{-1}), some textbooks use the quotient $\frac{\mu}{\rho}$ (in cm^2/g) under this name. This is the mass attenuation coefficient, an entity normalized to the mass density ρ, more suited for dealing with absorber materials containing mixtures of different elements. It is therefore prudent to check very carefully in which units μ is given!

The absorption law describes the number of photons retaining the primary energy $E_{\gamma 0}$ behind the absorbing layer. This number decreases exponentially. Some of the photons are removed totally by absorption via the photoelectric effect. In addition, photons are removed from the primary beam by Compton scattering or pair production, producing secondary radiation quanta of generally much lower energy and different propagation direction.

The interaction of electromagnetic radiation with matter therefore may produce several secondary quanta out of one primary photon, i.e. one or more Compton-scattered photons, Compton electrons and/or e^-e^+ pairs. Behind an absorbing layer this may *increase the total number of radiation quanta* ("radiation intensity buildup"), at the same time decreasing considerably the average energy of the secondary quanta.

When devising γ-ray shields (e.g. for ^{60}Co or ^{137}Cs sources), the possibility of radiation intensity buildup behind the shield has to be taken into account, especially with shielding materials of low average Z, such as water or concrete.

10.2.3 Interaction of Particle Radiation with Matter

During transit through matter, charged particles (e.g. α, β particles) lose their kinetic energy predominantly by multiple collision processes, i.e. ionization and excitation of atomic electron shells. The details depend strongly on particle energy, charge state, mass ratio of the interacting particles, mean atomic number and density of the absorber material's nuclei, etc. [10.1, 3, 4].

As long as the particle energy is still high compared to the electron shell binding energies, practically no decrease of radiation intensity (= number of particles/s) can be detected, and the specific energy loss $\frac{dE}{dx}$ of particles with energy E in an absorbing layer of thickness x shows only a slight energy dependence. For lower particle energy, the ion density increases along the path of the particle (*"Bragg-curve"*) [10.3].

In contrast to γ radiation, particle radiation in matter has a well-defined practical range R_p which can be described by empirical formulae.

An electron beam penetrating matter is widened by collisions with atomic shell electrons, as the particle rest masses are equal.

The practical range R_p of the electrons emitted by a β source is determined by the endpoint energy $E_{\beta,\mathrm{max}}$ of the β spectrum.

Compiled data tables frequently show, instead of R_p, the product of the mass density ρ with R_p, the so-called *mass range* ρR_p. This notation is better suited for handling mixed and/or diluted absorber materials, where the absorber shape or its element mixture are ill defined.

For $E_{\beta,\mathrm{max}} > 600\,\mathrm{keV}$ and an aluminium absorber, the empirical relation holds ($E_{\beta,\mathrm{max}} = E_\mathrm{e}$ in MeV) [10.1] (see Table 10.2)

$$\rho R_{\mathrm{p,Al}} = 0.526\,E_{\beta,\mathrm{max}} - 0.094\,\mathrm{g/cm^2}, \text{ or } \quad R_{\mathrm{p,Al}} = 1.95\,E_{\beta,\mathrm{max}} - 0.35\,\mathrm{mm}, \tag{10.4}$$

and for a lead absorber accordingly

$$\rho R_{\mathrm{p,Pb}} = 0.335\,E_{\beta,\mathrm{max}} - 0.091\,\mathrm{g/cm^2}, \text{ or } \quad R_{\mathrm{p,Pb}} = 0.29\,E_{\beta,\mathrm{max}} - 0.079\,\mathrm{mm}. \tag{10.5}$$

Table 10.2 Mass densities ρ of some common absorber materials in g/cm^3. C $\widehat{=}$ graphite, $(C_2H_3)_n$ $\widehat{=}$ polystyrene ($\approx$ nylon)

Material	Be	C	$(C_2H_3)_n$	Air	Al	Cu	Pb
Mass Density	1.84	2.2	1.04	$1.24 \cdot 10^{-3}$	2.7	8.8	11.35

10.2.4 Bremsstrahlung

The electromagnetic radiation emitted by an accelerated electrical charge is called *Bremsstrahlung.* It has to be taken into account when dealing with electron (or β) absorption processes.

Bremsstrahlung shows a continuous energy spectrum, with photon energies from 0 up to the total particle energy E_e. Its intensity is roughly proportional to E_e, and to the atomic number Z of the absorber material. The conversion efficiency η increases proportional to E_e from $\eta = 10^{-3}$ at $E_e = 14\,\text{keV}$ to $\eta = 0.74\%$ at $E_e = 100\,\text{keV}$. At higher energies the efficiency rises slightly less than proportional, e.g. to $\eta = 30\%$ at $E_e = 10\,\text{MeV}$ [10.3, 4].

10.2.5 X-Ray Fluorescence

Sufficiently energetic radiation, e.g. photons or electrons, may promote shell electrons of an atom into unoccupied excited states, or remove them totally, thus ionizing the atom.

Holes produced in an atomic shell may be filled by electrons from any less tightly bound states. The binding energy differences are emitted as X-ray photons with well-defined energies. All allowed electronic transitions between shell states will occur, producing a line spectrum characteristic for the absorber element. By transitions of L-, M-,...-shell electrons to holes in the K shell the *K series* (K_α-, K_β-,... -lines) of the absorber material is produced, which can be used for identification purposes (e.g., of impurities).

This process is called *X-ray fluorescence.* It is most intense when the incident radiation has only slightly more than the binding energy of the shell electrons. It becomes evident in the detector spectra – most of the time as an unwanted side effect – if for example the inner surface of shielding material is seen by source and detector simultaneously.

10.3 Detectors and Measuring Equipment

10.3.1 Scintillation Detectors for β and γ Spectrometry

In scintillation detectors the fluorescent light excited by ionizing radiation is used to measure the energy loss of single radiation quanta. There are two classes of substances with suitable properties:

a. The ***organic scintillators*** – special organic ring compounds dissolved in a transparent plastic material such as plexiglas. They are mechanically and chemically robust, nonhygroscopic, easy to shape conveniently, and they deliver signals with extremely well-defined timing. However, they interact with γ radiation mostly by Compton scattering, due to their very low effective $Z \approx 6$ (i.e. hydrocarbon compounds). As practically no photoelectric absorption occurs, no accurate γ-energy determination is possible from the spectra obtained with a plastic scintillator.
b. The *anorganic scintillators* – a heterogeneous group of anorganic metal salts with high Z of at least one of their components, and a high fluorescent light yield when excited by ionizing radiation. The most prominent substance in this group is NaI(Tl), produced as a machined single crystal of NaI, doped with 0.1 at% thallium. It can be produced in very large single crystals, showing fluorescent light yields of up to 20%. As iodine has the high $Z_I = 53$, it gives a high detection efficiency for γ rays via photoelectric effect. Therefore it is quite well suited for spectrometry purposes.

NaI is very brittle, very hygroscopic, and poisonous. It must be protected against shock, humidity and light by a hermetically sealed, opaque enclosure, which makes it unsuitable for α spectrometry. Its scintillations are short, they decay exponentially with a half life of

$$T_{1/2}(\text{NaI(Tl) scintillation light}) = 160 \text{ ns}.$$

They are so faint that the photo current of a photo cathode has to be amplified by at least a factor of 10^6, before it can be processed by electronic equipment. To achieve this high amplification factor with a tolerable signal/noise ratio, and with sufficient signal bandwidth to preserve the time and amplitude information of the detected events, a ***photomultiplier*** is used. It is a secondary-electron multiplier directly coupled to the photocathode looking at the detector crystal. For proper functioning it needs a stabilized voltage supply of 800 to 2500 V.

At the anode of the photomultiplier, small charge pulses are generated, whose amplitudes are proportional to the amount of radiation energy deposited in the detector crystal. A dedicated, pulse-shaping linear signal amplifier [10.5–7] is needed to prepare the detected events for registration by a multichannel analyser (MCA).

β-particle spectrometry preferentially is done using organic scintillation detectors with very thin entrance windows.
γ-radiation spectrometry detectors must have noticeable detection efficiency by photoelectric effect. This can only be achieved with detectors several cm thick, consisting of material with high effective Z, e.g. Ge or NaI.

10.3.2 Signal Recording Equipment; the Multichannel Analyser

After being processed by the linear pulse amplifier, the detector signals usually have rise times of 0.2 to 1 μs and total lengths of $< 8\mu$s, with amplitudes of 0..+10 V [10.7]. Their peak amplitude is a measure of the energy deposited in the detector by the radiation quantum.

The MCA sorts these pulse signals according to their amplitude [10.6]. The number of pulses found in each amplitude interval during the run time of the experiment is recorded in memory *channels*, i.e. in attached positions of a data array in memory. Each channel number can be attached to a radiation energy interval by calibration of the system with radiation sources of known energy.

The collected data are displayed as a histogram that shows relative frequency of absorption events versus their registered amplitude. As the latter corresponds to the energy loss in the detector, such an amplitude histogram is called *energy spectrum*, or in short, *spectrum*.

10.3.3 Energy Resolution of a Detector

Only if the radiation quantum is absorbed totally in the sensitive volume, can the detector signal be a measure of the primary radiation energy. Even for quanta with well-defined primary energy (α, γ), the total absorption peaks in the measured spectra have the shape of a gaussian amplitude distribution with non-negligible width arising from three main sources:

1. Unavoidably, the detector electronics adds noise to the input signal, deteriorating its amplitude definition.
2. The interaction processes within the detector involve 1 to 10^5 charge carriers, depending mainly on radiation energy and type of detector. In a scintillation detector only 1...3 photoelectrons per keV absorbed energy arise at the photomultiplier cathode. (A semiconductor detector generates $\approx$300 electron-hole pairs per keV absorbed radiation energy.) These numbers are so small, that the stochastical variance of the charge carrier numbers becomes visible. In a scintillation detector this factor mainly determines the energy resolution.
3. Particle radiation suffers an energy loss when penetrating the unavoidable covering layers of the source and the detector. The energy loss varies stochastically, causing an energy distribution and a decrease in average energy of the particles entering the sensitive part of the detector.

The *energy resolution* of a detector is defined as the *full width at half maximum* (FWHM) of the total absorption peak of a monoenergetic radiation (of energy E_0) in two different ways, depending on the properties of the detector.

In *semiconductor detectors*, the radiation energy-independent noise contribution of the electronic circuitry is the determining factor. Therefore, the energy resolution is given as *absolute width* Δ E (FWHM) in keV.

In *scintillation detectors*, only very few charge carriers per keV radiation energy loss are generated at the cathode of the photomultiplier, making the stochastical variance of the photoelectron number the prevailing influence, and causing an energy dependence of the resolution proportional to $\frac{1}{\sqrt{E_0}}$. Therefore, the energy resolution is given as *relative width* $\frac{\Delta E}{E_0}$ (FWHM) in %.

10.3.4 Radiation Detection Efficiency

a) α and β Particles In contrast to electromagnetic radiation, charged particles entering a detector of sufficient thickness will be totally absorbed with almost 100% efficiency. Their energy is transferred completely to the detector by ionization and excitation of atomic shells. For monoenergetic particles a spectrum with a nearly gaussian peak is produced. The amplitude corresponds to the primary energy of the particle minus its energy loss in the matter between source and detector not belonging to the sensitive volume, i.e. source and/or detector surface coatings, residual ambient gases (air!) etc. .

b) γ Radiation Here, the detection efficiency is usually considerably less than 100%, depending very strongly on detector volume, thickness, atomic number Z and the photon energy [10.2] (see Table 10.1)[6].

A γ detector always produces a continuous amplitude distribution in its spectrum. In addition, it may show a total absorption peak suitable for radiation energy determination (see Fig. 10.8).

Two mechanisms contribute to the total absorption peak – the photoelectric effect and multiple Compton scattering, or a combination of these two.

The continuum is caused mainly by single Compton interactions of incident photons, where the scattered photon leaves the detector unnoticed, while the Compton electron is detected. At amplitudes equivalent to $E_\gamma < 250\,\text{keV}$, additional peaks and continua appear in the spectrum, generated by X-ray fluorescence in source, detector, and shielding materials, and by photons Compton-scattered outside the detector. For γ energies below 1.5 MeV, e^+e^--pair production plays only a minor role.

---

[6] See also *Harshaw Radiation Detectors Catalogue*, Harshaw Chemical Comp.

10.4 Experimental Setup

10.4.1 Hardware Setup

The core of the experimental setup is a 100 MHz 4096 channel Wilkinson-ADC plug in board developed by the author. It is inserted into a free slot of Apple IIgs or IIe computers. Interfaced directly to the processor bus, it is running under immediate control of the 65C02 or 65C816 processor.

The hardware specifications of this system correspond closely to those of standard laboratory MCA systems. That is:

- input signal 0...+8.2 V into $1k\Omega$, DC coupled, unipolar or bipolar.
- input signal rise time 200 ns to $5\mu s$, fall time $\leq 5\mu s$.
- conversion nonlinearity $\leq \pm 0.2\%$ integral in upper 96% of range
- channel width variation (differential nonlinearity) $\leq 0.5\%$ (0.1%typ.).
- built-in fast lower input level discriminator (range 40...500 mV).
- coincidence/anticoincidence input, DC coupled, LS-TTL levels.
- hardware support provided for
  - dead time corrected run time,
  - elapsed net real time (net clock time of run),
  - momentary analysed event count rate of last elapsed second,
  - dead time corrected momentary event count rate of last second,
  - corresponding ADC live time in ms/s.

The quasiperiodic input event rate limit[7] is processor hardware dependent. The Apple IIe is able to analyse 23700 events/s maximum, the Apple IIgs yields 50300/s max., and the Æ-TransWarp[8] accelerated Apple IIee 68100/s maximum.

Correspondingly, the experimental front end usually consists of a conventional arrangement of radiation detector, high voltage supply, and linear signal amplifier with pulse shaping provisions (see Fig. 10.1).

In the practicals for first year natural science students, up to five experimental setups are run in parallel. To obtain compact, reliable, and easy-to-handle systems, a charge-sensitive variable gain pulse amplifier[9] [10.7] producing a charge symmetric bipolar output signal has been integrated into the voltage divider housing of the NaI(Tl) γ or β detectors.

The minimum necessary PC configuration consists of a 128 kB Apple IIe with two 140 kB 5.25" floppy disk drives.

---

[7] Note that due to Poisson statistics, the actual count rate input into the ADC has to be about 10 times higher in order to reach these count rate limits.
To avoid excessive spectral distortion by pulse pile up or preamplifier jamming, the actual detector input rates should never exceed the quasiperiodic limit rate.
Only with the accelerated Apple IIe, the intrinsic speed of the ADC limits the obtainable quasiperiodic input rate.

[8] Trade mark of Applied Engineering Inc., Carrollton, Texas 75006, US.

[9] Developed by the author.

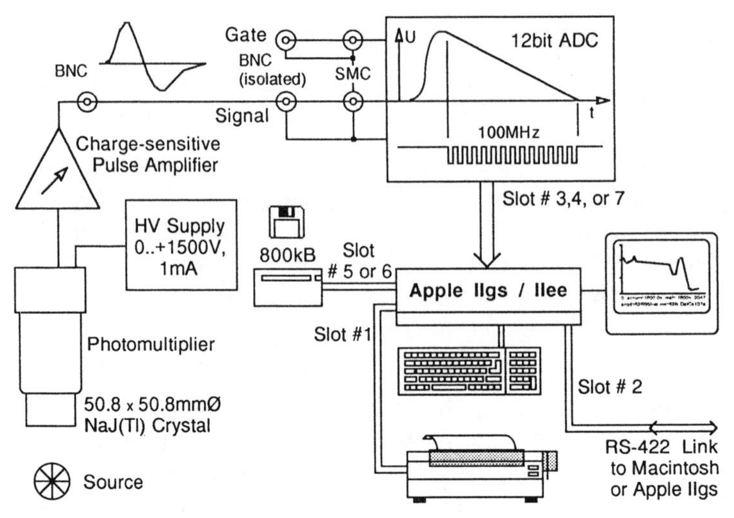

Fig. 10.1 Schematics of typical hardware setup

The preferred configuration is an Apple IIgs with at least one 800 kB 3.5" disk drive and $\geq$ 256 kB RAM-disk, or an Apple IIee with two 800 kB disk drives, ÆTransWarp accelerator board, and an additional RS-232/C serial I/O card to establish a data communications link with other computers.

A dot matrix printer, connected to a 9600 Baud RS-232/C or RS-422 serial interface port, provides graphic display hard copies and/or data listings.

10.4.2 General Structure of the MCA Program; Program Kernel

The MCA data memory is organized either as one range of 4096 channels, or as two ranges of 2048 channels each, with a capacity of $2^{24} - 1 = 16777215$ counts/channel.

The program kernel consists of an I/O error handler supervising two independent but communicating procedures for data collection and data display, i.e. the ADC event handler and the MCA display manager. Arranged around this kernel is a shell of data I/O routines and general MCA system administration utilities.

a) The I/O Error Handler Considerable effort has been invested into the program's internal error handler. As far as possible, it provides a "bomb-proof" user I/O interface, a defined recovery path within the program which includes diagnostic messages after I/O error conditions, and provisions to avoid accidental data losses. At strategic points in the program, and every half hour during experiment runs, a bit image of all essential program parameters and the MCA data arrays is saved automatically to reserved files on disk (files MCA.pref [2 blocks] and MCA.bkup [24 blocks]).

As a consequence, the user may exit the program at any time without data loss. At a later restart, the original program status at exit time is restored automatically, enabling the user to continue the experiment where he left off. In the worst case, hardware failures during an experiment will destroy the data of the last half hour run time only.

b) The ADC Event Handler This is an optimized assembly language [10.9–11] procedure. It manages the data collection into the preselected active memory range, and the various timers of the MCA system. Every half second during the experimental run, it scans the keyboard buffer register to detect display manager calls (see Table 10.3).

Like almost all PCs, the Apple IIgs hardware includes only a rather slow serial system clock. The Apple IIe has no system clock at all. For MCA timing requirements, the use of such a serial clock is very inefficient, as it introduces too much administrative overhead.

The 6522 VIA circuit[10] interfacing the ADC to the 65C02/65C816 processor bus provides a 16-bit timer register (T1), and a 16-bit timer/event counter register (T2). While T1 can be used to generate periodic real time interrupts (every 62.5 ms) in the VIA circuit, register T2 is employed in time multiplex for both ADC event rate counting (R_{acc}), and ADC live time measurement (t_{live}). The latter is needed for dead time correction of the experimental run time (t_{run}), and to compute the true ADC input rate R_{input}.

To obtain the fastest possible system response, and to avoid operating system management overheads, the hardware timer and ADC events do not generate processor interrupts. Instead, the VIA interrupt flag register is monitored by a program loop comprising only seven processor clock cycles.

Table 10.3. ADC event handler structure diagram. When active, the ADC handler can be in five different states. It idles if neither a timer event nor an ADC event is true. Different timer events generated by the VIA 6522 T1 timer occur every 62.5 ms, 500 ms, and full seconds. They have higher priority than the randomly occurring ADC events

t_{clock}	random	ΔT1=62.5ms	ΔT1=(n+0.5)s	ΔT1=1s
idle	handle ADC event, update MCA memory	update t_{real}	update t_{real}, T2 $\rightarrow R_{acc}$, update t_{live}, update t_{run}. if Keybd event then call Display Manager	update t_{real}, toggle T2 $\rightarrow t_{live}$, update R_{acc}. if $t_{run} = t_{preset}$ or t_{real} mod 16=0 or Keybd event then call Display Manager

[10] Rockwell International Corp.

This structure, however, has disadvantages. It slows down the response of the MCA display manager during the experimental run, and, in addition, precludes a real live display of the MCA data histogram.

Therefore, only a pseudo live display can be generated by periodically updating the MCA data histogram, say, every 16 seconds. Every time the update condition becomes true, the ADC event handler calls the MCA display manager. During this update, activities of the ADC event handler and its hardware timers are suspended, thereby excluding display rebuild times from the registered real times and experimental run times. For experimental conditions where these interspersed system dead times cannot be tolerated, the MCA display manager functions may be switched off, or reduced to short timer display updates during the experimental run.

c) The MCA Display Manager The MCA data histogram display is controlled by the MCA display manager, an Apple UCSD Pascal[11] procedure. To improve display processing speed, it is backed up by a number of specialized arithmetic functions written in assembly language [10.9–11].

For the pseudo live display of the MCA data, a compromise has to be found between graphic resolution and build-up speed. With the 280*192 pixel graphic display mode of the Apple IIe, histogram rebuild times of 2...5 s can be achieved. With proper display organisation and a good monochrome monitor, this gives an acceptable presentation of the MCA data.

A region of interest for display is defined, which is always an integer multiple of 256 MCA memory channels wide. It may be changed or shifted at any time.

The data from the selected region of interest are automatically compressed and scaled into a histogram of 256 data points. Vertical scaling is determined by the maximum channel content in the region of interest, rounded to the next higher power of 2. It may be modified by a zooming option (2*, 4*,..., 64*). Display modes selectable are either

1. linear display,
2. logarithmic display with optional 1 to 8 octades value range between top and bottom, or
3. a simplified Kurie plot ($\sqrt{\text{channel content}}$) for use in β spectrometry.

Ten cursor marks may be positioned on the histogram display, numbered 0 to 9. Their attached content windows show either the channel content, or the energy (or time of flight) calibration value of the cursor position. As with the cursors, the content windows may be hidden, or moved to suitable positions where they do not interfere with the histogram display (see Fig. 10.2).

For pre-evaluation purposes, three additional cursor content options have been implemented:

---

[11] Apple Pascal is a trademark of Apple Computer Inc., Cupertino CA., UCSD Pascal is a trademark of The Regents of The University of California.

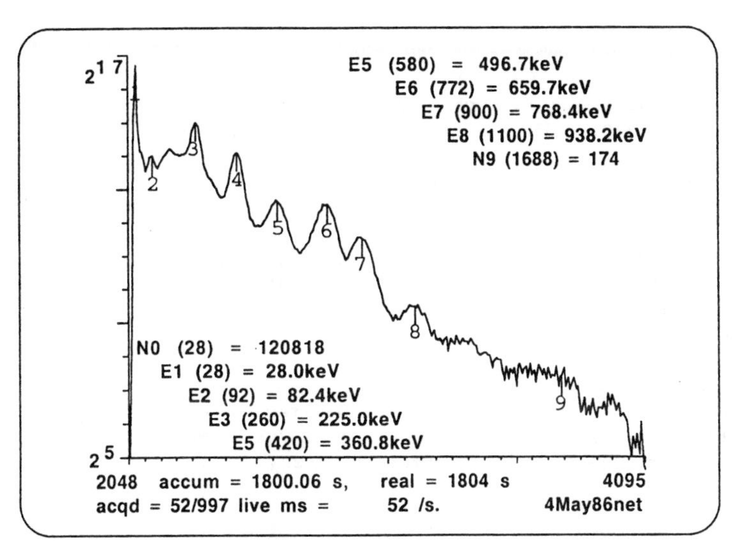

Fig. 10.2. Hardcopy of typical MCA data histogram display. The content fields of the cursors may optionally display different information. Instead of the corresponding energy, alternatively the delay time, the numeric content of the selected channel or channel range, i.e. the number of events stored for the corresponding energy or time interval, or Compton edge or photopeak energies can be shown

1. channel content summation between the positions of the last two cursors activated,
2. the display of the Compton edge energy, assuming the selected cursor's position to be that of a corresponding photopeak, and
3. the inverse of (2) – the display of the photopeak energy, assuming the selected cursor's position to be that of a corresponding Compton edge.

The latter two options have been introduced as evaluation auxiliaries when working with organic scintillation detectors, or for Compton scattering experiments with low-Z materials.

10.4.3 MCA Program Menues

To make optimum use of the limited amount of information the Apple II monitor screen can hold, either the MCA program menu or the MCA data histogram is displayed on the screen.

10.4.3.1 MCA Main Menu During start up the MCA program displays the results of a check of proper installation of an ADC board in one of the slots no. 2 to 7 of the PC, a printer interface (in slot no. 1 and/or 2), and a communications interface (in slot no. 2). If the ADC board or these peripherals are missing, the main menu is reconfigured correspondingly.

Optionally, the default dot density/font settings of the printer driver may be redefined at this point. Drivers are provided for Apple, C.Itoh, NEC, Epson and Star printers.

```
MCA Menu:        today is the 30-Jun-91
                 active Memory (1): ch 0..2047 with Fallout4.Data
new Mem Range:  <0> ch 0..4095   <1> ch 0..2047    <2> ch 2048..4096

Utilities        Submenus          Data I/O          Graphics I/O

I(nfo MCA        A(cquire          L(oad from Disk   F(oto from Disk
K(eysInfo        D(isplay          W(rite to Disk    G(raph to Disk
N(ew Date        M(ath.Ops.        P(rint            H(ardcopy
S(urvey Disk     C(alibrate        R(eceive
Q(uit Program                      T(ransmit     M

Math.Operations:..into Mem1 copy
1) Mem1-const.          2) Mem1*const.         3) N(0)*exp(-k/const)
4) Mem1-const*Mem2      5) Mem2-const*Mem1     6) dN(E)/dE of Mem2 (>0 only)
7)|dN(E)/dE| of Mem2    8) 4k Range compressed to 2k    E(xit
```

Fig. 10.3. MCA menu display screen: The upper half of the screen showing the MCA main menu, the lower half showing the selection of one of the options, the "Math.Ops." submenu. As far as possible, the program is controlled by two sets of single key commands, one set for the main menu, the other for the data acquisition and display modes (these include combinations with the open-Apple and solid-Apple special keys)

Then, to define the initial state of the program, the saved backup files MCA.pref and MCA.bkup from the previous run are loaded from disk. If they are not found, new backup files are generated with a default set of parameters.

The non-hierarchical structure of the MCA main menu provides direct access to all program functions. Restricted program modes have been avoided as far as possible. The two main working modes of the program, i.e. real time data acquisition with concurrent pseudo-live display, and display and/or manipulation of stored data, have been designed to present almost identical user interfaces and command lists.

Available program functions of the MCA menu are arranged in columns according to their type. Program utilities comprise (see Fig. 10.3)

- "Info MCA": a screen display with short hardware specifications, and disk data array declarations,
- "KeysInfo": a screen display of all key commands codes important for histogram display manipulation. It may also be called at any time when the histogram display is active. During data acquisition, this screen remains invisible in the background until called. Then it exchanges places

with the histogram display. It does not interfere with data acquisition,
- "New Date": redefines the date stored in the Pascal system, on the boot disk, and in data files written to disk, and
- "Survey Disk": a display of the directory of a selectable disk.

The dialogue entered under "Acquire" defines – or reconfirms – a data file name, and the preset time for the intended experimental run. It provides options for both – start a new run, or resume an interrupted run with new preset run time. "Acquire" – like "Display" – then calls the MCA display manager.

b) Mathematical Operations Submenu The active memory range Mem0 (4096 ch.), Mem1 (lower 2048 ch.) or Mem2 (upper 2048 ch.) selected beforehand in the main menu, serves as destination buffer for the arithmetic data manipulation to be performed (see Fig. 10.3). The Mem indices shown in the display change according to the selected range. Options 1 to 5 are auxiliary functions for spectrum stripping, and option 3 generates an exponential function with selectable start value N(0) and negative decay constant k ($\frac{1}{\text{channel}}$) for the separate stripping of exponential spectral components.

If Mem0 is selected, options 4 to 8 are not possible, and therefore are suppressed. Equivalents to options 4 and 5 for a 4k memory range are available under the "Load Data from Disk" menu.

Options 6 and 7 are auxiliary functions for the analysis of Compton continua in γ spectra. They first smooth the histogram data over a selectable number of channels to reduce the statistical variance, and then compute the differences ΔN(E) of adjacent channel contents. The result is roughly equivalent to the first derivative of the data histogram N(E) (E = energy, or time of flight, depending on the spectral calibration unit chosen).

There are two reasons why these options are implemented for 2 k memory ranges only:

1. For the unequivocal identification of spectral details on Compton continua, one should have the opportunity to view the differentiated and the original spectrum simultaneously in overlay mode. This is only possible if both spectra can be held in MCA memory simultaneously.
2. The differentiation algorithm dramatically increases the unavoidable statistical variance of the experimental data points. Although smoothing means sacrificing part of the spectral resolution of the measured data, it generally improves significantly the legibility of the spectra.

c) Calibrate Submenu (see Fig. 10.4) "Define Calibration" allows one to explicitely enter calibration values. "Calibrate with Cursor" switches the display region of interest back to the first 256 channel group of the histogram display. There, one of the cursors may be placed on a spectral detail of known energy (or time of flight, whichever is appropriate), if necessary, after shifting the region of interest window. The cursor being positioned, the calibration routine is called again, this time presenting a dialogue for input

```
place Cursor in turn on 1..10 Reference Points, then C(alibrate,
or  D(efine Calibration [keV/ch.] and Zero Offset [ch.].

Between Calibration steps, Display Window, Data Set {L(oad from Disk}
 and {2k} Memory Range may be changed.

Calibration   0.8491 keV/ch., with Offset -5 ch.

  Unit is   1) eV   2) keV   3) ps   4) ns
C(alibrate with Cursor,   D(efine Calibration,   - E(xit █
```

Fig. 10.4 Upper half of screen, showing the "Calibrate" submenu on entering a new calibration run. Once a calibration run is under way, instead of this display the coordinates of the reference points already defined are tabulated. If only the calibration units are to be changed, the program returns to this calibration start screen

of the appropriate energy or time value. After entering the last reference point, a linear regression is computed automatically. In the special case that only one reference value was entered, channel 0 with E=0 is implied as a second reference point.

d) Disk I/O Submenues (see Fig. 10.5) The options offered here correspond to those of the "Math. Ops." submenu. Instead of the other MCA memory half, the specified disk file is involved as the second operand data source, thus extending the range of spectrum stripping operations to the handling of 4 k spectra.

```
Q(uit Program                      T(ransmit  L

Delete old Fallout4.Data? <Y/N> y
Load from Vol.#<4,5,11,12>: 4
FileName: " Fallout4 " - ok? ..Y/N - E(xit N
FileName? :Tshernob .Data

Get Disk File Data:..into Mem1 copy
1) Diskfile           2) Disk-const.        3) Disk*const.
4) Mem1-const*Disk  5) Disk-const*Mem1  6) 4k compressed to 2k  E(xit █
```

Fig. 10.5 Lower half of screen, showing the "Load Data from Disk" submenu. Of the four Disk I/O submenues, only this one offers a submenu after the initial dialogue to define the file transfer source or destination

e) Receive/Transmit Serial Data Link Submenues (see Fig. 10.6) The main menu offers "Receive Data" and "Transmit Data" only, if the corresponding hardware is implemented, otherwise these menu options are suppressed. The same applies to "Acquire Data" and "Print Data".

```
Q(uit Program                    T(ransmit  T

Now ready to send Fallout4.Data as
B(itmap  T(ext or text compressed to  1(  2(  4( *256 channels - E(xit
```

Fig. 10.6 Segment of screen, showing the "Transmit" submenu

"Bitmap" transfers a bit image of the MCA data array and the complete set of experimental parameters, including file name and memory range, between two Apple IIgs/IIee computers both running the MCA program.

9600 and 19200 Baud rate settings of the RS-422 or RS-232/C interface result in almost identical transfer times.

Alternatively, the MCA data from the chosen region of interest can be transmitted in ASCII text format either uncompressed, or compressed to 256, 512 or 1024 numbers. Again, the text string additionally includes all essential experimental parameters, i.e. data file name, and all timer, count rate and dead time information.

As it makes little sense for the MCA program to receive compressed ASCII-formatted data from foreign sources, a "Text Receive" option has not been implemented.

10.5. Experiments

10.5.1 General Considerations

The spectrometry experiments discussed here are examples from a group of practical experiments implemented during the last five years. They are intended for natural science students (medicine, biology, geology, etc.), for first year physics, and for advanced (third year) physics students. These three groups each require a quite specific approach to practical organisation.

a) Experiment Organisation Second or third year physics students in advanced practicals will be provided with general information about the experiment to be performed, i.e. the kind of data to be taken, the extent of data evaluation expected, the available hardware, a 50 page user manual of the MCA program, and some hints to textbooks and supplementary literature useful for the experiment. It is then left mainly to their initiative to find their way through the experiments. Evaluation steps exceeding the features implemented in the MCA program itself are usually executed after transferring the MCA data over the serial link to other computers. The target computer chosen depends on the experimental requirements, and the preferences of the students.

A Macintosh 512e and a Macintosh Plus[12] have been available for this purpose during the last three years. Equipped with standard graphics and text processing software, they are the preferred choice if advanced graphics and/or combination with text is required. Alternatively, and preferentially, if students want to write their own special evaluation routines, Atari 1040ST or MS/DOS[13] computers are used.

The other extreme of practicals organisation requirements is represented by the first year non-physics natural science students. Here, very tight guidance and supervision by the instructor is necessary. Physics knowledge levels, and motivation for physics, are generally lower. The time available for the experiments and their evaluation is shorter (at most 4 hours). First year physics students should be treated with an intermediate approach, leaving more room for their individual initiative.

b) Detectors for β or γ Spectrometry In these practical experiments, well-type 50.8 mm*50.8 mm⌀ NaI(Tl) detectors[14] are used. This is by no means a necessary condition. For β spectrometry, dedicated β detectors with thin windows would yield even more accurate results, and would enhance the range of experimental options. At the time being, however, the 5 setups in service under supervision of the author are almost exclusively used with first year natural science students. Consequently, technical considerations such as mechanical robustness and easy handling have to take precedence. The advanced physics practical and research groups in the house employ this MCA system with a wide variety of detectors and electronics configurations.

Well-type detectors have been chosen, as inside the well the entrance windows are only $\approx$ 0.3 mm thick, in contrast to ordinary crystal mounts, where the entrance window usually is much thicker. In addition, they offer the opportunity to perform some simple, but very instructive experiments on positron annihilation, which otherwise would require two detectors and coincidence circuitry.

c) Calibration of the MCA Independent of the variant of spectrometry experiment to be performed in the practical, the first step always is an MCA system calibration run. Sources used are [10.8]

1. ^{237}Np, 59.6 keV γ from ^{241}Am-α source (Buchler/Amersham),
2. Pb-K X-ray fluorescence (76.7 keV), exited by ^{22}Na or ^{137}Cs,
3. ^{22}Na, 511 keV positron annihilation photons, and 1276 keV γ, (Buchler/Amersham source),
4. ^{137}Cs, 661.7 keV γ, (Buchler/Amersham source), and
5. ^{40}K, 1460.7 keV γ, (1 kg crystalline KOH in plastic container).

---

[12] Macintosh Plus and 512e are Trademarks of Apple Computers Inc. .

[13] Trademarks of Atari Computer Corp. and Microsoft Inc. .

[14] *IntegralLine* detectors from Harshaw Chemical Company, and *MonoLine* detectors from Bicron Corp.

10.5.2 γ-Ray Absorption; Radiation Intensity Buildup by Compton Interaction

Several variations of this experiment may easily be implemented, using high-Z materials such as Pb and Sn, or in contrast, low-Z materials such as water, polystyrene, brick or marble. Water or plastics would yield the biggest effects. However, their low absorption coefficients would lead to very large source–detector distances, which in turn causes prolonged run times inconvenient for a practical experiment (see Fig. 10.7).

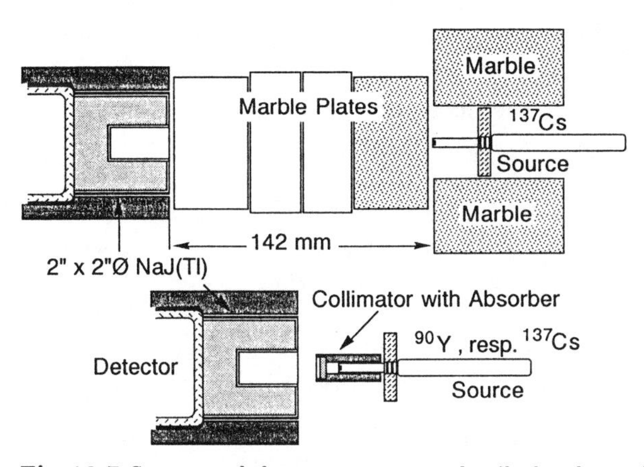

Fig. 10.7 Source and detector geometry details for the γ-absorption and β-energy loss practical experiments respectively (proportions approx. to scale). The detector crystals are shielded by removable lead cylinders. *Top* Marble plates of different thickness may be inserted between ^{137}Cs source and detector. *Bottom* Either a ^{137}Cs source with Pb collimator and Pb absorber foils is used for a γ-absorption experiment, or, for β-energy loss experiments, a ^{90}Sr/^{90}Y source is used, with polystyrene, or aluminium absorbers and collimators, respectively

As an example, the complete itinerary for the γ-absorption experiment employing marble absorbers is given below. The strict guidance rules given apply to first year students, who, in groups of three, will spend about 4 hours doing selected subsets of the complete experiment under close supervision by the instructor. As mentioned above, advanced students (third year physics) will only be provided with general information, and then left to their own initiative. After the introductory discussion of the planned experiment, the instructor will be available on request, should any problems arise.

(1) Run a calibration of the MCA system on a 2048 channel range, using the appropriate combination of reference sources, in this case ^{137}Cs and ^{237}Np. If necessary, remove the lead shielding of the detector during this part of the calibration run, in order to avoid interference of the 59.6 keV peak with the 76.7 keV Pb X-ray peak.

Adjust the distances of the calibration sources, so that the total event rate is ≈ 2000/s, therefore avoiding undesirable photomultiplier gain shifts with count rate. A run time of 500 to 1000 s should suffice for good statistical accuracy.

Use this time span to have a look into the MCA program manual. Try to acquaint yourself with the user surface of the MCA program, especially of the MCA display manager, by trying various display options during the run.

When the run has ended, chose (C)alibrate to start the calibration algorithm (see MCA manual). After entering the last reference energy value, the program's built-in linear regression algorithm computes the calibration in $\frac{\text{keV}}{\text{channel}}$ with a zero offset in *channels*. Reconfirm this result by testing whether cursors placed on the reference peaks in the calibration spectra display correct energy values. Otherwise either an error has been made during input of the reference values, or the linear amplifier has been damaged by improper handling.

If the calibration is not in the range of $0.8\frac{\text{keV}}{\text{channel}} \pm 10\%$, adjust the amplifier gain accordingly, and repeat the calibration. Plot the result into a diagram showing reference energy versus channel number of the corresponding peaks, make a cross check of the calibration, and determine its accuracy.

For first year students, this sequence should be cut short by the instructor by making sure beforehand that the calibration is in the correct range.

(2) Define the experiment geometry. During the whole experiment, both detector and source may not be moved! (see Fig. 10.7)

If the lead shield of the detector was removed during calibration, put it back in place. Fix the ^{137}Cs source at the proper distance from the detector front, so that the thickest absorber layer to be investigated just fits snugly between source and detector. Record all relevant settings of the hardware.

(3) Preset the run time to 800s (200s min.), name the data set, e.g. "CsMa000" (the trailing three digits denoting the absorber thickness), and start the first run. Store the measured spectrum on disk using the main menu option "Write Data to Disk".

Make sure that the data file was stored correctly (16 Blocks long) by selecting "Survey Data" from the main menu. This is a precaution in case the data file did not fit completely into the remaining free space of the storage disk, an error not detected by the system.

Place 6 cursors on the spectrum at positions corresponding to energies of about (see Fig.10.8)

(1) $E_0 = 60\,\text{keV}$ (low energy cut off),
(2) $E_1 = 175\,\text{keV}$ (low end of Compton backscatter peak),
(3) $E_2 = 235\,\text{keV}$ (high end of Compton backscatter peak),
(4) $E_3 = 477\,\text{keV}$ (Compton edge of ^{137}Cs),
(5) $E_4 = 620\,\text{keV}$ (low end of ^{137}Cs photo peak), and
(6) $E_5 = 705\,\text{keV}$ (high end of ^{137}Cs photo peak).

As a matter of precaution, these cursor settings should be kept constant during the experiment! Instead of moving one of them, use one of the remaining 4 cursors.

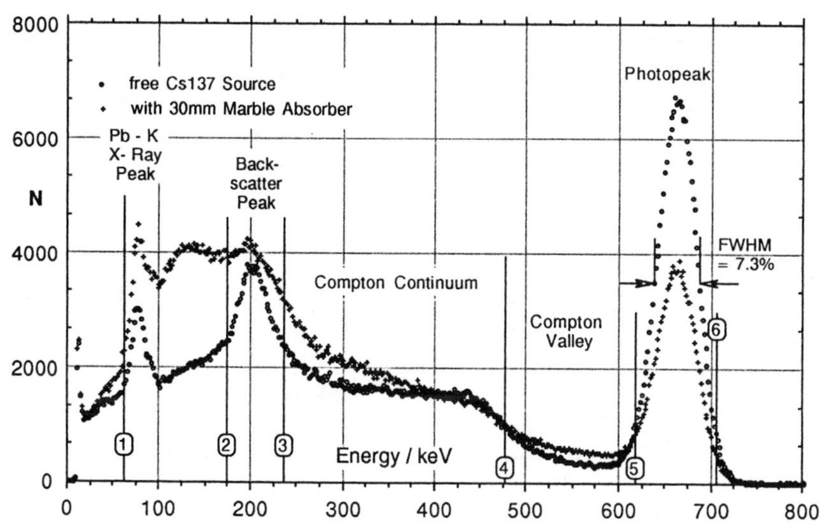

Fig. 10.8 Overlay of two ^{137}Cs γ spectra obtained with and without a 30 mm marble absorber layer between source and detector. Cursor positions numbered (1) through (6) mark the spectral ranges used for the evaluation. In addition, markers illustrating the definition of energy resolution of a scintillation detector are shown. The Pb–K X-ray peak is caused by X-ray fluorescence in the lead shielding of the detector crystal

From the recorded spectrum, get the event sums between the different cursor positions by applying the "Add" option of the MCA Display Manager, i.e. press the number keys attached to the two cursors delimiting the region of interest, then press "A". Read the result from the content window of the second cursor, and record it. Take the sums between cursors

1. (1) and (6), $\sum_{\Delta E_0} N(x)$ for the total spectrum, (x = absorber thickness),
2. (1) and (2), $\sum_{\Delta E_1} N(x)$ the region below the backscatter peak,
3. (2) and (3), $\sum_{\Delta E_2} N(x)$ the backscatter peak,
4. (3) and (4), $\sum_{\Delta E_3} N(x)$ the Compton continuum,
5. (4) and (5), $\sum_{\Delta E_4} N(x)$ the Compton valley below the photo peak, and
6. (5) and (6), $\sum_{\Delta E_5} N(x)$ for the photopeak of the γ spectrum.

(4) Place two 40 mm marble plates on both sides of the source (see Fig.10.7). Combine the various marble plates to produce different absorber layers. Place these absorbers as near to the source as possible. (Placing the absorber near to the detector, or not placing the marble plates on both sides of the source, would make a different variant of the experiment (see Fig. 10.9)). Then measure γ spectra for 3 to 6 evenly spaced absorber thicknesses up to 140 mm. Use the identical run time setting as before. Store the spectra on disk, naming them appropriately, e.g. "CsMa140" for 140 mm absorber layer. For each spectrum, take the same summations as described under step 3.

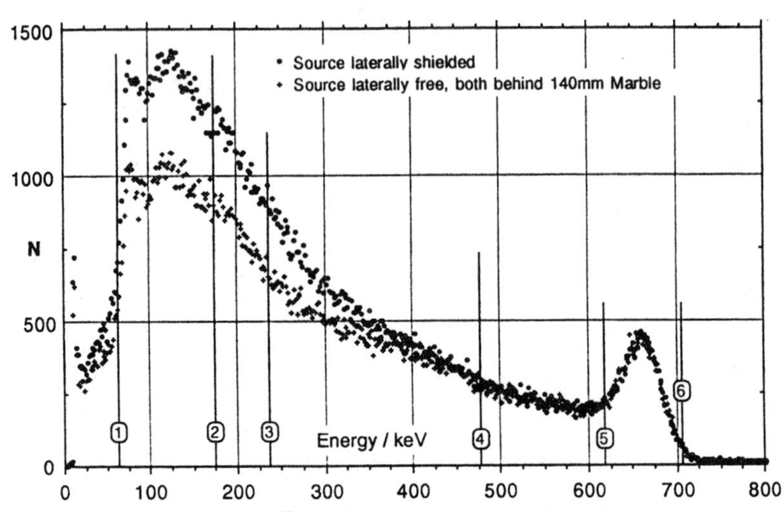

Fig. 10.9. Overlay of two ^{137}Cs γ spectra both obtained with a 140 mm marble absorber layer between source and detector, with otherwise identical settings. A more than 50% increase in radiation intensity around and below the Compton backscatter peak is caused by additional lateral marble shielding of the source, covering an additional solid angle of approximately $\frac{\pi}{2}$. The cursor positions marked are identical to those of the preceding figure

(5) For the energy intervals ΔE_k (k = 0,..,5) defined in step 3, and all absorber layer thicknesses x, compute the logarithms of the event sum ratios

$$R(k,x,\Delta E_k) = \ln \frac{\sum_{\Delta E_k} N(x)}{\sum_{\Delta E_k} N(0)}.$$

Plot $R(k,x,\Delta E_k)$ versus absorber layer thickness (see Figs. 10.10 and 10.11). It is up to the students to do this either by hand, or to use a computer of their choice.

Compare the results with γ-absorption values given in literature [10.2, 3].

Advanced students are expected to do this experiment for 5 to 7 absorber thicknesses, and at least two different absorber materials, e.g. marble (low-Z), and lead (high-Z), including a thorough calibration run of the MCA system. They usually work in groups of two on a flexible time frame.

First year students usually work in groups of three. They will do their experiments for 4 to 5 absorber depths, and one absorber material only. Step 5 will be done by pocket calculator. The results are drawn by hand on graph paper. In order to fit the experiment into the given time frame of $\approx$4 hours, the calibration run will be condensed to reconfirming an established value.

The measured spectra do not arise exclusively from the effect of Compton scattering of γ radiation in a low-Z absorber layer, but are a convolution

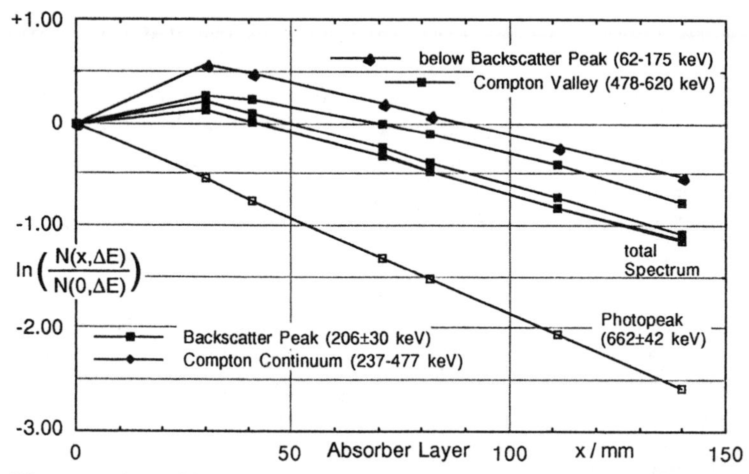

Fig. 10.10 Logarithmic plot of radiation attenuation in marble versus absorber thickness, for the different energy ranges defined above, with the single line γ source ^{137}Cs. Note the intensity *rise* below 175 keV of up to 80% for an intermediate increase in absorber thickness, demonstrating considerable radiation intensity buildup by Compton interaction in the low-Z absorber material

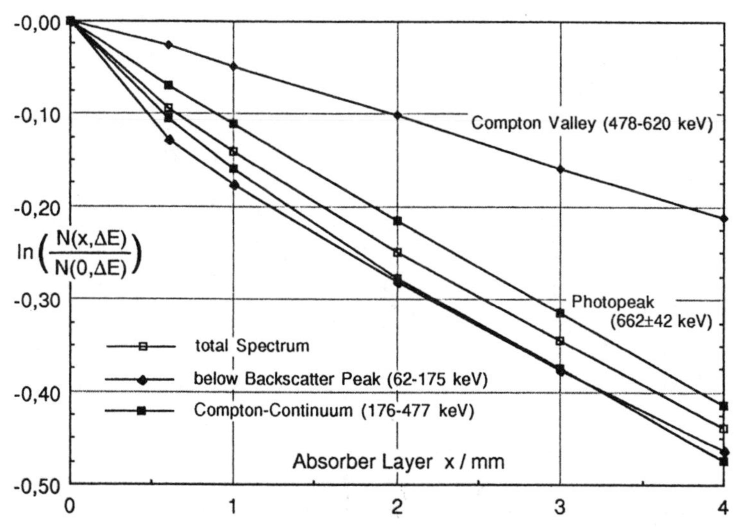

Fig. 10.11 Logarithmic plot of radiation attenuation in lead as a function of absorber thickness, for the different energy ranges defined above, with the single line γ source ^{137}Cs. Note the difference in attenuation behaviour compared to a low-Z material. Here in Pb, the absorption by photoelectric effect prevails almost totally. Radiation intensity buildup by Compton interaction is almost negligible

of this effect with photoelectric absorption and Compton scattering processes in the detector. A quantitative analysis would exceed the scope of a practicals experiment. The qualitative evaluation gives, however, a vivid illustration of the effects of radiation intensity buildup by Compton scattering. The essential information content of this experiment is difficult to arrive at by other means.

Applying the spectrum stripping options implemented in the "Math. OPs." submenu of the MCA program, the analysis of the spectra may be extended further. For example, subtracting the spectral contribution of γ rays which had no interaction with the absorber would make the effects of Compton scattering in the absorber even more visible. This step has intentionally not been included in the evaluation.

Using larger NaI(Tl) detector crystals, having nearly 100%γ-detection efficiency by photoelectric effect below 350 keV, would even allow semi-quantitative investigations on Compton-scattering cross sections. Evaluation steps in this direction should therefore be part of another variant of advanced practicals experiment.

10.5.3 β Spectrum; Energy Loss of Electrons in Matter

This experiment deals with the shape of the high energy end of β spectra, and the absorption behaviour of high energy electrons from a ^{90}Y β-radiation source in polystyrene and aluminium absorbers.

Other absorber materials, or other β sources with $\geq$2 MeV end energy, may easily be substituted. For sources with lower end energy the use of a dedicated thin window β detector would be required.

While the experiments discussed in (10.5.2) relied heavily on arithmetical tools provided by the MCA display manager, these experiments predominantly make use of spectrum stripping features implemented under the "Math. Ops." menu of the MCA Program.

Using the source–detector geometry shown in Fig. 10.7, the complete experiment itinerary to be performed by advanced students looks like this:

After calibrating the MCA system, β spectra are to be taken for a set of 6 to 8 absorber thicknesses, and for two different absorber materials. The spectra are to be stored on disk for later analysis. Again, first year students are expected to do a reduced set of experiments, measuring on one absorber material only.

(1) Like in step 1 of (10.5.2), this is a calibration run on a 2048 channel MCA range. The same procedure is to be followed. However, this time the preferred calibration value is 1.15 $\frac{\text{keV}}{\text{channel}} \pm 10\%$. Calibration sources to be used are ^{40}K, and ^{22}Na or ^{137}Cs (see (10.5.1.c)).

(2) Define the experimental geometry according to Fig. 10.7. During the whole experiment, make sure that after every change of the absorber foil, the source position is reproduced exactly. The detector window is thin and fragile, therefore be very careful while manipulating the source geometry or the detector! For the first run, use the ^{90}Y Buchler-Amersham source with an empty collimator made from the same material as the absorber foils to be used.

(3) Preset the run time to 400s (120s min.), name the data set, e.g. “YPs00”, or "YAl00" (the trailing two digits denoting the absorber thickness), and start the first run. Afterwards, store the measured spectrum on disk[15] using the main menu option “Write Data to Disk”.

(4) Place polystyrene absorbers of 1, 2, 3, 4, 5, 6, 8 and 12 mm thickness into the appropriate recess of the source collimator, reposition the source carefully, and run the corresponding set of measurements, each time storing the data on disk after naming them. (In first year practicals, only a subset of 5 absorbers is to be measured).

(5) Repeat step 4 using an Al collimator, and 0.3, 0.6, 1.0, 1.5, 2.0, 2.5, 3.0, and 4.0 mm Al absorbers. In first year practicals, only a subset of 5 absorbers is measured. In addition, either step 4 or step 5 is done, not both of them.

(6) First data evaluation step, i.e. spectrum stripping.

The thickest absorber to be chosen will stop all β particles from the source. Therefore, the corresponding spectra contain only bremsstrahlung, mainly arising in the source. Subtracting this contribution reduces the X-ray background in the spectra, thus enhancing the accuracy of the intended determination of the end-point energy of the β spectra in the next evaluation step.

Use the “Math. Ops.” or the “Load Data from Disk” submenu options of the MCA program to subtract the spectrum obtained with the thickest absorber from each of the other spectra. Beforehand, make an intelligent estimate of the proper normalization factor for the spectrum to be subtracted. Consider the influences of ADC deadtime and average analyzed event count rates as displayed by the MCA program, as well as X-ray absorption and production in the absorber foils. Store the resulting net spectra on disk after renaming them.

(7) Determination of the end-point energy of the measured β spectra. This next step may be taken in two different ways, either by means of the MCA display manager's “Kurie Plot” option, followed by a hard copy of the graphic screen, and a graphic "fit by eye" to determine the end-point energy, or by transferring the spectra to another computer for handling with more sophisticated plotting and/or fitting software (see Fig. 10.12).

[15] See remarks under step 3 of previous experiment.

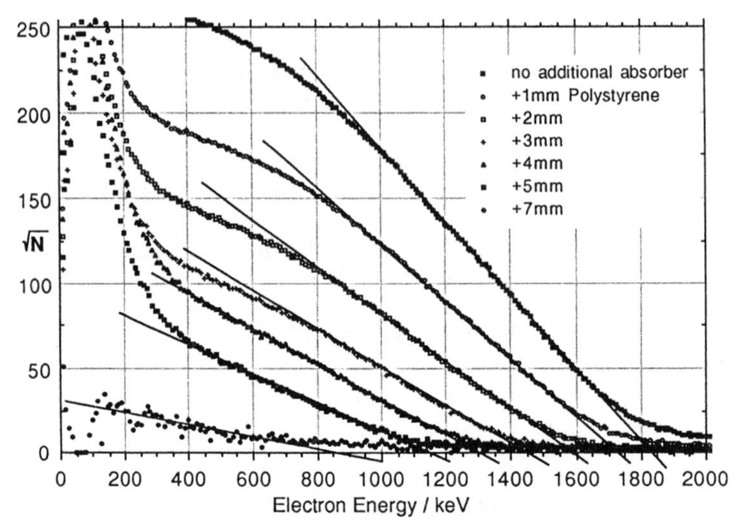

Fig. 10.12 Simplified Kurie plots of ^{90}Y β spectra, for polystyrene absorbers of different thickness. The high energy tails in the spectra are caused by pulse pile up in the detector electronics. In the Kurie plot, these tails can be distinguished clearly. In a logarithmic plot of the data, e.g., they would invariably lead to a systematic overestimation of the effective β end energies, leading to values 100...200 keV too high

The second possibility is encouraged with advanced students. In beginners practicals, the first method has emerged to be a very convenient way to evaluate the spectra, yielding results with only slightly larger error margins. As a side effect, advanced students here may try the different available methods to prepare and present data for graphic and/or numeric evaluation, assessing the relative merits.

(8) Plot the energy loss $\Delta \mathrm{E} = E_{\beta,\mathbf{max}}\ (^{90}\mathrm{Y},0) - E_{\beta,\mathbf{max}}\ (^{90}\mathrm{Y},x)$ versus absorber layer thickness x, using the known β-decay end-point energy for zero absorber thickness $E_{\beta,\mathbf{max}}\ (^{90}\mathrm{Y},0) = 2260\,\mathrm{keV}$. To obtain the unknown effective absorber thickness of the source enclosure and of the detector front window, fit the data assuming a linear dependence of the energy loss in matter for electrons of more than 600 keV energy (Fig. 10.13).

This evaluation step is skipped in first year practicals, where only the β end energies versus absorber layer thickness are to be plotted.

Variants of this experiment may include a comparison of the integral event rates from the measured spectra with the known β-ray absorption law. Also, with a thin source and a dedicated thin window β detector, the shape of allowed or forbidden spectra may be studied, etc.

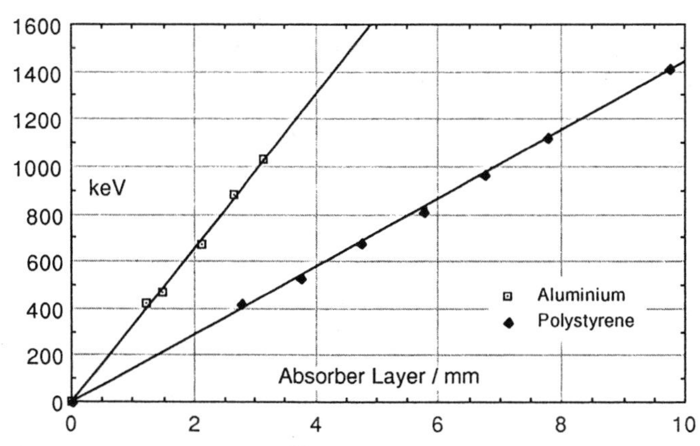

Fig. 10.13 Energy loss of ^{90}Y β particles in polystyrene and aluminium absorbers, derived from the spectra shown in Fig. 10.12 and from an analoguous experiment with aluminium absorbers. As the exact absorption properties of the source enclosure and the detector entrance window are not known, the zero absorption values have been fitted, using the known end energy $E_{\beta,\max} = 2260\,\mathrm{keV}$ of the ^{90}Y β spectrum, and assuming a linear energy dependence of the electron energy loss on layer thickness in the 2 MeV region

10.6 Student Reactions

The following remarks apply mainly to first year natural science students (medicine, biology and geology). Since the introduction of this setup as a standard practical experiments in September 1988, more than 2500 students have been seen through a 4 hour variant of one of these experiments.

Compared to this, the author's experiences with advanced students and the complete experiments are rather scarce. They come mainly from about 30 third year physics students, who, in preparation for working as assistant instructors for this experiment in the practical course, had to acquaint themselves with the apparatus and the various experiments implemented. At that stage of their studies, many of them were just completing the last experiments of their regular advanced physics practicals.

Students confronted with the experimental setups discussed here may generally be divided into four groups according to their response pattern:

1. Students already acquainted with computers. In general, students belonging to this group will immediately begin to scan the MCA program menu, try out the MCA display manager, service utilities and some I/O options, and acquaint themselves very quickly with the MCA program. Experiencing no technical difficulties with the MCA system, they can – and generally do – concentrate on the physics of the experiment. About 20% of the participants belong to this category.
2. Students afraid of touching a computer keyboard, or even any keyboard at all. These students appear to be emotionally blocked against com-

puter use for some reason external to the practical. Another $\approx 20\%$ of the participants, they are the main problem group, and apparently have severe difficulties to grasp what is going on during the experiment. Consequently, they usually are not able to gain access to the physical contents of the topics discussed.

3. Students not interested in physics at all. This group, containing up to 10% of the participants, is a phenomenon special to non-physics student groups having to pass physics practicals as a precondition for their exams. From the author's experience, considering the time frame and manpower given and the total number of students to be tutored, there is little a practicals instructor can do to motivate this group without neglecting the majority of students willing to perform their experiments properly.
4. Students unexperienced with keyboards and computers, but without emotional blocks against computer use. After initial uneasiness, probably caused mainly by difficulties in finding the appropriate keys on the keyboard, most of them develop a fair working knowledge of the MCA program during the four hours of experimenting with the MCA system. They need more or less continous supervision by the instructor, as they often appear to be helpless when unusual error conditions occur.

There is a typical error condition that occurs time and again. Desktop space in the practicals room is in notoriously short supply. Thus, every so often some student inadvertently places his notebook on top of one of the computer keyboards, thereby continuously pressing some keys. As the MCA program is protected against data loss by adverse input handling like this, nothing serious happens. However, any experiment running is stopped, or any data display cancelled, and all sensible screen displays are scrolled far off screen. In their first encounter, even most of the assistant instructors have difficulties to handle this situation properly, i.e. to get the idea to find an input key appropriate to bring back the MCA program's menu screen.

All in all, confronting the students with state-of-the-art equipment, and doing experiments previously considered to be out of range for a beginner's practical, appears in itself to have a strong motivating effect on the students.

When introducing this experiment, the author had expected a lot more complaints than actually came, as the preparational work the students are required to do for these experiments is distinctly above average. These days, nuclear physics appears to be a very unpopular – or even avoided – topic in many schools.

References

10.1 G. Knop and W. Paul: *Interaction of Electrons and α Particles with Matter*, in: K. Siegbahn (ed.): *Alpha-, Beta- and Gamma-Ray Spectroscopy*, (North Holland, Amsterdam 1965) p. 1–36. (and Appendices)

10.2 C.M. Davisson: *Interaction of γ Radiation with Matter*, in: K. Siegbahn (ed.): *Alpha-, Beta- and Gamma-Ray Spectroscopy* (North Holland, Amsterdam 1965) p. 37–78. (and Appendices)

10.3 G.F. Knoll: *Radiation Detection and Measurement* (Wiley, New York 1979)

10.4 D. Kamke und W. Walcher: *Physik für Mediziner* (Teubner, Stuttgart 1982)

10.5 Horowitz and Hill: *The Art of Electronics* (Cambridge University Press, Cambridge 1980)

10.6 H.J. Stuckenberg: *Detektor- und Experimentierelektronik* (Braun Verlag, Karlsruhe 1973)

10.7 E. Fairstein and J. Hahn: *Nuclear Pulse Amplifiers, Fundamentals and Design Practice* I–III, Nucleonics **23**, No. 7, 8, 9 (1965), and IV–V, Nucleonics **24**, No. 1, 2 (1966)

10.8 Nuclear Data Group (ORNL), *Nuclear Data Sheets* **B1–B12**, No. 2 (1966–1974), (Academic, New York)

10.9 Apple Pascal: *Language Reference Manual*, Apple Computer Inc., Cupertino, CA., USA, Apple Product # A2L0027 (1980)

10.10 Apple Pascal: *Operating System Reference Manual*, Apple Computer Inc., Cupertino, CA., USA, Apple Product # A2L0028 (1980)

10.11 L. A. Leventhal: *6502 Assembly Language Programming* (OSBORNE/Mc Graw-Hill, Berkeley, CA. 1979)

11. Parity Violation in the Weak Interaction

E. Kankeleit, H. Jäger, C. Müntz, M.D. Rhein, and P. Schwalbach

11.1 Introduction

Until 1956, it seemed to be a matter of course that experiments performed in force-free space should lead to the same result if performed in a shifted or rotated position, but a symmetric condition. The physical processes we observe in a movie can proceed the same way with the film left-to-right inverted. The reasons for these effects are borne out in the physical laws, which stay the same in the mirror symmetric world by exchange of $(x, y, z) \rightarrow (-x, y, z)$, which is the same as the parity operation $(x, y, z) \rightarrow (-x, -y, -z)$ with rotation arround the x axis. Because of these symmetries it seemed to be against common sense that in a "Gedankenexperiment", a foil covered on one side with a β source and mounted perpendicular to a very thin wire should be able to start rotating in only one direction. That this can be the case due to the *parity non-conserving weak interaction* in β decay is the subject of our laboratory course.

As predicted by *Lee* and *Yang* in 1956 [11.1] and experimentally confirmed in 1957 by *Wu* [11.2], in weak interactions (and only in these) mirror symmetry no longer holds. As a consequence, leptons (electrons and neutrinos) emitted in β decay show a longitudinal polarization with respect to their momentum vector: electrons and neutrinos are fully left-circular polarized, the antiparticles right polarized. It is the electrons circular polarization which is to be determined in this experiment as a clear indication for parity violation.

Unfortunately a direct study of the circular polarization of electrons from β decay [11.3] is too difficult for a simple course. Therefore we make use of the fact that the bremsstrahlung photons generated by decelerating electrons when stopped in an absorber, take over the polarization of the electrons to a great extent due to conservation of angular momentum. The circular polarization of the photons has thus to be measured, which is much easier than that of electrons. For this, the fact is used that Compton scattering of a circular-polarized photon occurs more strongly when both spins are parallel rather than opposite. We find polarized electrons in magnetized iron. Reversing the magnetization will lead to a difference in scattering events, which depends of course on the photon polarization. A difference of count rate for the two electron polarizations becomes evidence for the parity violation.

Counting rates for sources admitted for this course are low and because of the small analysing power of the Compton scatterer, the effect is small.

It may be clear by now that this experiment is not simple, and years ago a considerable amount of man power and money for the electronics was needed to study this effect. We will see with the help of a simple and (relativly) cheap PC and some of our own soft- and hardware developments, that we are able to confirm this most important and fundamental effect.

The manifold physics involved in this experiment makes it impossible to touch all the necessary points without exceeding the framework of this chapter.

11.2 Basic Physics

Parity was shown not to be a conserved quantity in the weak interaction [11.2]. It results in a polarization of the electrons emitted in the β decay of nuclei. Polarization of the electrons means a fixed correlation between the spin and the momentum vector. Electrons exhibit a left-handed behaviour (spin opposite to momentum), while positrons are right-handed particles. The fact that electrons originating from a β decay are always left-handed is a clear sign for symmetry breaking in the weak interaction.

Figure 11.1 shows schematically the basic ideas of the experiment, carried out by *Goldhaber* [11.4]. Firstly the electrons from a β source are stopped in an absorber. In the energy range of electrons available with β sources, the energy loss is dominated by ionisation processes,

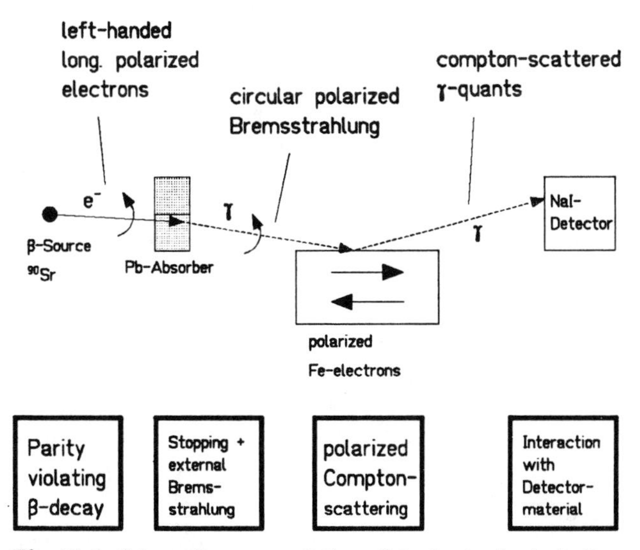

Fig. 11.1. Schematic representation of the basic physical effects used to examine parity violation in the weak interaction

while usually a small fraction of their kinetic energy is carried away by external bremsstrahlung. If the primary electrons were originally polarized, the bremsstrahlung will also be circular polarized. The polarization of the bremsstrahlung [11.4, 5] and the fraction of energy loss due to bremsstrahlung depends on the energy of the electrons. From this it is advisable to use high-energy electrons in the experiment. The polarization is defined as

$$P = \frac{n^+ - n^-}{n^+ + n^-} , \tag{11.1}$$

with n^+ and n^- are the number of γ quanta with spin in the same and in the opposite direction to the momentum, respectively. The polarization is measured with a magnetic γ polarimeter, where the γ quanta are Compton scattered with polarized electrons. The scattering cross section depends on the relative orientation of the electron and the photon spin, and can be calculated with the Klein-Nishina formula to be

$$\frac{d\sigma}{d\Omega} = \frac{r_0^2}{2} \left(\frac{k}{k_0} \right)^2 \cdot (\Phi_0 + f \cdot P \cdot \Phi_H) , \tag{11.2}$$

with k_0 being the initial photon momentum, k the photon momentum after scattering, P the degree of the polarization of the γ rays and f the fraction of oriented electrons in the scattering material. The quantity Φ_0 gives the part of the Klein-Nishina cross section which is independent from the polarization:

$$\Phi_0 = 1 + \cos^2 \vartheta + (k_0 - k)(1 - \cos \vartheta) , \tag{11.3}$$

and Φ_H the part which depends on the relative orientation of the spins:

$$\Phi_H = -(1 - \cos \vartheta) \{(k_0 + k) \cos \vartheta \cos \psi + k \sin \vartheta \sin \psi \cos \varphi\} . \tag{11.4}$$

φ is the angle between the direction of the incident photon k_0 and the direction of the electron spin $\boldsymbol{s}$, ϑ is the scattering angle and φ the angle between the $(\boldsymbol{k}_0, \boldsymbol{s})$ and the $(\boldsymbol{k}_0, \boldsymbol{k})$ plane. After being scattered, the photons are detected in a $3'' \times 3''$ NaI crystal.

If the primary γ quanta are polarized, the number of detected photons Z, which is proportional to (11.2), can be influenced by changing the direction of $\boldsymbol{s}$ from ψ to $\psi + \pi$. Using a cylindrical formed scattering magnet, the angle φ becomes zero, and with the two different counting rates $Z^+(\psi)$ and $Z^-(\psi + \pi)$ the measurable effect can be defined as

$$\eta_{exp.} = \frac{z^+ - z^-}{z^+ + z^-} = P \cdot A; \quad A = f \cdot \frac{\Phi_H}{\Phi_0} , \tag{11.5}$$

with A being the analyzing power of the experiment. From this equality the polarization P can be deduced.

Performing an energy differential analysis of the measured γ spectra one can obtain the quantity $P(E_\gamma)$, and with the crude approximation $P(E_\gamma) =$

$P(E_{e^-})$ the energy differential degree of the polarization of the primary electrons can be found. Whether a parity violation was detected or not is best seen in a plot of P_{e^-} versus $\beta = v/c$ of the electrons. Even if the theoretical prediction of $|P| = v/c$ [11.6,7] will not be obtained by the experimental results, an increasing degree of the polarization with increasing energy of the electrons cannot be explained with systematic errors, thus showing the parity violation in the weak interaction.

11.3 Experimental Setup

Three different experimental arrangements are possible to detect the polarization of the γ rays [11.8]. The transmission geometry chosen here [11.9] is schematically shown in Fig. 11.2.

The electrons emitted by a ^{90}Sr source are stopped in a Pb absorber of 1 mm thickness. The longitudinal polarized electrons thereby produce circular polarized γ quanta. The source is mounted 10 mm above a cylindrical magnet (200 mm long). 31 mm below the cylinder, a NaI detector is placed. To keep the stray field of the magnet from influencing the amplification of the photomultiplier tube, the crystal is coupled to a 60 cm-long light guide to allow a safe distance for the multiplier.

In the centre of the magnets cylinder, a lead cone shields the direct radiation from source to detector. The geometry source–magnet–lead cone detector is chosen such that the mean scattering angle of detected γ quanta is $\approx 45°$. The length of the lead cone (15 cm) is sufficient to reduce the intensity of primary 2.27 MeV photons to 1/760. The shape of this absorber

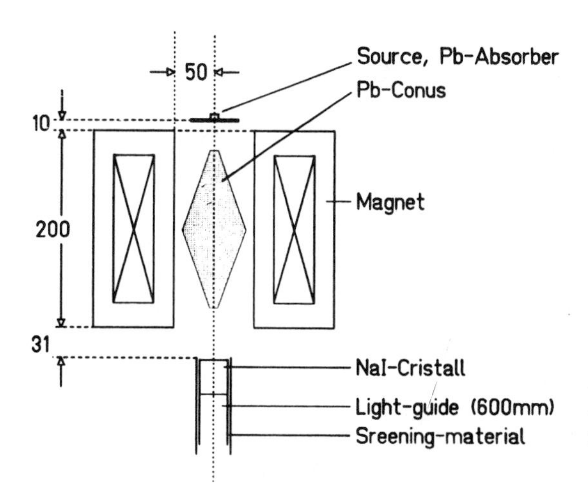

Fig. 11.2. Experimental setup to measure the polarization of γ rays. All quantities are given in mm

suppresses unwanted scattering angles. The photons scattered in the walls of the magnet could also interact with the magnetizing copper coils. To take care that these contributions are smaller than the statistical accuracy of the data, a wall thickness of 2 cm was chosen. This reduces the contribution of 2.27 MeV quanta scattered at 46° to 6%, the contribution at 1 MeV to 3%.

11.3.1 Electronics

The electronic set up is shown in Fig. 11.3. The γ quanta are detected with a NaI crystal. The light produced in the scintillator is transferred to the photomultiplier and its output signal amplified by a preamplifier, followed by a standard main amplifier. The main amplifiers analogue output signal is digitized by an ADC (10 bits, 1024 channels). The ADC is in turn coupled to a memory module (MANY86, 16k channels (14 bits)). This module is interfaced to a standard IBM-compatible PC running under DOS. Both memory module and interface have been developed in our electronics workshop. PC, MANY86 and ADC, combined with a program VKS for data acquisition and analysis, are a cheap and flexible replacement for a conventional multichannel analyzer. VKS is a MS WINDOWS application. It enables the user to start and stop the measurement, display and analyze the spectra, clear the memory or dump its contents into a file on a disc.

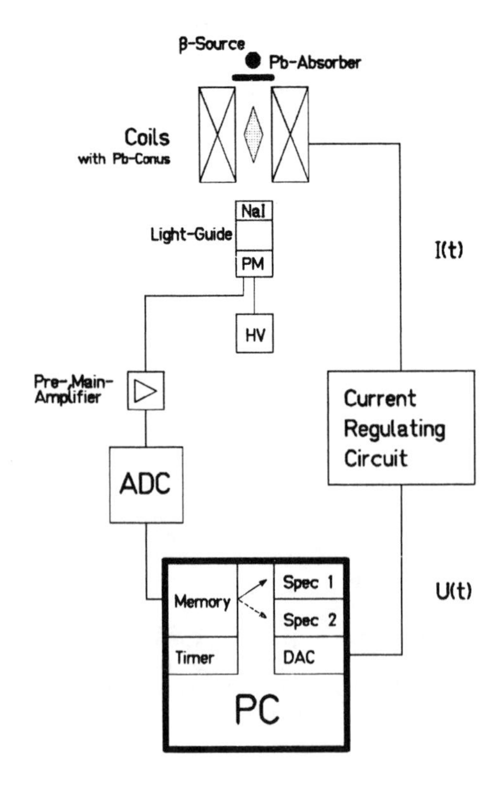

Fig. 11.3. Electronic setup

The magnetic cylinder serves as an analyzer with which we can measure the difference in photon flux for two different directions of the magnetic field. Typical overall measurement times are of the order of 1-2 hours for each direction. It turns out that this time is too long to keep all conditions stable enough, especially in a laboratory classroom. Room temperature fluctuations change the performance of the photomultiplier, the electronics and the magnet, and result in small drifts affecting the results. Since the measurable effect is small, one has to take care that such unavoidable drifts affect the measurements for the two magnetizations in the same way, and thus cancel for short measuring times.

With the help of the PC the problem could be solved easily: The PC controls the current in the magnetizing coils. The sign of the current is being changed periodically in periods of 60 s and the γ quanta are registered in two corresponding memory areas of the MANY86. It is crucial that the intervals of the measurement times for the two field directions are of exactly the same length. The PC controls the magnet via a DAC to which it is interfaced. The DAC output is used as a reference value for the operational amplifier whose output is the magnet current. Due to hysteresis effects, the periodic signal is a little more complicated than just a square wave function. At each change of the sign of the magnetic current, it is necessary to go beyond the wanted value I to I_{max} and then back to I (see Fig. 11.4) to ensure a reproducible polarization of the iron electrons.

The PC also synchronizes the measurement to the current: To distinguish between field up and field down, an extra bit is generated by the PC for field down and fed to the MANY86 as bit 11, thus routing the pulses into a

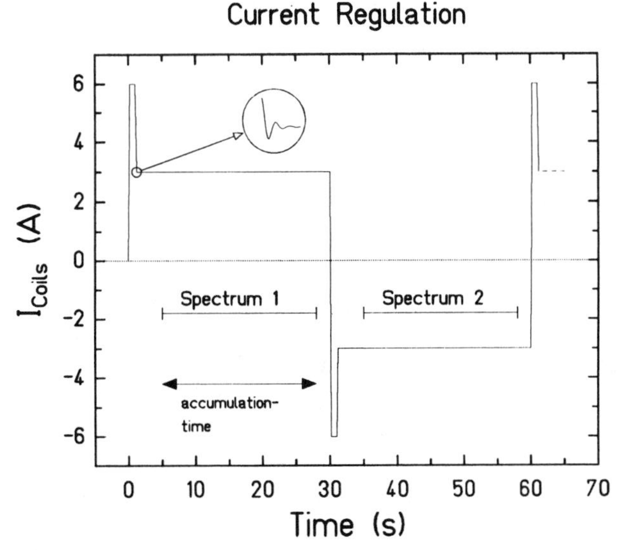

Fig. 11.4. Diagram to visualize the direction of the magnetic current and the corresponding measuring times of the ADC

second memory area (channels 1024 to 2047) as compared to field up, when channels 0 to 1023 are open. The PC stops the measurement whenever the current is changed to I_{max} and starts it again when I is reached. The length of a period, the height (I) of the plateau and the number of periods to be run can be chosen by the student when starting the measurement.

11.3.2 Software

VKS is part of a data acquisition system which was developed for research work in the laboratory and is applied in combination with various, sometimes large, electronic setups in our nuclear and solid-state physics experiments.

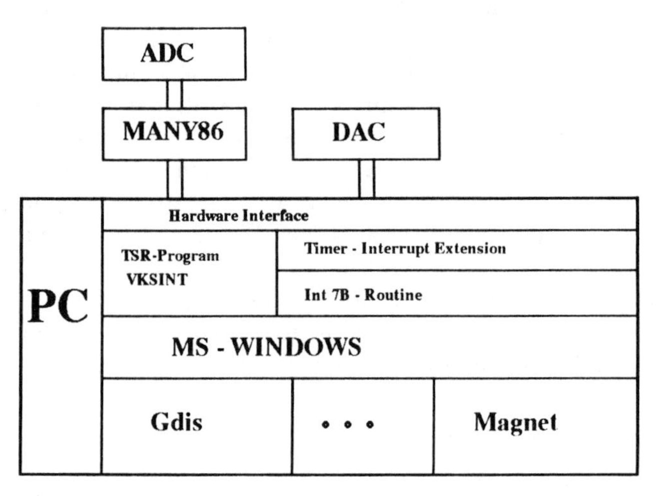

Fig. 11.5. Block diagram of the soft- and hardware components used to run the experiment described

Figure 11.5 shows the block diagram of the version discussed here. The software consists mainly of our standard WINDOWS application GDIS, which, e.g., displays spectra, starts and stops measurements and clears or dumps memories. GDIS can be run independently in several windows and enables the user to view several spectra at the same time. MAGNET is a small WINDOWS application, especially written for the parity experiment, and allows one to choose the amplitude of the magnetic current and the cycle time. To reach the required timing precision in the control of the current, a "terminate and stay resident" (TSR) program VKSINT is used. It is loaded before starting MS WINDOWS and applies the timer interrupt of the PC for exact timing. The communication between VKSINT and the WINDOWS applications GDIS and MAGNET is done with interrupt 7B, described in detail in the source file on the program disk.

11.4 Measurements and Results

11.4.1 General Remarks

All measurements and the discussion of the details of the experimental setup and physical aspects of the experiment should be done within one laboratory course day. Data analysis and the interpretation of the experimental results has to be done by the students at home and summarized in a protocol. The limitation in time means one has to accept short measuring times leading to large statistical errors, compared to the physical effect to be examined. As consequence the error treatment is the most important part of the data analysis.

11.4.2 Energy Calibration

The energy calibration of the NaI detector setup is done with a ^{22}Na source. The β^+ decay of Na ends up (branching ratio 90.4%) in an excited level of ^{22}Ne, which decays into the ground state by emission of a 1275 keV γ quant. The positron annihilates with an electron in two 511 keV photons. These two γ energies (511, 1275 keV) are used for a linear energy calibration. The error of this calibration determines the uncertainty of the velocity β of the polarized electron. In an extended version of this experiment, the unpolarized 511 keV quanta can be used to test the symmetry of the γ polarimeter concerning the two accumulation modes (normal and inversed magnetic field).

11.4.3 Background Measurement

As shown by a simple calculation, it is not useful to measure the background longer than the time used for the source measurement. The background measurement can be only used for a qualitative discussion. The measuring time needed for a reasonable statistical error compared to the error of the polarization measurement would exceed the limit of one day. Neglecting the background leads to a systematic underestimation of the γ polarization discussed below, but will in principle not effect the tendency of its energy dependence.

11.4.4 Measurement of the γ Polarization

The measurement of the γ polarization is done with the γ polarimeter described above. The accumulation time is typically 150 cycles (1 cycle = 60 s). The data presented in this article were accumulated within 24 hours. The two γ-energy spectra related to both orientations of the magnetic field are analyzed in bins with an adequate bin size of between 40 and 50 keV and within a γ-energy window of 430 and 820 keV. The position of this energy window is determined by the β-end point energy of ^{90}Y decay and the fact that the effect, which is to be verified, is proportional to the original velocity of the β decay electrons. In the following, the polarization of the electrons is analyzed in the electron energy interval of 500 to 2200 keV. It results the γ-energy window given above from two assumptions: (i) that the Compton scattering angle is 45 degrees (averaged over the setup geometry), (ii) that the energy of the γ bremsstrahlung is equal to the energy of the stopped electron. The last assumption means a systematic but indispensable underestimation of the electron velocity.

To computate the γ polarization, the following steps have to be considered:

- The energy of the incoming γ quant has to be evaluated, taking into account the energy of the Compton-scattered γ quant and the mean scattering angle θ=45°, averaged over all scattered photons reaching the detector.
- By comparing the γ-energy-dependent counting rates Z^+ and Z^- for normal and inversed magnetic field orientation, the counting rate effect η can be derived.
- The degree of polarization f of the Fe electrons can be determined with the help of the magnetization curve of the applied iron (see Fig. 11.6). It results in a degree of polarization of 4.6%: on average only one electron per Fe atom is polarized for a current of 3 A. The magnet is not operated in saturation to avoid heating of the experimental setup, resulting in a temperature drift of the electronics.
- The γ-energy-dependent analyzing power $A = f \times \Phi_H / \Phi_0$ can be derived with the formulae given above by taking into account the cylindrical geometry of the setup: (i) the scattering angle is on average 45 degrees, (ii) the relative angle between the spin of the Fe electrons and the polarization vector of the incident γ quanta is on average 22.5 (202.5) degrees, for normal (inversed) magnetic field. The result is given in Fig. 11.6.

The γ polarization, P, is then given by

$$P = \frac{\eta}{A} = \frac{\eta}{f} \left(\frac{\Phi_H}{\Phi_0} \right)^{-1} . \tag{11.6}$$

11.4.5 Results and Discussion

If one assumes that the bremsstrahlung excited by stopping the polarized electrons in the Pb absorber has the same polarization as the electrons, and that the electron energy is totally converted in the energy of one γ quant,

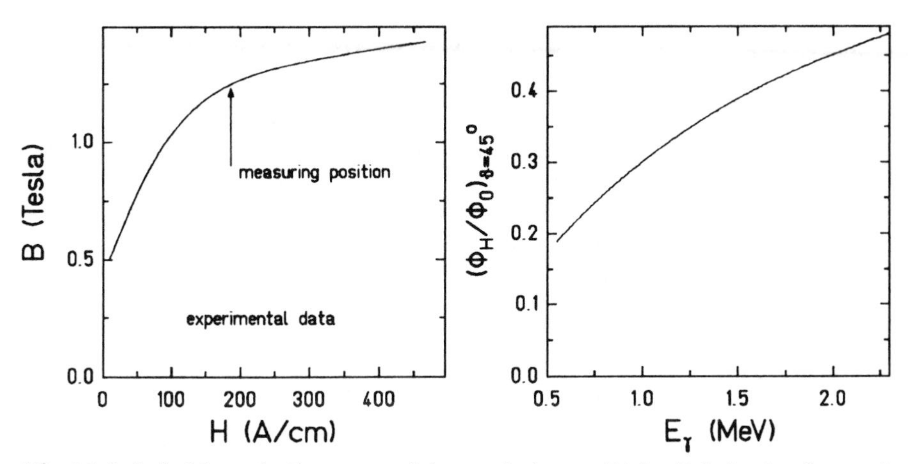

Fig. 11.6. Left: Magnetization curve of the γ polarimeter. Right: Polarization factor of Compton scattering

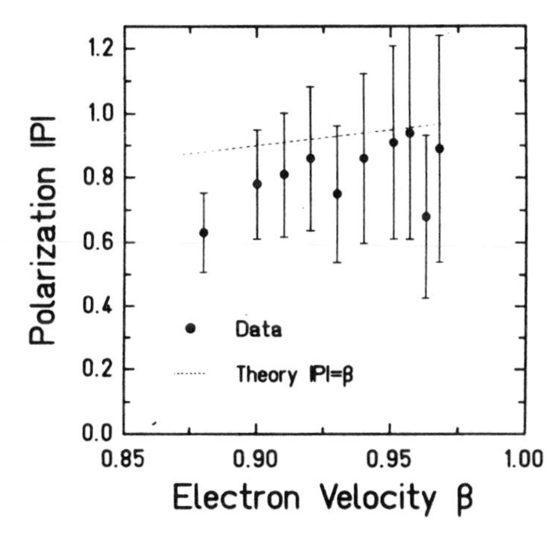

Fig. 11.7. Electron polarization as function of the electron velocity

one can calculate the electron polarization as a function of the electron velocity and compare the experimental data with the theoretical prediction $P = -\beta$. This is done in Fig. 11.7.

As expected, neglecting the background leads to a systematic underestimation of the polarization relative to the theory of β decay. The tendency of the data points as a function of the electron velocity however, exhibits a clear hint at the violation of the parity in β decay, even with respect to the large error bars, including statistical errors of Z^+ and Z^- and the error in determination the analyzing power A.

11.5 Didactic and Pedagogical Aspects

The measurable effect is rather small, and it is only possible to extract qualitative results within a reasonable measuring time. Furthermore, a large amount of assumptions enter into the analysis of the experimental data, leading to large error bars. However, a detailed physical discussion can be triggered by this experiment, giving insight into basic physical questions and experimental techniques.

The following list should give a short overview of what should be discussed during the experiment:

1. General remarks about parity and parity violation.
2. Parity violation in β decay.
3. Interaction of electrons with matter.
4. Production and energy spectrum of internal and external bremsstrahlung.
5. Interaction of γ radiation with matter.
6. Dependence of the scattering amplitudes on the polarization of the γ rays and the medium to be scattered on.
7. Polarisation of matter and fraction of polarized electrons.
8. Measurement of γ radiation with scintillation counters.
9. Coming back to the Gedankenexperiment in the introduction: Why should the foil with the β-ray activity on one side start to rotate?

References

11.1 T.D. Lee and C.N. Yang: Question of Parity Conservation in Weak Interaction, Phys. Rev. **104** 254 (1956)
11.2 S. Wu et al.: Experimental Test of Parity Conservation in Beta Decay, Phys. Rev. **105** 1413 (1957)
11.3 H. Frauenfelder et al.: Parity and the Polarization of Electrons from ^{60}Co, Phys. Rev. **106** 386 (1957)
11.4 M. Goldhaber et al.: Evidence for Circular Polarisation of Bremsstrahlung Produced by Beta Rays, Phys. Rev. **106** 826 (1957)
11.5 A. Bisi and L. Zappa: Relation between Circular Polarisation and Energy of External Bremsstrahlung from β-rays, Nucl. Phys. **10** 331 (1959)
11.6 T.D. Lee and C.N. Yang: Parity Nonconservation and a Two-Component Theory of the Neutrino, Phys. Rev. **105** 1671 (1957)
11.7 L. Landau: On the Conservation Laws for Weak Interaction, Nucl. Phys. **3** 127 (1957)
11.8 H. Schopper: Measurement of the Circular Polarisation of γ-rays, Nucl. Inst. Meth. **3** 158 (1958)
11.9 D. Schnitger: Aufbau eines Polarimeters zur Messung der zirkularen Polarisation von γ-Strahlung, Diplomarbeit Institut für Kernphysik, TH-Darmstadt 1962

12. Receiving and Interpreting Orbital Satellite Data. A Computer Experiment for Educational Purposes

T. Kessler, S.M. Rüger, and W.-D. Woidt

12.1 Introduction

The Physics Department of the FU Berlin operates a computer laboratory for the education of students of physics and related disciplines. Special attention is paid to the application of computers in a real time environment, because this is the main use of computers in today's laboratories.

Of course the students need a source of real time data in order to test their programs. While physical experiments normally are quite expensive data sources, artificially generated data are boring and do not give much motivation to the students. The data source we found delivers useful data, is inexpensive to use and at the same time stems from one of the technologically most advanced devices operated today: satellites.

The University of Surrey (UK) is building and operating experimental satellites for educational purposes and for getting experience in building low-cost satellites. One important aspect of this satellite experimental program is that the satellite's engineering and experimental data are transmitted in such a manner that they are readily received by simple, low-cost amateur ground stations.

One of the satellites (UoSAT-B) seemed to us quite usable as a data source for a computer experiment. The techniques of receiving the satellite's data are simple and cheap. The transmitted data contains information which can be interpreted in a physically relevant manner.

Part of the data are transmitted in real time, so there is a challange of real time data acquisition and display. Since the satellite's orbit is only 700 km above the earth's surface, the satellite's data signal can be received only for periods of about ten minutes several times a day. During these periods the data link quality changes rapidly, so that one has to develop a method to discriminate valid data from noise or interference.

Another part of the transmitted data are whole orbit data (WOD); that means a method by which values from selected telemetry channels spanning several orbits are collected in the on-board computer memory for subsequent transmission to the ground. These data show a more complete picture of the satellite's operation than can be gathered from the real time data of a ten minute pass.

12.2 The UoSAT Satellites

UoSAT-A was the first of two low-cost satellites, constructed by a team of research engineers at the University of Surrey, launched successfully in 1981 into a sun-synchronous, circular polar orbit (i.e. the same geographical area is crossed at the same time of day) with an initial altitude of 550 km. Because of the atmospheric drag the satellite's orbit continually decayed and UoSAT-A burned up on re-entry into the atmosphere.

UoSAT-B was launched on March 1st 1984 into a 9 am/pm sun-synchronous, circular polar orbit at an approximate altitude of 700 km (Fig. 12.1). An advantage of the higher orbit is that there is much less atmosphere at this altitude, so the effect of drag on the orbital parameters is greatly reduced. The orbital lifetime is estimated to be around 50 years.

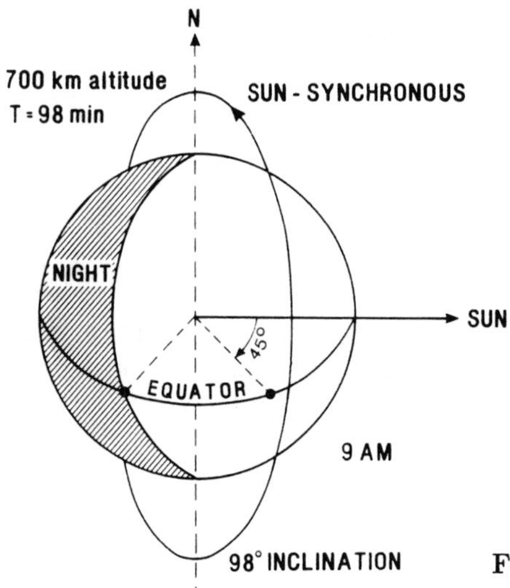

Fig. 12.1 Orbital geometry of UoSAT-B

With an orbital period of about 98 minutes, UoSAT-B is accessible in the mid-morning and again during the late evening. A typical pass lasts around 12 minutes and a minimum of two passes will occur during one day, with up to six passes for temperate latitudes. The most important data of UoSAT-B are summarized in Table 12.1.

The simplest way to find the position of the satellite is to use published tables of its equator crossing (AMSAT Journal) or to phone the special message service at the University of Surrey. From the prediction tables for UoSAT-B, one can see for example an ascending equator crossing at 20:12:32 GMT at 351 degree west (20:12:32 351 >). These data can be used to position the ground track of UoSAT-B marked off in minutes on a polar map of the world (Fig. 12.2).

Table 12.1 Properties of UoSAT-B

Orbit:	inclination: 98 degrees period: 98 minutes altitude: 700 km
Mechanical:	dimension: 33.5 cm x 35.5 cm x 58.5 cm mass: 60.5 kg
Communication (downlink):	145.825 MHz, 0.4 Watt, 1200 Baud FKS, Digitalker 435.025 MHz, 1.0 Watt, 1200 Baud FKS, Digitalker
Payload:	Digital Communication Experiment (DCE) Digitalker (synthesized voice output) particle detector electron spectrometer 3-axes magnetometer attitude control system CCD camera

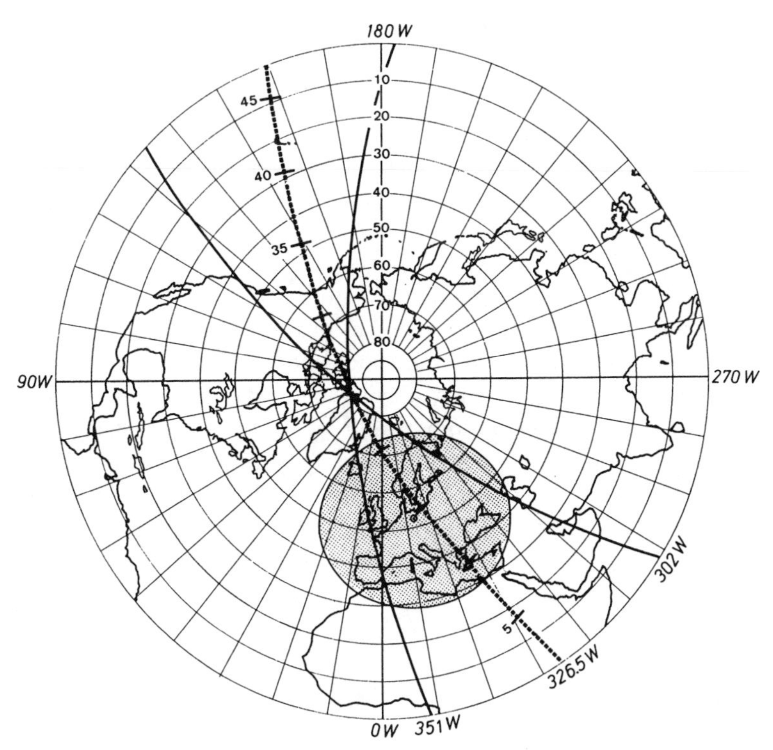

Fig. 12.2 Polar map of the Northern hemisphere showing ground tracks of UoSAT-B intersecting the radio horizon of Berlin. The number of minutes marked off on the intermediate track plus the time of the equator crossing is the time the satellite comes into view for a ground station at this position

If the path of the satellite intersects the radio horizon or range circle of the ground station, the satellite comes into view at a time calculated by adding the equator crossing time to the number of minutes marked off on the track at that intersection. The next equator crossing can be calculated from the given longitude plus the amount the earth rotates in one orbital period. This method will prove fairly accurate for short term predictions.

12.3 The Receiving System

UoSAT's communication system is constructed in such a way that receiving stations can be set up with a minimum of technical and financial effort (Fig. 12.3).

The downlink frequencies lie in the amateur radio bands and frequency modulation is used. Therefore one can use readily available and inexpensive amateur FM receivers.

In our case, the receiving section of an old crystal-controlled 145 MHz walky talky is used. The self-made antenna consists of two crossed dipoles with reflectors.

UoSAT's downlink uses the 'Kansas City' data format, which comprises one period of 1200 Hz for a logical '0' and two periods of 2400 Hz for a logical '1' . This signal is used to modulate the frequency of the satellite's transmitter.

A suitable data demodulator which was published in [12.2] has been built up on a perforated eurocard and extended by a V.24 interface, and worked without any problem. It's data output is connected to a serial input line of our computer system.

Data are transmitted as a modulated audio subcarrier, which can be simply recorded by an ordinary audio tape or cassette recorder. The data clock frequency is recovered by the data demodulator, so there arises no difficulty from tape speed variations when playing back the data signal. Incoming satellite data were recorded automatically by a cassette recorder prior to demodulation for our experiment. Thus we had always 'canned' data for testing of our real time software. This facility proved to be very useful since good satellite passes are too scarce for this purpose.

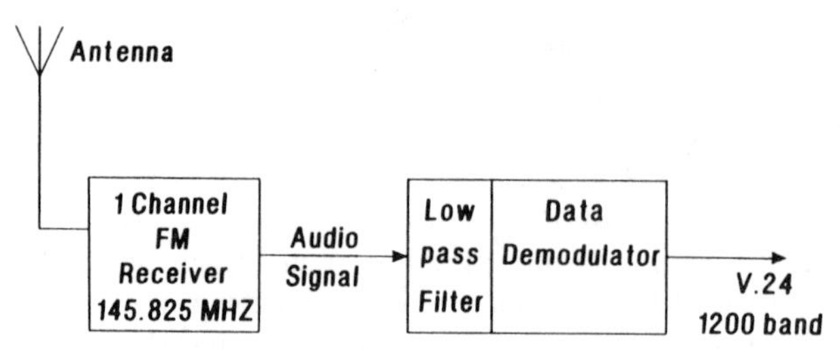

Fig. 12.3 Block diagram of the receiving system

12.4 Discriminating Valid Data from Noise and Interference

The data decoder of our receiving system outputs a bit stream of 1200 bits per second. Therefore in order to process the satellite's data, we need some means to determine if the received bits represent valid data or not. Though in principle this can be accomplished by some additional hardware, we realized this task purely by software. If we suppose that the receiving hardware (UART) of the computer is in synchronization, then any single bit error in the 7 data bits or in the parity bit will cause a parity error. Furthermore, an error of at least one of the two stop bits will cause a framing error.

Both kinds of error are recognized by the computer hardware. By counting the parity and framing errors we can get information about the bit error rate and therefore information about the quality of the data link between satellite and ground station.

Because these data errors are occurring statistically, we must describe them by the laws of probability [12.4]. Let's assume a given and constant bit error probability P_b. The probability for receiving a bit correctly is 1 - P_b. If a data word consists of N bits, then the probability for receiving a data word correctly is $(1 - P_b)^N$ and thus the word error probability $\mathrm{P}_w = 1 - (1 - P_b)^N$. An odd number of bit errors in the n bits representing $n - 1$ data bit plus parity bit will produce a parity error. Any wrong stop bit results in a framing error.

A false start bit will also bring the dataflow out of synchronization, which produces most probably a parity or framing error. In a few cycles, maybe immediately, the data is brought back to synchronization[1] by finding a correct start bit. In sum, a bit error leads to a word error, so that the error rates (i.e. the errors per second) for bits and words are equal, as long as their values are small.

However, there is a small chance of undetected transmission errors; an even number of bit errors of the n data and parity bits cannot be seen by the hardware

[1]The data transfer is asynchronous, i.e. there may be an arbitrary long pause between data words. This pause appears as a series of stop bits, which are coded as logic '1'. The start bit is coded as '0', which means that the first signal change after a received data word is taken as the beginning of a new data word. If a start bit could not be transferred correctly, the next '0' in the data field will trigger erroneously a new data word. Any long enough pause betweeen data words will bring the data flow back to synchronization. If there are never pauses between data words, the synchronization depends on the transfered data; and there are possible scenarios where the data will *never* come back to synchronization. In the following we assume that we had a correct start bit. False start bits typically result in a more or less severe loss of data.

$$P(\text{undetected errors}) = (1-P_b)^{N-n-1} \underbrace{\left(\sum_{\nu=2,4,6,\ldots}^{n} \binom{n}{\nu} P_b^{\nu}(1-P_b)^{n-\nu} \right)}_{P(2,4,6,\ldots \text{ errors})}$$

$$= (1-P_b)^{N-n-1}(\frac{1}{2} + (1-2P_b)^n - (1-P_b)^n) \approx \binom{n}{2} P_b^2 \quad \text{for small} \quad NP_b \quad .$$

The factor $(1-P_b)^{N-n-1}$ is due to the $N-n-1$ correctly transferred stop bits. In our case N=11 and $N-n-1$=2 (7 data bits, 1 parity bit , 2 stop bits, 1 start bit). The above approximation holds for reasonable small bit error probabilities characteristic for usable data. Then the fraction of undetected errors in data words $\binom{n}{2}P_b^2$ is negligible compared to the word error rate $P_w \approx NP_b$.

If the input to the data decoder is only noise, as is the case if the satellite is not accessible, we get a pure random bit stream at the output with an error probability $P_b = 0.5$ for each bit. There are 3 bits checked per data word. The probability of all three to be correct is $P_b^3 = 0.125$ and thus the probability of at least one to be incorrect is $P_w = 1 - P_b^3 = 0.875$, which is the probability of hardware error detection.

In our computer experiment the statistical properties of the occurrence of errors has been used to derive a figure of merit for the data link quality directly from the received data stream. For determining the error rate one has to count the detected errors over l consecutive data words. The larger l is, the more precise is the result. However l cannot be made arbitrarily large because the error rate changes with time and we are not only interested in the mean error rate. A good compromise has been found experimentally with l = 16, which corresponds to a time of 0.15 seconds at 1200 Baud transmission rate.

The probability P(l,k) for having k errors in l data words is characterized by a binomial distribution

$$P(l,k) = \binom{l}{k} P_w^k (1-P_w)^{l-k} \quad .$$

Figure 12.4 shows computed values for the probability of having k errors in $l = 16$ data words.

The parameter for the different curves is the bit error probability P_b. The discrete probability distributions have been drawn as continuous curves for easier readability. As can be seen from Fig. 12.4 the statistical properties of usable data (in terms of error probability) and pure noise are quite different, so their evaluation is a good method to discriminate between good and bad signals. In our experiment, data were accepted if there were not more than two errors in 16 words.

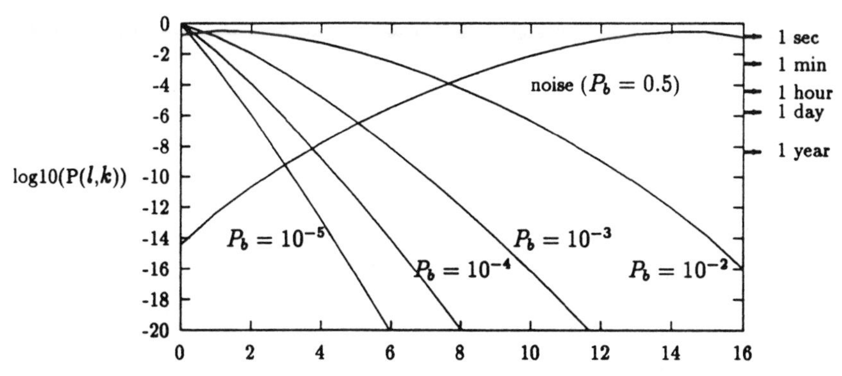

Fig. 12.4 log10(P(l,k)) versus the number k of errors in l = 16 data words. The right ordinate shows time values for $16P(1,k)^{-1}$ word transmissions at a rate of 1200 bits per second. In the case of noise, for example, one has to wait about 1 sec to expect 12 errors in a 16 word group; once a year one can expect to detect only four errors

12.5 The Real Time Data Acquisition System

We changed the low-level input routine of the serial device to add a high bit when finding a framing error and to create an underscore when running into a parity error. By this means it was possible to detect phases of valid data transfer. There are different types of satellite data, e.g. plain English text with schedules for the coming orbits, whole orbit data or real time data of the on-board experiments. The latter are coded in 60 analogue channels (each of which has values between 0 and 999) and 96 digital channels with 0 and 1 as possible values. Each channel is associated with a certain meaning: the analogue channel 1 contains a number n, which when calibrated by $0.1485 * n - 68$ means the measured magnetic field of the navigation magnetometer x axis in μTesla. The digital channel 29, for example, tells whether the CCD camera experiment is currently working or not.

When operating in real time data transmission, the satellite sends a complete dump of all channels in ASCII decimal digits along with additional checksum bytes, which guarantee with a high probability that no corrupted data are taken by mistake.

However, the major problem of real time evaluation is to accomplish many tasks at the same time:

- read data when the satellite is on duty and write the original data to a backup file,
- display all data in a readable form,
- evaluate all aspects of particular interest, and

- depending on the status of the satellite, give new commands during the uplink session. Although we had neither the hardware nor the allowance to accomplish this, this point should be considered in our concept.

A solution for all these requirements to be fulfilled at the same time was to use the abilities of a multiuser real time operating system.

Tasks 1–4 cannot work independently from each other. The first task produces the data and the others rely on this information. In a typical multiuser environment there are some methods of interprocess communication, namely:

- sharing information in a file
- pipelines
- message queues
- signals
- shared memory

We decided to use the real time operating system OS-9 with shared memory as interprocess communication, which is quick compared to shared files and which can be re-read many times (as compared to pipelines and message queues). The operating system allows many processes to allocate the same piece of memory, which is to be written to or read by any involved process. The idea was to export the knowledge of the actual data read by process 1 (read process). All other processes may read this data. Anyway, there is a problem of synchronization: Suppose the only information which is updated by the read process is the actual time stored in 3 bytes for hour, minutes and seconds. If the time is to be updated from 00:00:59 to 00:01:00, it might happen that this procedure is interrupted; the process which is scheduled next might read a wrong time of e.g. 00:00:00 or 00:01:59. Semaphores are used to overcome this situation. Like the motel guest who is allocated a room by getting a room key from the motel manager, all processes are expected to allocate access of the memory by setting a semaphore: If the semaphore was already set (the room key is not present in the motel office) somebody else has the resource under control and the process is expected to wait until the semaphore is cleared.

In our situation the shared memory was divided in three parts, each of which is associated with a semaphore:

- a bulletin with a structure containing the satellite data
- a screen queue with commands for a central screen-mapping process **curses** (the name was taken from a similar UNIX library)
- a queue containing the keystrokes done at the terminal

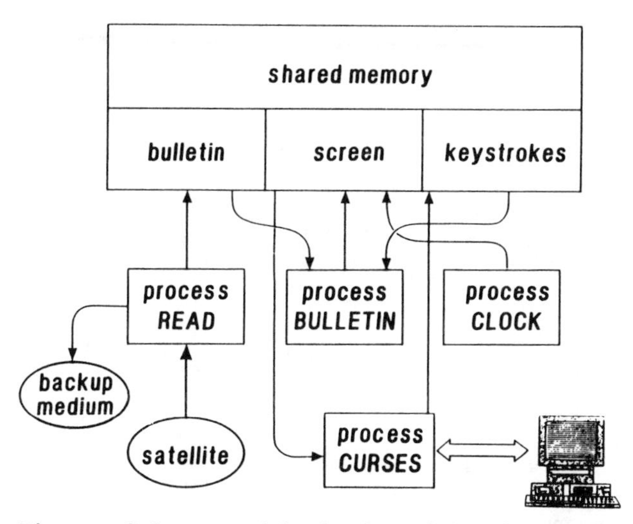

Fig. 12.5 Software modules for the real time data acquisition system

We arrive at the following principal structure for our software (Fig. 12.5).

The process **CURSES** controls the screen and the terminal. Each keystroke of the user is placed into the shared memory and may be read by any other process. Many processes are allowed to change the screen display (Fig. 12.6) by placing appropriate commands into the screen queue. In the above-shown minimal configuration of the software two other processes, **CLOCK** and **BULLETIN**, write onto the screen. The process **CLOCK** updates a running headline including a time display; a system load is also calculated and displayed as load. Load 0 means that no process is active at the time.

The smallest, but nevertheless important process is **READ**. This process analyzes the bit stream of the satellite, places the status of the satellite (noise, sending) along with the actual channel information into the bulletin and backs up valid data lines along with the time stamps into a data file for later processing. During phases of valid real time data transfer, the bulletin in the shared memory contains the actual image of the analogue and digital channels of the satellite. By this structure it is simple for anyone to link to the shared memory and analyze, plot or print the data of interest. This, for example, is done by the process **BULLETIN**, which has the knowledge about the meaning of all the channels and their calibration functions. The process **BULLETIN** displays a list of all accessible data onto the main screen along with the number of valid and invalid data lines detected so far by the **READ** process.

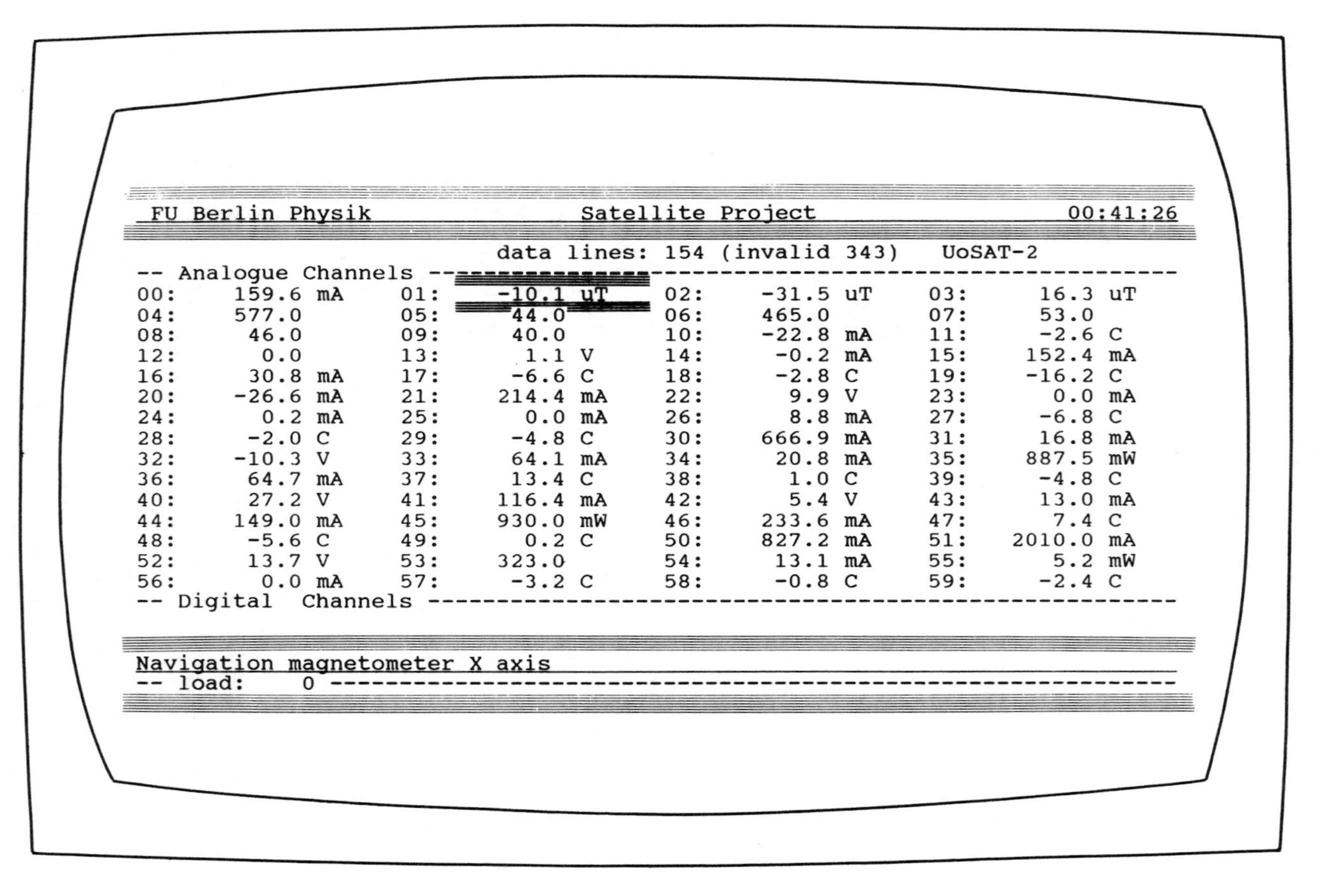

Fig. 12.6 Screen layout of the real time data acquisition system

The user of the main terminal might choose a certain channel by moving the cursor, the accorded meaning of this channel is then highlighted in the last line of the display. This highly flexible concept of our software easily allows the implementation of many independent real time data evaluation processes.

12.6 Whole Orbit Data Analysis

A regular part of UoSAT-B data transmission are the whole orbit data (WOD). The on-board computer has direct access to the telemetry system and can store selected data channels into a 32 Kbyte memory over the course of one or several orbits. Typically 4 telemetry channels are selected for a WOD survey, i.e. about 3 complete orbits can be read into the free memory. This facility enables data of interest to be collected and dumped later when the satellite is in range of ground stations. Normally WOD are interspersed with other downlink data such as real time telemetry and computer status messages. In the status messages (Fig. 12.7) information about the channels included in the WOD surveys can be found besides other useful data such as the satellites's spin period etc..

```
***UOSAT 2 COMPUTER STATUS INFORMATION***
COMMAND DIARY V4.6 IN OPERATION
UNIVERSAL TIME IS 20:06:30
DATE 28/08/87
AUTO MODE IS SELECTED
0F RAM ERRORS AT E8A3H
SPACECRAFT SPIN PERIOD IS -0107H SECONDS
LAST CMD SENT BY COMPUTER WAS 40H TO 1
LAST CMD RECD BY COMPUTER WAS 0EH TO 0 WITH DATA 00H
CURRENT WOD COMMENCED AT 00:00:00
DATE 28/08/87
SURVEY INCLUDES CHANS 01,02,03,61,
MAGNETORQUER HAS BEEN ACTIVATED 00 TIMES
LIBRATION CONTROL INITIATED
```

Fig. 12.7 Computer status message

The WOD format comprises a line number or sample address followed by a checksum (Table 12.2). The first line (number 0) is used to record the selected telemetry channels in the WOD survey.

Table 12.2 Whole orbit data format

	NNNNXYZ......CC
NNNN:	4 hex digits sample address coding the time of measurement
XYZ:	decimal telemetry channel value as ASCII digits
CC:	one byte checksum $NN + NN + 0X + YZ + \ldots + CC = BB_{hex}$ (8 bit binary addition)

Another important mechanism of error reduction is to minimize the effect of burst errors on the received data values. The data are transmitted with an interleave of 8, i.e. first the data values with the address 0, 8, 16, 24, ... are transmitted followed by 1, 9, 17, 25, ... in the second run etc.. The sample period for UoSAT-B is 4.84 seconds. Figure 12.8 shows the beginning of the WOD survey from 05/08/86 including the three axis flux gate magnetometer (channels 1, 2 and 3) and the battery voltage (channel 52).

000000100200305263 ← channels 1,2,3,52
000834962736769535
00103376283776952E
00183276323896941B
002032063539969407
002831564041069487
003031364541969473
00383156524296944C
00403186584366942E
004832366444369303
0050331671449693DA
0058341677455693B0
00603516844626937E

Fig. 12.8 Beginning of WOD survey as received from UoSAT-B

Together with the appropriate calibration equation (published in [12.1]) the telemetry data values are transformed into physically relevant quantities. The telemetry data value for the z axis of the magnetic field, for example N = 627, put into the equation $H_z = 0.1523*N-69.3$ results in a vertical magnetic field intensity of 26.2 μTesla. Figure 12.9 shows a plot of the complete WOD survey of the magnetic z axis for about three orbits.

In a first approximation one can regard the earth's magnetic field as the result of a magnetic dipole with a momentum of $M_E = 8 \cdot 10^{21}$Tesla · cm ^{3}. The vertical intensity of the magnetic field can be calculated to

$$Z = 62 \cdot \left(\frac{R_E}{R_E + h}\right)^3 \cdot \sin\beta \text{ in } \mu\text{Tesla} \quad ,$$

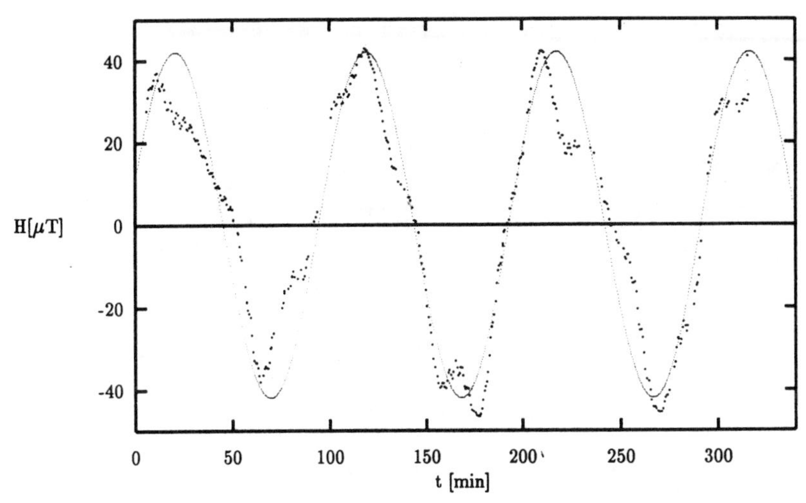

Fig. 12.9 WOD of navigation magnetic z axis

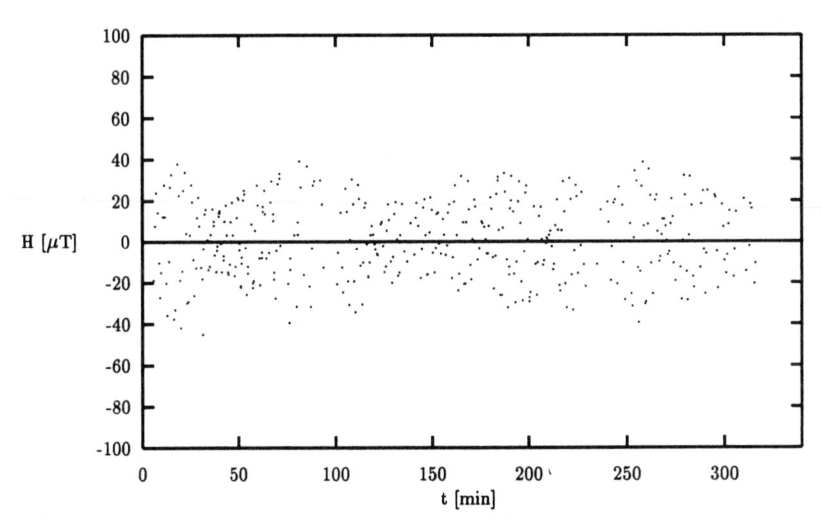

Fig. 12.10 WOD of navigation magnetic x axis

with β as the geomagnetic latitude, R_E as the earth's radius and h as the satellite's altitude.

Computed values from the equation of the vertical intensity given above are included into the WOD plot of Fig. 12.9.

In the case of x-axis data of the magnetometer (Fig. 12.10), one must take into consideration that the satellite is gravity gradient stabilized and therefore spins slowly around its z axis, which is directed towards the centre of the earth. For details on the method of gravity density stabilization, refer to the chapter "Attitude Determination, Control and Stabilisation" in [12.1].

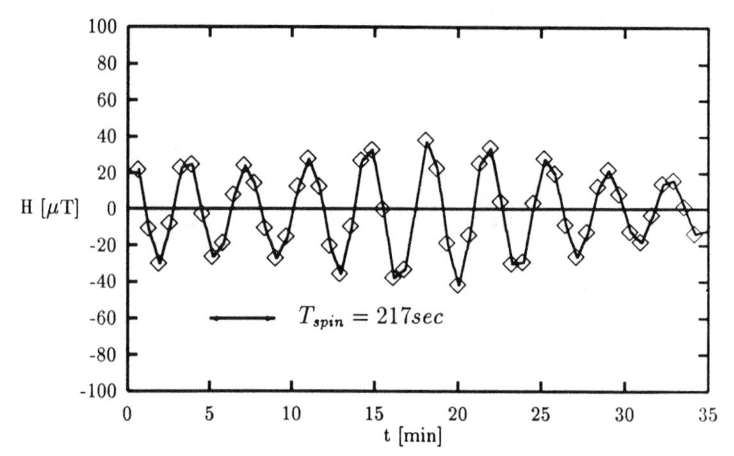

Fig. 12.11 WOD of navigation magnetometer x axis (enlarged)

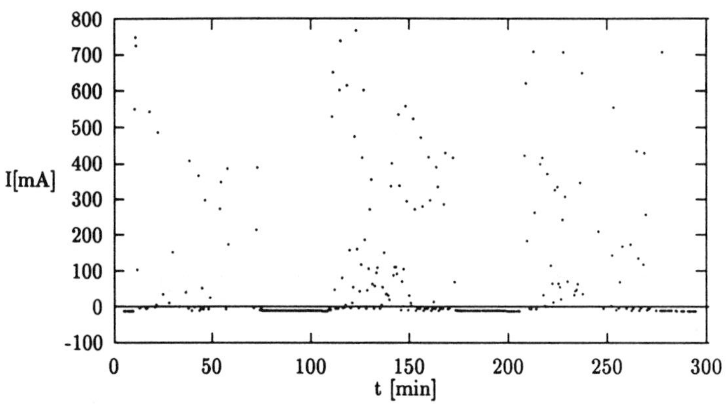

Fig. 12.12 WOD of solar array current -y

Because of the satellite's spin, the angle between the horizontal component of the magnetic field and the x axis of the magnetometer changes continuously with time.

As a consequence, the data values of the magnetic x axis include the information of the satellite's spin period. As long as the sample period is at least half of the satellite's spin period (sample theorem), the spin period can be directly derived to 217 sec from Fig. 12.11, which is an enlarged part of Fig. 12.10. The same argument holds for the solar current values of the four solar arrays (channels 0, 10, 20, 30) when the satellite is in sunlight.

As an example, the -y solar array current[2] is plotted in Fig. 12.12. When the satellite is over the night side of the earth, there is no solar array current

[2]Because the satellite is a cube, the two solar arrays in every direction are distinguished by the corresponding sign.

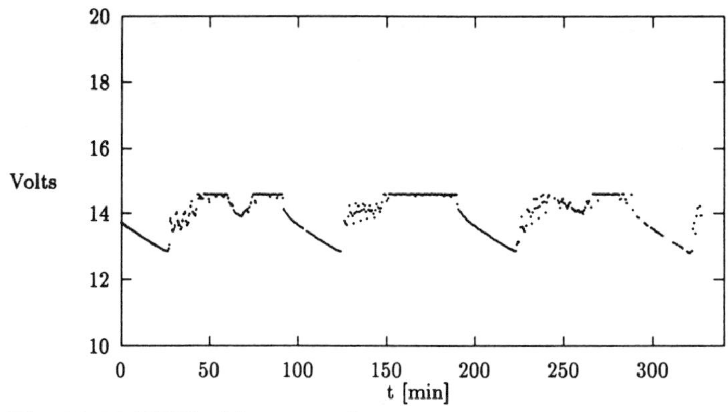

Fig. 12.13 WOD of battery voltage

for about half of an orbital period. As a consequence of the absence of the solar current, the battery system is not being charged during this period, which can be seen in the typical decrease of the battery voltage (Fig. 12.13).

12.7 Practical Experience and Further Aspects

Receiving, decoding and analyzing the data derived from the UoSAT-B spacecraft turned out to be a very stimulating computer experiment for the participants of the computer laboratory. The students learned important aspects about data integrity, satellite tracking, analysis and interpretation of satellite data and got deeper insight into a real time operating system, especially into interprocess communication via shared memory and semaphores. Although many ideas have been contributed and integrated into this project, some problems are still unsolved.

Electromagnetic compatibility (EMC) proved to be a most serious problem for our experiment. The noise level due to high-frequency radiation of computers is very high at our site. The situation could be improved by using a directional antenna, which must be controlled as the satellite moves over the ground station. The use of a directional antenna means that the signal-to-noise ratio (SNR) will be enhanced significantly, so that the satellite passes could fully utilized from horizon to horizon, which is impossible under the momentary cicumstances. A specially designed receiver with matched bandwidth and automatic frequency control to compensate for doppler shift will also enhance the SNR. Writing a program for real time tracking can be an interesting task solved by the students as an exercise.

Another improvement of the experiment concerns the satellite tracking. Instead of extrapolating the position of the satellite from tables of equator crossing, the times of the satellite passes can be calculated from published Keplerian elements [12.5] (describing the orbits's parameters), which give results with a maximal deviation of a couple of minutes for a forecast of a few weeks. Orbital data could be corrected by exactly recording the times of acquisition of satellite (AOS) and loss of satellite (LOS) together with the doppler shift of the received VHF signal. A few doppler measurements have been made by recording the DC component of the receiver output, which varies linearly with the frequency of the received signal, by an analogue chart recorder. Consequently this task could be done by the computer via an analogue-to-digital converter.

Unfortunately it was impossible to receive 2D pictures of the earth beneath the spacecraft due to malfunction of the CCD-camera experiment on UoSAT-B.

Acknowledgements (from the third author)

Thanks to T. Kessler, who had the initial idea for this computer experiment and spent many hours on building and testing the receiving system. Thanks also for his permanent interest in the progress of the project. S.M. Rüger has to be mentioned for his excellent knowledge of the OS-9 operating system and the C language. He contributed most of the software for the real time data acquisition system and a remarkable training script on the C language. We thank A. Koffinke for her sovereign control of the text setting system LATEX. Last, but not least, some words about Malte (three years old), who permanently interrupted the writing of this paper with his never-ending questions, especially concerning the computer mouse.

References

[12.1.] UoSAT Spacecraft Data Booklet, University of Surrey, Guildford, Surrey, GU2 5XH, England

[12.2.] J.R. Miller: Data decoder for UoSAT (Wireless World, May 1983) p.28

[12.3.] AMSAT – DL Journal: AMSAT-DL e.V. Holdenstrauch 10, 3550 Marburg 1

[12.4.] Seymour Lipschutz: *Theory and Problems of Probability* (Mc Graw-Hill, New York 1974)

[12.5.] M. Davidoff: The Satellite Experimenter's Handbook AMSAT-UK, 1984

Subject Index

Abbe theory of microscopes 219
ADC 3, 113, 138, 261
ATR (attenuated total reflection) 175, 182
Absorption
– spectroscopy 200
– spectrum 199
– strength 205
Acceleration 30
Acorn BBC model B microcomputer 138
Air table 34
Amplitude modulation 9
Analyzer 287
Anharmonic oscillator 196
Anharmonicity constant 197
Aperture ratio 200
Apple II PC 251
Assignment of quantum numbers 209
Autocorrelation 225

BASIC 17
BBC Basic 142
Ballistic motion 29
Beat period 9
Beats 6
β radiation energy loss 276f.
Binominal distribution 298
Birge-Sponer extrapolation 197, 210
Boltzmann distribution 131, 205
Boundary conditions 174
Bremsstrahlung 284
– , external 292
– , internal 292
Burst errors 304

CAMAC interface 114
CCD (Charge Coupled Device) 202
CGA graphics 16
Carrier 10
Centrifugal force 33
Chaotic motion 36
Chopper stabilized operational amplifier 136
Circular polarization 282
Cluster physics 45
Coherent
– optics 223
– illumination 241
Collisions 34
– , elastic 37
– , inelastic 38
Compton effect 254
Computer Assisted Instruction Laboratory (CAIL) 114
Constant pressure specific heat 130
Constant current source 113
Constant volume specific heat 130
Constraint force 32
Contact resistance 119
Cooling
– by radiation 123
– curves 123
– law 122
Copper resistance thermometer 134
Coriolis force 33
Cracks 119
Critical temperature T_c 116
Curie temperature 134
Cut-off frequency 133

Data
– acquisition 28
– acquisition system 288
– demodulator 296
– processing 28
Debye
– model for the specific heat 133
– temperature 133
Defect of focus 225
Deslandre table 197, 210
Detector slit 238
Dielectric constant 173
Diffraction 239
Dispersion
– curves 175
– equation 174

Dissociation energy (D_e) 197
Distribution function 29, 34
Drift 63
Duffieux Integral 225
Dysprosium 146

Elastic continuum 133
Electromagnetic
– compatibility 307
– wave 172
Electron diffraction 45
Energy
– level 198
– resolution 259
Equilibrium distance 197
Equipartition of energy 130
Error
– analysis 209, 213
– probability 297
Excitation of surface plasmon-polaritons 176
Experimental lectures 16
Explicit finite difference approximation 85
Extrapolation process 141

FWHM (full width half maximum) 119
Fe-Constantant thermocouple 111
Fictitious force 33
Filter equation 223
Focal length 200
Forbes bar experiment 71
Four-f setup 224
Four point probe technique 111
Four terminal device 136
Fourier
– analysis 3
– integral 14
– optics 245
– pairs 221
– series 3
– spectrum 29
–'s equation of heat flow 69
– transform 220
Frame of reference 33
Framing error 297
Frank-Condon Factors (FCF) 216
Fraunhofer diffraction 223
Free path 34
Frequency counter 118
Fresnel diffraction 245
Fundamental frequency 5
γ polarimeter 284, 289
γ radiation intensity buildup 274f.

Gas handling system 135
Gibbs phenomenon 14

H-bridge switch 189
Harmonic wave, simple 5
He-filling gas 123
HeNe laser 201, 206
Heat
– transport 123
– conduction 123
– conductivity of gases 123
– losses 139
Heater
– circuit 137
– coil 134
Helium exchange gas 135
High-T_c superconductors 111
Homogeneous nucleation 45

IBM compatible PC microcomputer 138
IEEE-488 bus 76
Imaging optics 228
Impurity scattering 116
Inert-gas-aggregation 45f.
– source 47
Inhomogenous granular-like model 119
Instrumentation amplifier 113
Interaction
– of electrons with matter 292
– of radiation with matter 255
Interface 187
Interfacing course 14
Interference image 232
Interprocess communication 300
Iodine molecule I_2 195

JOD (program) 208

Kurie plot 276

LAmDA (Licht Analyse mit Dioden Array) 202, 207
– Data manipulations 208
– Filtering 208
– Fourier Transformation 208
– Smoothing 208
LC-oscillator 113
LIF (Laser Induced Fluorescence) 201, 206
– spectrum 199, 201, 206, 207
Least squares fit 144
Light line 176
Liquid nitrogen 116

MCA
- calibration sources 270
- design 252f.
- display manager 263
- event handler 263
- program menu 265f.
- program structure 262f.
MINUIT program 116
Macintosh PC 270
Mass range 256
Matthiessen's rule 116
Maxwell's equations 173
Meissner effect 114
Microphone 3
Microsoft QuickBasic 142
Model dielectric function 174
Modulation transfer function (MTF) 222
Molecular
- constants 210
- spectroscopy 195
Monochromator 200
Morse potential 197
Motion data 23
Multiplexing circuit 75

NILFIT (program) 215
Néel temperature 134
NaI detector 285
NaI(TI) 258
- scintillation detectors 258f., 270
Newton's law of cooling 122

OMA (Optical Multichannel Analyser) 202
ORVICO 23
Offset balance 116
Optical transfer function (OFT) 219
Optional exercises 214
Opto-isolators 137
Outdoor recordings 30

PID control 54
Paramagnetic susceptibility 113
Parasitic thermal voltages 119
Parity 283, 292, 297
- violation 282, 285, 292
Pendula, coupled 37
Pendulum
—, rigid 31
—, spheric 35
Periodic
- function, strictly 6
- signals 3
Peripheral Interface Adapter 138
Phase
- transition 111
- transfer function 222
Physics teachers 16
Plasmon
- -based devices 178
- -polaritons 172
Point mechanics 23
Polarization 284
- of matter 292
p-polarized light 177
Polarized electrons 282
Population 205
Position-time function 29, 37
Potential (curve) 198
Power supply circuit 137
Prism geometry 179
Progression 210
Proportional
- plus derivative (PD) control 53
- plus integral (PI) control 51
- control 50
Pupil function 224

Quantum number 213
Quartz tube 111
"Quasi-equilibrium" technique (QE) 116
Quasiperiodic motion 36

RKR (program) 214
Radiation 139
- shielding 135
Radio horizon 296
Real time clock 114
Real-time applications 124
Rectangles, Fourier analysis of 12
Resistance measurements 113
Resonant frequency 113
Rotating table 33
Rotational
- quantum number 198
- structure 198
- constant 198
Rydberg, Klein, Rees 214

Surface plasmon experiment 184
Sampled-data PID algorithm 55
Sampling interval 63
Satellite data 293
Scanning 191
Scattering amplitudes 292
Schrödinger equation 216
Scintillation
- counters 292
- detectors 258f., 270

Screening currents 114
Searle's apparatus 69
Sector star 230
Semaphore 300
Shared memory as interprocess communication 300
Side band 10
Signal generator 3
Signal-to-noise ratio 219, 307
Simple harmonic oscillator 131
Sine grating 223
Solid-state sinter reaction 111
Specific heat anomalies 134
Spline-fit program SPLFIT 114
Square wave 12
Start bit 297
Statistical motion 34
Steady-state-error 63
Stop bit 297
Stroboscopic flashlight 28
Student response 279
Surface
– modes 176
– plasma oscillations 172
– plasmon 172, 191
– plasmon-polariton 173
– plasmon-polaritons dispersion curve 175
– waves 174
Symmetry 282

T^4 power law 149
TDO-measurements 113
TM wave 173
TURBO PASCAL 3, 17
Telemetry channel 303
Temperature variation of specific heat 142
Termscheme 198
Thermal conductivity 67f.
Thermocouples 72
– amplifier 73
Thermometer
– calibration 138
– circuit 136
Time derivative 25
Total reflection, attenuated (ATR) 172
Trajectory 30
Transmission errors 297
Tuning forks, pair 3
Tunnel diode oscillator (TDO) 111, 113
Turbo Pascal 142
Two-dip structure 122
Two-peak structure 119

Ultrafine particles 45

Vibration
– rotator 198
– quantum number 196
– constant 196
Video camera 23

Wave functions 216
Weak interaction 282ff.
Whole orbit data 303
Wilkinson - ADC 261
Wind-up 57

$YBa_2Cu_3O_7$ 111

Zero resistivity 111
Ziegler-Nichols' methods 60